BACK INJURY *among* HEALTHCARE WORKERS

Causes, Solutions, and Impacts

BACK INJURY *among* HEALTHCARE WORKERS

Causes, Solutions, and Impacts

Edited by
William Charney
Anne Hudson

LEWIS PUBLISHERS

A CRC Press Company
Boca Raton London New York Washington, D.C.

Library of Congress Cataloging-in-Publication Data

Back injury among healthcare workers : causes, solutions, and impacts / edited by William Charney and Anne Hudson
p. ; cm.
Includes bibliographical references and index.
ISBN 1-56670-631-9 (alk. paper)
1. Backache—Prevention. 2. Nursing. 3. Industrial safety. I. Charney, William, 1947- II. Hudson, Anne (Mary Anne)
[DNLM: 1. Back Injuries—prevention & control. 2. Nursing. 3. Occupational Health.
WE 720 B1262 2003]
RD771.B217B295 2003
617.5′64—dc21 2003047547

Lewis Publishers is an imprint of CRC Press LLC

International Standard Book Number 1-56670-631-9
Library of Congress Card Number 2003047547
Printed in the United States of America 1 2 3 4 5 6 7 8 9 0
Printed on acid-free paper

Dedication

This book is dedicated to the thousands of back-injured healthcare workers who have sacrificed their well-being, and often their careers, to painful injuries from manually lifting patients. It is the authors' hope that this book will lead to the implementation of no manual lifting of patients in hospitals, nursing homes, and home health through the use of technology by nursing staff or specially trained lift teams. We also hope that states will correlate the national nursing shortage with nursing injury and will pass "zero lift" for healthcare legislation to halt the unnecessary loss of healthcare workers to preventable disabling injuries. Finally, we look forward to the day when nursing organizations will negotiate for retention of back-injured nurses, including, when necessary, provision by employers of permanent light duty nursing work.

The Editors

William Charney, D.O.H., is currently the safety coordinator for Washington Hospital Services based in Seattle. He has served as a director of environmental health at San Francisco General Hospital and a safety officer at Jewish General Hospital in Montreal. He is a diplomate in occupational hygiene from the University of Montreal.

Charney is the author and editor of three volumes on *Essentials of Modern Hospital Safety,* published by CRC Press and *The Epidemic of Hospital Worker Injury: an Epidemiology,* also published by CRC Press. Charney has written many articles on hospital safety systems. He has done research on lifting teams which was published in *The American Journal of Occupational Health Nursing*. His research also included needle safety devices, published in *The Journal of Healthcare Safety, Compliance and Infection Control*; as well as ribavirin, published in *The Journal of Respiratory Care*. Charney has given lectures on healthcare safety issues at major conferences and received the 1998 Environmental Health Award by the California State Department of Health. Charney is a member of the American Conference of Governmental Industrial Hygienists and the Canadian Society of Safety Engineers. He has designed many safety devices currently in use in hospitals around the country, such as the HEPA filtration unit for tuberculosis control and a portable patient lifting device for home healthcare.

Anne Hudson, R.N., B.S.N., received the degree of Associate of Science in Nursing from Southwestern Oregon Community College, Coos Bay, in 1990, and a Bachelor of Science in Nursing from Oregon Health Sciences University, Portland, in 1998. She is a member of the American Nurses Association, Oregon Nurses Association, Physicians for Compassionate Care, and Sigma Theta Tau International Honor Society of Nursing. She maintains certification in medical/surgical nursing and advanced cardiac life support.

Hudson worked for a community hospital in medical/surgical, telemetry, and intermediate care units from September 1990 until a work-related spinal injury disabled her from lifting patients and bedside nursing in June 2000.

Becoming aware that most patient-handling injuries to nurses could be prevented led Hudson to become an activist for safe patient handling and an advocate for back-injured nurses. She is the founder of Work Injured Nurses' Group U.S.A. (WING USA).

Through writing and speaking opportunities, Hudson strives to increase awareness of the needless loss of nurses to preventable disabling injuries and to bring the plight of injured nurses before the public.

Contributors

Cynthia Barnes, R.N., B.S.N., C.C.R.N.
Staff Nurse
United Hospital, JNHH
St. Paul, Minnesota

William Charney, D.O.H.
Safety Coordinator
Washington Hospital Services
Seattle, Washington

Rahul Chhokar, B.Sc.
Project Coordinator
Occupational Health and Safety Agency for Healthcare (OHSAH)
Vancouver, British Columbia, Canada

Chris Engst, B.Sc.
Ergonomics Program Manager
Occupational Health and Safety Agency for Healthcare (OHSAH)
Vancouver, British Columbia, Canada

Maggie Flanagan, R.N.
NICU Staff Nurse
Fox Island, Washington

Guy Fragala, Ph.D., P.E., C.S.P.
Director of Environmental Health and Safety
University of Massachusetts Medical School
Worcester, Massachusetts

Susan Gallagher, Ph.D., R.N., C.W.O.C.N
Clinical Affairs Coordinator
Size Wise
Sierra Madre, California

Anne Hudson, R.N., B.S.N.
Public Health Nurse
Coos County Public Health Dept.
and
Founder
Work-Injured Nurses' Group U.S.A.
Coos Bay, Oregon

Elizabeth Y. Langford, R.N., R.M., B.N., Grad. Dip. (Adv. Nsg.)
Coordinator
Injured Nurses' Support Group (ANF Victorian Branch)
Preston, Victoria, Australia

John D. Lloyd, Ph.D., M.Erg.S., C.P.E.
Associate Director Technology Division
Patient Safety Center of Inquiry
James A. Haley VAMC
Tampa, Florida

Aaron Miller, B.Sc.
Project Coordinator
Occupational Health and Safety Agency for Healthcare (OHSAH)
Vancouver, British Columbia, Canada

Audrey Nelson, Ph.D., R.N., F.A.A.N.
Director
Patient Safety Center of Inquiry
James A. Haley VAMC
Tampa, Florida

Bernice D. Owen, Ph.D., R.N.
Professor Emeritus, Nursing
University of Wisconsin
Madison, Wisconsin

Beth DeWees Piknick, R.N.
Staff Nurse
Cape Cod Hospital
Hyannis, Massachusetts

Elizabeth Shogren, R.N.
Staff Specialist Labor Education
Occupational Health and Safety
Minnesota Nurses Association
St. Paul, Minnesota

Judy Sims, B.S.N., R.N., C.C.R.N., C.L.N.C.
Nurse Case Manager
Standard Insurance Company
Portland, Oregon

Bernadette Stringer, Ph.D.
Assistant Professor
Department of Epidemiology and Biostatistics
Faculty of Medicine and Dentistry
University of Western Ontario,
London, Ontario, Canada

Susie Lyons Toftum, former C.N.A.
Manager
AYA Copy Center
Coos Bay, Oregon

Jocelyn Villeneuve, D.E.S.S.
Ergonomist
Association for Health and Safety in Social Affaires
Montreal, Quebec, Canada

Annalee Yassi, M.D., M.Sc., F.R.C.P.C.
Founding Executive Director
Occupational Health and Safety Agency for Healthcare (OHSAH)
Vancouver, British Columbia, Canada

Contents

1 History and Vision for Work-Injured Nurses' Group USA

Anne Hudson

CONTENTS

INTRODUCTION

In 1999, the U.S. Bureau of Labor Statistics (BLS) identified healthcare patients as the source of 72,363 injuries and illnesses requiring time off work, including 59,002 musculoskeletal disorders (MSDs) with the trunk and back as the body parts most often affected. In the year 2000, healthcare patients caused time away from work for 10,983 registered nurses and 44,854 nursing aides, orderlies, and attendants with "overexertion" and "overexertion in lifting" identified as leading events for time lost. Also in 2000, of 129 occupations with time away from work due to MSDs, the BLS ranked nursing aides, orderlies, and attendants second with 44,660 MSDs, registered nurses sixth with 12,074 MSDs, and licensed practical nurses eighteenth with 5598 MSDs. A footnote listing inclusion of sprains, strains, tears, back pain, etc., specifies that, while herniated spinal discs may be considered MSDs, they are classified elsewhere, indicating that the actual number of musculoskeletal injuries to healthcare workers may be greater than shown by these figures (http://www.bls.gov).

The numbers give an idea of the magnitude of the problem concerning patient handling injuries. Behind the numbers are individuals, each with a story of service to others, injury, pain, and loss. It is unknown where these injured healthcare workers are today, whether any organization tracks them, and whether any type of practical assistance is available to them. It is likely that, if unable to resume heavy lifting, most were not allowed to return to work by their employers.

While back-injured RNs and LPNs may, at least theoretically, be employable elsewhere, CNAs often have more severely limited employment options following disability. Any category of healthcare worker is subject to disability and potential loss of career from patient handling injuries.

LACK OF SUPPORT FOR INJURED NURSES

Disabled nurses are generally ushered out the door and have traditionally just "gone away." The mention of intolerance of injured and ill nurses in the literature is curious because it seems that if rejection of nurses with infirmities has been recognized, then the nursing community should have responded to the needs of injured and ill nurses. There should be enough nurses to effect change. Healthcare workers represent the largest female workforce in the country and also suffer more disabling injuries than any other population of working women. These injured women have traditionally more or less accepted disability as a risk associated with work requiring the physical lifting and moving of human beings.

1-56670-631-9/03/$0.00+$1.50

This is not meant to diminish in any way the tragic physical, emotional, and financial losses suffered by male nurses following disabling injuries. Male nurses are noted to be at increased risk of disabling injuries because they may be called upon more often to lift and move patients. A recent example is of a male CNA, about 6 feet 2 inches and 30 years old, sent by a nursing supervisor to a different unit to help manually transfer and transport a 500 lb patient. Though transfer equipment was available in the hospital, the unit said it was not going to use it. The CNA did as he was instructed. He helped pull the 500 lb. patient onto a stretcher, and pushed the stretcher and patient to another unit. Lack of enforced policy mandating use of available equipment, and the CNA's compliance at the unreasonable request, could have resulted in permanent disability from this single transfer. In addition, the day-in and day-out repetitive lifting and moving of patients takes its cumulative toll. In "The First to Go" (Injured Nurse Story #5) Kmak speaks of the profound impact of back injuries to a male nurse from physically lifting patients.

Disabling back injuries are no respecters of gender but some of the long-standing societal attitudes may be influenced by gender. It remains that nurses are 89.9 to 94.6% women, depending on the data source, and it is primarily female nurses who are losing their health and careers from largely preventable injuries. Consider whether any college-educated, male-dominated profession would tolerate preventable injuries to inflict lifetime disability and destroy their means of livelihood. Is it possible that both unsafe patient handling practices, which result in multiple thousands of preventable injuries annually, and the historical intolerance of injured and ill nurses within healthcare, are related to nursing being predominately female? Consider whether the innate caring nature of nurses could possibly contribute to a type of codependent enabling and toleration of the exploitation of fellow nurses.

Once disabled and discarded, there appears to be no safe haven for injured nurses who have faithfully provided care for others. They are now in pain, rejected by their employers, and effectively shunned by fellow nurses who, one might think, would rebel at such treatment of the injured, demand change, and build a support system prepared to care for the victims.

RESPONDING TO THE NEED

While it appears there is no organized nationwide assistance, wonderful progress is occurring within some states. The Massachusetts Nurses Association promotes strong contract language for no-lift policies and lift teams, and has also established the Injured Nurses Network. The Minnesota Nurses Association has worked tirelessly for many years on health and safety issues and contract language. In "An Advocate for the Ill, Injured, or Disabled Nurse" (Injured Nurse Story #17), Cynthia Barnes describes her advocacy role within her local Minnesota Nurses Association bargaining unit, backed by contract language protecting seniority and providing priority hiring rights to nurses with work-related injuries — a huge step forward. Cynthia Barnes looks to a future with "no more disappearing nurses." Other states may be making strides forward as well.

On the national level, however, it appears that there is no place for back-injured nurses to turn. One can only speculate why this is so, why nurses have not prioritized providing personal support to their injured coworkers and assistance in returning to work as nurses. If such assistance is available, it has not yet been located by countless disabled nurses who ask, "Is there any help for me?"

Other countries are responding to the need. WING, Work Injured Nurses' Group, part of England's Royal College of Nursing, offers practical support to all injured and ill nurses, whether their infirmities are work-related or not — newsletters, support groups, a telephone advice line, attorneys to represent injured nurses with their court cases, and practical assistance in continuing to work as nurses. INSG, Injured Nurses' Support Group, with the Australian Nursing Federation (Vic Branch) offers support and informational meetings, guest speakers, and telephone support service, addressing the physical, personal, employment, and legal issues faced by injured nurses. These groups, and perhaps similar others around the world, recognize and address the very real needs of injured nurses. In addition, in the way of prevention, the Royal College of Nursing and the Australian Nursing Federation have put forth no lifting policies, which ban the manual lifting of patients in all but exceptional or life threatening situations.

My response to being injured in June 2000 was to seek information on back injuries among nurses and contact other back-injured nurses. Attempts to locate a designated chat line for back-injured nurses were not successful. This was perplexing because I knew the injured nurses were out there. I discovered the enormous body of evidence on patient handling and methods proven to prevent back injuries to nurses. I had been lifting and moving patients every hour on the job but had never before heard the phrase "patient handling" and was totally unaware of the work done on prevention of back injuries with mechanical patient lift equipment, friction-reducing devices, lift teams, and no-lift policies. My injury, and the injuries of countless others, could have been *prevented?!* I now wanted to share some of this information with other nurses.

With the help from good friends, a web page called "B.I.N. There — Back Injured Nurses" was begun as a way to reach injured nurses. B.I.N. There is receiving approximately 550 hits each day. Injured nurses and interested others have sent e-mails from across the country and beyond. They are grateful for the source of information on patient handling issues and for validation of their experiences.

Stories from the injured are each unique, yet, the same elements come through time after time: "I am a nurse. I was injured lifting patients at the job I love. I'm in physical and emotional pain and may be facing financial ruin. My employer will not allow me back because of permanent limitations from my injury. Is there any help for me?"

Though many could continue working, apart from heavy lifting, injured nurses across the country report the same response to their pleas to work: "There isn't, we don't, you can't. There is no light duty. We don't create jobs. You can't return unless you resume full physical performance, including lifting requirements." Even during the current nursing shortage, many of the readily available population of experienced nurses are not permitted back to work, and are often treated as though their value is directly related to the strength of their backs.

BIN There's focus has been to provide emotional support and encouragement, to be a source of information and available research on patient handling, to acknowledge the needs of the injured and validate their experience, and to let injured nurses know that they are not alone. Thousands of disabled nurses are at home, in pain, and out of work. Though isolated, they are not forgotten. BIN There has successfully begun a network of injured nurses who exchange information and draw support from each other.

It is vital to share available information with nurses injured by manual patient handling. Of particular note is Section E. "Disorders of the Low Back," pages 68469 to 68483, of the *Ergonomics Program Final Rule*, Federal Register, Tuesday Nov. 14, 2000, Vol. 65, No. 220, Part II, Department of Labor, Occupational Safety and Health Administration, 29 CFR Part 1910, which "...summarizes and discusses the evidence that physical work-related risk factors contribute to the pathogenesis of specific disorders of the back." The thorough review of research includes studies specific to nurses and the proven risk of back injury with lifting patients, cadaver studies providing tolerance limits of spinal structures to compressive forces, and the pathophysiology of cumulative trauma to spinal structures from repetitive heavy lifting. Though the Ergo Standard was rescinded, the research stands, making the document a powerful tool for proving the relationship between cumulative trauma spinal injuries and the repetitive lifting of patients. (http://www.access.gpo.gov/su_docs/fedreg/a001114c.html/. Scroll down to Occupational Safety and Health Administration, Rules. Select 68461–68510. Scroll down to page 68469 and Section E on "Disorders of the Low Back.")

I have been privileged to have had the opportunity of speaking at a number of healthcare conferences on behalf of back-injured nurses. I feel the need to speak out and draw attention to the exploitation and unnecessary loss of nurses from out-dated manual patient handling, to inform nurses that the injuries they may have considered "just part of the job" are largely preventable, to describe potential ramifications of back injuries, and to inform of available research, which may be helpful to back-injured nurses.

B.I.N. There - Back Injured Nurses is undergoing change to become "Work Injured Nurses' Group USA," with the expanded vision of joining the international community of nurses working to protect its own from preventable disabling injuries and needless loss of career and to see practical support and assistance offered American nurses such as provided by England's WING and Australia's INSG.

WING USA is being launched with goals, through collaborating with others, to become a united voice for work-injured nurses; to build a network of mutual support; to provide information on medical, legal, and other issues faced by injured nurses; to develop informational materials; to endorse establishment of priority hiring rights and permanent light duty for work-injured nurses; to pursue reduction of preventable injuries and unnecessary loss of nurses by promotion of lift teams, zero lift policies, and safe patient lift and transfer equipment; and to campaign for zero lift for healthcare legislation, which, when enacted, will protect from needless injuries, reduce medical and compensation costs, and facilitate retention of nurses.

Input and involvement from others dedicated to assisting injured nurses are most welcome. A broad base of financial and practical support will be necessary to fulfill the goals of WING USA. One hundred percent of the proceeds from this book are going directly to WING USA with the hope that one day work-injured nurses will no longer need to ask, "Is there any help for me?" Please visit WING USA at: www.wingusa.org.

2 Magnitude of the Problem

Bernice D. Owen

CONTENTS

INTRODUCTION

Back injuries have been a significant long-term problem for those nursing professionals providing direct patient and/or resident care. Research studies, dating back to the 1960s, indicate the incidence of back injuries to be high in nursing personnel as compared to other workers. Jensen et al. (1989) cites 90 studies prior to 1988 that deal with this problem. Many of these studies identified patient lifting and transferring tasks as major factors associated with overexertion injuries; these tasks involved risk factors including awkward and twisting postures, heavy weights, and forceful exertions.

Klein et al. (1984) found, through analysis of national worker's compensation claims, that nursing personnel ranked fifth among all workers for occupationally related back problems. Nursing personnel were surpassed only by occupations involving heavy physical labor — miscellaneous laborers, sanitation workers, warehouse workers, and mechanics.

The back injury rate among nursing personnel remains high. In 1993, the back injury rate for nursing assistants in nursing homes led all other occupations (Bureau of Labor Statistics [BLS], 1995); their rate was four times higher than the average rate for private industry. In 1995, the back injury rate for nursing assistants was still four times higher than the rate for workers in private industry (BLS, 1997a). That same year, the rate for home healthcare workers was more than double that of private industry; this rate even surpassed the rate of overexertion injuries for hospital nursing personnel (BLS, 1997b). The average lost work days per injury in home healthcare workers was 7; this was 2 days more than the median absence from work for other workers (BLS, 1997c).

The goal, established for nursing personnel in *Healthy People 2000: National Health Promotion and Disease Prevention Objectives* issued by the U.S. Department of Health and Human Services (USDHHS) in 1990, was far from reached. The goal was to reduce the rate of back injuries from 12.7 per 100 full-time (FT) workers to 9, instead the rate actually increased by midcourse to 17.8 per 100 FT workers (USDHHS, 1997b). (These nonfatal injuries include more than just back injuries, but these latter injuries make up the greatest number of injuries.) In the 2010 goals, a specific goal for back injuries is not specified, but there is a citation that indicates low back disorders remains a high priority for research (USDHHS, 2000).

1-56670-631-9/03/$0.00+$1.50

In 2000, the BLS (BLS, 2000) reported on the incidence rates of nonfatal occupational injuries for private sector industries with 100,000 or more cases; the incidence rate for workers in nursing and personal care facilities was second highest in the nation. Only scheduled air transportation workers surpassed the rate for nursing.

The statistics presented above on work-related back injuries represent those that have been reported. There is some indication that these statistics may be revealing only the tip of the iceberg. In a random sample of 503 nurses, Owen (1989) found that 38% stated they had suffered at least three consecutive days of occupationally related back pain, but only 33% of that number ever reported this. This group also averaged 6.5 days of its own sick days for unreported back pain perceived to be occupationally related. Of these nurses, 20% had made at least one employment transfer to decrease the amount of patient lifting. For example, they transferred to a different unit (from intensive care to obstetrics), changed employment settings (from hospital to clinics), or changed positions (from hospital staff nurse to drug company representative). Another 12% said they were considering making an employment transfer and 12% were thinking about leaving the profession of nursing because of occupationally related back pain.

In England, Stubbs et al. (1986) found 12% of all nurses intending to leave nursing permanently cited back pain as either a main or contributory factor. In the Netherlands, Knibbe and Friele (1995) found 91% of the home care personnel continued to work even though they had significant back pain; of these, 16% exchanged assignments with colleagues while 9% withdrew from the heavy part of work.

By analyzing answers to questionnaires given to 3548 nurses, McGuire and Dewar (1995) found 33.4% sustained an occupationally related injury but only 51.9% of them completed an accident form. The reasons given by these nurses for not completing the form included not realizing until the next day the level of pain involved, not thinking it was important, and feeling the process of reporting was too time-consuming.

Researchers (Klaber-Moffert et al., 1993) have even found the back pain problem to exist in student nurses. Of the 199 students studied over a 30-month period, 64% reported at least a one-day episode of back pain related to heavy work on the units; 37% reported occupationally related back pain that lasted for at least 3 consecutive days.

So it is easy to see that nursing personnel have a significant back injury problem. The following changes in the delivery of healthcare may certainly have had an impact on the occurrence of these overexertion injuries: (1) patients are staying fewer days in the hospital, so their care becomes more intense; (2) more acutely ill patients are transferred from hospital to long-term care facilities and rehabilitation settings where they require more patient care; (3) and many patients are discharged to their homes while still in need of much care. At the same time that these changes occurred, many hospitals have downsized the number of professional staff, increased the number of unlicensed assistive personnel, and added patient care duties to some employees who have other types of duties (e.g., housekeeping staff being trained to help feed patients). These changes may have a direct effect on the overexertion rates. The shortage of nurses and the decrease in staffing ratios are important to the problem.

There are many reasons for addressing this occupational injury problem in healthcare settings. They include the impact of the disability on the worker and his or her family, the impact of the lost worker to patient care and loss of continuity of care to the patients, and of course the ever-increasing financial burden to everyone. As the healthcare industry continues to strive for cost-saving opportunities, many are viewing the worker compensation costs, medical costs, and associated indirect costs of occupational injuries as an opportunity for significant cost savings. In addition, improving this situation can also result in enhanced quality of life for workers and quality of care for patients.

CONTRIBUTING FACTORS

Biomechanical and postural stressors are the most likely triggers for back pain and back injury. Variables such as the weight of the load, distance of the load (patient) from the lifter's center of gravity, duration of the lift, awkward lifting position, confined work space, unpredictable patient behavior, and the amount of stooping, reaching, and bending involved lead to excessive force in the spinal area. Incident reports, worker compensation claims, and nursing personnel perception all implicate the stressors involved in lifting and transferring patients as causative agents in back pain (Bell, 1984; Nelson and Olson, 1996; Owen and Garg, 1993; Smedley et al. 1995; Villeneuve, 1998). The frequency of carrying out these tasks, combined with the cumulative nature of the insults, is important to the problem of back injuries.

The tasks perceived to be most stressful by nursing personnel in a nursing home were: transferring residents on and off the toilet, in and out of bed and chairs, bathing, and weighing the resident (Owen and Garg, 1989). Tasks perceived most stressful by nursing personnel in hospitals were: lifting patients from the floor, transferring patients on and off stretchers and cardiac chairs, in and out of bed on and off commode, and lifting patient up in bed (Owen et al., 2000). Tasks perceived most stressful in home care were: lifting up in bed, putting on antiembolism stockings, transferring chair to chair, giving tub bath, repositioning in chair, and toileting patient (Owen and Staehler, 2003).

The method used most frequently for these patient handling tasks has been found to be a major problem. Owen and Garg (1993); Owen et al. (1995) found the "under-axilla" method was used 98% of the time for lifting and transferring residents and patients. This method is one in which two nursing personnel stand facing the patient or resident, each grasps the patient under the axilla and vertically lifts the patient to a standing position or carries the patient to a new location (see Figure 2.1). Many body mechanic principles are violated: the torso is rotated, the lift is asymmetric, and the weight to be lifted exceeds NIOSH recommendations. Owen et al. (1999) even found that 83% of nursing educators taught nursing students this under-axilla method for transferring patients in and out of bed.

This under-axilla method has been studied by Garg and Owen (1992) and Knibbe and Knibbe (1990). Through the use of computerized static-loading biomechanical models, they found the compressive force to L5–S1 of the workers exceeded the level recommended as maximum by the U.S. Department of Health and Human Services (1981). The recommended maximum level is 3400 Newtons (N) of force; the average found by Garg and Owen (1992) was 4751 N (SD = 106). Findings by Knibbe and Knibbe were comparable. In addition, subjects perceived physical stress to their backs as "high" when using the under-axilla method of transfer.

Owen et al. (1995) studied patient handling tasks in a hospital setting. Again, nursing personnel perceived physical stress to the back as "high" while using the under-axilla method. Patients also found this method uncomfortable.

Marras et al. (1999) studied the under-axilla method of transfer using a computer model that took into account the coactive nature of multiple muscle groups used for lifting and transferring.

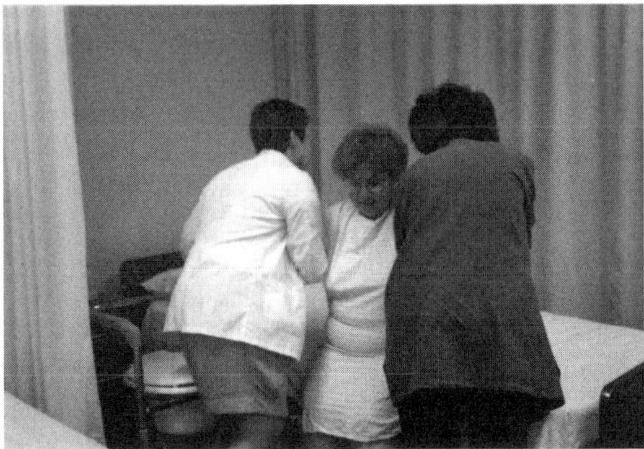

FIGURE 2.1 Manual lift. Note back torque. (Photo courtesy of Bernice Owen, University of Wisconsin, School of Nursing.).

They found the compressive force to L5–S1 was greater than the recommended maximal level in about 62% of the transfers from bed to chair and even exceeded 6400 N in about 17% of these transfers. (They were using a healthy, cooperative 115 pound subject as a patient). With the task of lifting up in bed using the under-axilla technique, they found 60% of the tasks above the 3400 N level and 35% were even above the 6400 N level of force.

In addition to this under-axilla lift being hazardous to nursing personnel, it is also hazardous to the patient. The brachial plexus is located in the axilla area and pressure against this plexus can affect nerve involvement in the neck/shoulder/arm/hands of the patient. Also, pressure to the musculature under the arm can subluxate the shoulder.

In England, this under-axilla method is called the "drag" and is considered inappropriate for use (Corlett et al., 1992). By law, nurses and employers must work toward no-lift policies for patient care and the "drag" technique must be eliminated (Royal College of Nursing, 1996). In addition, the British have stated the drag is "deplorable... inefficient, dangerous to the nurses, and often painful and brutal for the patient" (Hardicre, 1992).

APPROACHES TO DECREASING THE PROBLEM

It has been difficult to change the attitudes and work patterns of nursing personnel in relation to approaches to back injury prevention. The nursing culture tends to be one of accepting these injuries as part of the job. An editorial written by Huey in the *American Journal of Nursing* in 1993 stated: "For years we've blamed the victims of back injury for not using proper body mechanics as if the size, shape, and resistance of the loads we expect nurses to haul doesn't matter. We've acted as if teaching nurses the best-known lifting techniques and exercises to strengthen the abdominal muscles connected to the back would be enough to prevent back problems. And when they weren't, we have been too eager to accept back injury as an occupational hazard and to believe that if the nurse can't take it, she needs to get out." It is certainly an understatement to say that much more work needs to be accomplished in this area.

BODY MECHANICS

Various approaches have been tried to decrease this problem of back injuries. Emphasis has primarily been on education and training with a definite focus on body mechanics. However, these approaches have had little impact on the problem (Daltroy, 1997; Langerstrom and Hagberg, 1997; Personick, 1990; Fragala, 1996). With these approaches, the aim has been to change the worker instead of changing the job or the task. Though body mechanics are important, many times workers are limited in their ability to apply them well.

Pheasant (1991) aptly stated, "Many people (both within the nursing profession and elsewhere) take the view that nurses have back problems because they are undertrained. The reality is that they are physically overloaded by their work activities. *In situations of this kind, training is necessary but not sufficient.* To make further progress, we need to identity the features of the working system which are responsible for the physical overload." This author states some programs experienced a decrease in back injuries after a training program, but within a short time, the injury rates returned to pretraining levels or even higher.

St. Vincent et al. (1987) instituted a training program for nursing personnel that taught basic principles of body mechanics. They found that nursing personnel could not use many of these principles in their clinical practice; they lifted and transferred patients with force on their backs in 90% of the observed patient handling tasks. They concluded that the manual techniques taught during training did not integrate all the factors important to carrying out the tasks required in clinical practice. The variables that needed to be included in addition to nursing knowledge and skill were patient factors (weight, combativeness, ability to bear weight, and predictability) and environmental factors (height of bed, confined work space, flush surfaces for transferring, and wheelchairs with adjustable arms).

Friele and Knibbe (1995) hypothesized that the more the emphasis is placed on teaching body mechanics, the greater will be the back injury rate. They state that emphasis only on training in body mechanics implies that a person can do anything with their bodies as long as they use correct body mechanics. This approach negates the need to consider, and work with, the patient and environmental variables.

Ergonomics

The Occupational Safety and Health Administration's general duty clause states clearly that all nursing personnel have the right to a safe and healthy workplace. According to the statistics previously discussed, however, something is very wrong with the present workplace of nursing professionals. General industry has been able to improve many work settings by applying ergonomic programs. These programs are based on the principle that the job must fit the capabilities of the worker; when it does not, then the job must be changed.

According to Fragala (1996), ergonomics is defined as "the science or discipline of matching jobs and job tasks to the specific capabilities of the work force. Ideally, this involves designing jobs to avoid unacceptable risk factors such as lifting excessive weights or performing the same task over and over again. Realistically, it more often involves redesigning jobs to minimize or eliminate such risk factors." In other words, ergonomic programs aim to change the work environment and work practices to prevent physical stress and injuries.

There are healthcare settings that have been successful in implementing ergonomic programs. Several of these were cited in the *Draft Guidelines For Nursing Homes* (USDHHS, 2002, p. 3) that was developed by the Occupational Safety and Health Administration (OSHA). According to OSHA, "Facilities that have implemented ergonomics-based injury prevention programs using effective engineering and work practice controls have achieved considerable success in reducing work-related injuries and workers' compensation costs. In addition, some institutions have experienced additional benefits, including reduced staff turnover and associated training and administrative costs, reduced absenteeism, increased productivity, improved morale, reduced resident injury, and increased resident comfort. Many examples exist of effective ergonomics programs in nursing homes...."

Important elements of the ergonomic process include: the identification of stressful jobs and tasks, the evaluation or assessment of these jobs and tasks to determine the magnitude, force, and duration of the stress, the study of work organization, the assessment and application of various controls, and the evaluation of the results. This process is described in a number of articles (USDHHS, 2002; Owen, B.D., 1999; Cohen et al. 1997; Feletto and Graze, 1997; Fragala, 1996; USDHHS, 1997b). A short synopsis of the ergonomic process follows.

Management support and involvement in the full ergonomic program is essential. Some of management's roles are learning about ergonomics, helping to identify stressors, encouraging reporting of stressors, providing adequate resources and training time, establishing and carrying out policies, supporting a medical management program for those injured, and evaluation of the total ergonomic program. Employee support and involvement are also key to success. Employees should learn about ergonomics, understand and report risk factors for back pain and injury, help to identify stressors, participate in decision-making and goal formation for control of stressors, be skilled in the use of controls (e.g., equipment and devices), and help with the evaluation of the program.

Because patient handling tasks have been found to be major stressors, it is important that a thorough assessment of patient and resident abilities be conducted. This assessment should be a combined effort of those involved in the direct care of the patients and/or residents. Variables important to this assessment are mobility needs, patient weight and ability to bear own weight, predictability and cooperativeness of patient, cognitive ability and medical condition of patient. Results of this assessment are then used to determine control measures for decreasing risk of injury.

Adequate controls need to be in place. Examples include appropriate lifts, slings, toileting devices, transfer belts, etc. These need to be well maintained, adequate in number, and readily accessible to personnel for patient use. As discussed previously, staff must be well trained in the use of controls. Oral and written communication between care giving staff (and also communication with the patient) will help to assure consistent use of controls.

Good medical management of injured workers is important. Incidents should be investigated in a timely manner, case management of injured worker provided, transitional work duties provided and supervised, and prevention of re-injury emphasized. Monitoring and evaluation of the program are also essential elements of the program. Management and staff need to work together with this process.

Research — Applying the Ergonomic Process

In an effort to decrease physical stress due to use of the under-axilla method of transfer, Garg and Owen (1992) studied the following tasks identified as most stressful by nursing home personnel: transferring residents in and out of bed, on and off the toilet, on and off the weight scale chair, and bathing the residents. The above ergonomic program was implemented. After a study of tasks, resident needs, training of staff, etc., the following policies were instituted for most residents: a full mechanical lift was to be used for residents who could not bear weight and a transfer belt with handles was used for those who needed assistance but could bear some weight. Through use of the transfer belt with handles, the compressive force to L5–S1 of the nursing personnel decreased to an average of 2030 N of force (maximum for safety is 3400 N). The perceived stress ratings decreased from an average of hard and very hard to very light. In addition, the incidence of reportable back injuries decreased from 83 per 200,000 work hours to 47 per 200,000 work hours over an 18-month post-intervention period.

The ergonomic process was studied in the hospital setting by Owen et al. (1995). The tasks identified by nursing personnel as most stressful were: transfers in and out of bed, on and off the commode, on and off stretchers, lifting up in bed, lifting up from the floor, and toileting in bed. The controls selected, based on patient assessment, were: a full mechanical lift for those patients who could not bear weight or were not predictable in bearing weight, a stand up assist mechanical lift or a walking belt with handles for those patients who needed assistance but could bear some weight, a friction reducer was used under the draw sheet for repositioning up in bed and for transferring on and off stretchers and cardiac chairs, and a toileting apparatus designed specifically for toileting in bed. Data were collected in the control hospital following the usual annual back injury prevention in-service; data were collected in the experimental hospital following implementation of the full ergonomic program. Ratings of perceived exertion (physical stress) to the low back were completed after each task was carried out in both the control and experimental sites. (The scale used was 0 = no stress, 10 = extreme, maximal stress). Ratings for all the tasks were significantly lower for the nursing personnel from the experimental hospital than from the control site.

The patients also rated their feelings of comfort and security immediately after the tasks were performed. (The scale used was 0 = extremely comfortable and 7 = extremely uncomfortable and 0 = extremely secure and 7 = extremely insecure). For all tasks, the patients in the experimental setting rated their feelings of comfort and security significantly higher than the patients in the control setting.

Eighteen months prior to the intervention in the experimental hospital, there were 20 back injuries, 64 lost work days, and 15 restricted/transitional days. Eighteen months after the intervention, there were 12 back injuries, three lost work days, and 12 restricted/transitional days. The number of injuries decreased as well as the severity of these injuries.

A 5-year follow-up study showed that the ergonomic program could be sustained (Owen et al., 2001). During the total 5 years, 20 back/shoulder injuries were reported. Sixty-one percent ($n = 16$)

occurred in year 3. Many of the reports for year 3 indicated the injuries occurred when working with heavy patients. There were 87 lost work days (LWDs) during these 5 years. Seventy-seven percent ($n = 67$) occurred in year 3; of these, 64 were attributed to one nurse. There were no LWDs recorded in year 1 and 5 of the follow-up period. There were no restricted days recorded for years 2, 4, and 5. In year 1, all 42 restricted days were taken by one nurse.

In another study, Charney (1997) and his team applied the ergonomic process in 10 hospitals using lift teams as major controls. They found a significant decrease in injuries and also in worker compensation costs.

Ronald et al. (2002) used overhead ceiling lifts as controls for lifting and transferring patients. The musculoskeletal injuries were significantly reduced. In this pre- and poststudy design, the residents indicated an increase in their level of comfort and their satisfaction with the new controls postintervention.

The Evanoff team (Evanoff et al., 1999) implemented an ergonomic program in a hospital setting. The 2-year postintervention period showed a significant drop in lost work days and worker compensation costs. In addition, there were significant improvements in job satisfaction, perceived psychosocial stressors, and social support among the subjects.

A hospital surgical area was the setting for implementing an ergonomic program by Garb and Dockery (1995). There was a 25% reduction in the rate of back injuries in operating room personnel 18 months postintervention by the ergonomic program.

At present, there are a number of studies being conducted by researchers seeking methods that may help prevent back injuries in healthcare workers. Examples include the National Institute for Occupational Safety and Health, the Patient Safety Center at the Veterans Hospital in Tampa, Florida, the University of Maryland, School of Nursing, and the University of Wisconsin-Milwaukee.

The Need For More Work

The problem of back injuries continues to exist; this is especially true in long-term care settings, hospitals, home care, and hospice centers. These injuries are particularly prevalent in nursing personnel who provide direct patient and/or resident care. Continued research is needed in the abatement of this problem. Greater emphasis needs to be placed on *intervention studies* related to the total ergonomic process. For instance, what are the best ways to get management involved in the process, what are effective methods for securing worker involvement, what education and training models are most effective in healthcare settings and for various levels of personnel, what technology is best suited for what patient population. The emphasis must be on intervention studies.

The under-axilla technique for lifting and transferring patients and residents must be eliminated from nursing practice. Faculties in schools of nursing, as well as the entire nursing profession, need to learn the ergonomic process and be adamant about the application of these principles for lifting, transferring, and repositioning patients and residents.

Continued study into the nursing culture needs to be highlighted. Acceptance of back injuries as an occupational hazard needs to be changed. What will it take for the attitudes of nursing personnel to be directed toward safety, prevention, and protection?

McGuire and Dewar (1996) found nursing staff tended to blame themselves when injured, did not think any changes were needed, and many preferred to work alone even when lifting patients. Furthermore, some comments from nurses in cited in this study were: "It seems to be part of the nurses culture to manage alone," and "You are thought of more highly if you just get on with the job." Kane and Parahoo (1994) found nurses would use unsafe techniques rather than receive negative reactions from the staff.

The Healthy People 2010 (USDHHS, 2000) does not state a specific goal for reduction of back injuries in healthcare personnel. We hope this does not mean there will be no emphasis on this significant problem. Funding for research and training from NIOSH and OSHA needs to continue at an accelerated pace. Even funding from the National Center for Nursing Research should contribute to the goals of back injury prevention because these injuries have a great impact on the profession of nursing and patient care. The shortage of nurses is well known and this problem of back injuries is not helping the cause. As discussed earlier, nurses leave the profession because of back injuries.

In Europe, the Royal College of Nursing showed great leadership in taking political action to make the European Directive become a reality. This legislation mandated that lifting should be avoided as far as is reasonably practicable (Health Service Executive, 1992). The aim of the legislation was to decrease significantly the back injury rate of nursing personnel through the establishment of a code as close to no-lift as possible. It is time now for the professional nursing organizations of the U.S. to come together in political action to help draw up legislation on the national and state level for the zero-lift policy. Nurses have the largest membership in the health services and therefore have much power in the political arena as shown by the role played by the American Nurses' Association (ANA) in enactment of legislation on needle stick injuries. The ANA is looking at staffing problems and must also give considerable time and effort to preventing back injuries in nursing personnel. Consider the political strength the nursing profession would have if all professional nursing organizations teamed with the ANA to prevent back injuries.

An ergonomic standard for all of industry is probably not realistic at this time. However, an OSHA standard specific for healthcare settings should be reasonable. The draft of *Ergonomics for the Prevention of Musculoskeletal Disorders: Guidelines for Nursing Homes*, published in 2002, is a start. There continues to be much work to do and now is the time for researchers and nursing to move ahead at an accelerated rate.

REFERENCES

Bell, F., *Patient Lifting Devices In Hospitals*, Groom Helm, London, 1984.

Bureau of Labor Statistics, *News* (USDL-95–142), United States Department of Labor, Washington, D.C., April, 1995.

Bureau of Labor Statistics, *News* (USDL-97–188), United States/Department of Labor, Washington, D.C., 1997a.

Bureau of Labor Statistics, Injuries to caregivers working in patients' homes, in *Issues in Labor Statistics* (Summary 97–4), United States Department of Labor, Washington, D.C., 1997b.

Bureau of Labor Statistics, Home health care services: injuries resulting in absences from work lost worktime injuries and work hazards, in *Supplement of News Release* (USDL-96–163), United States Department of Labor, Washington, D.C., 1997c.

Bureau of Labor Statistics, *News* (USDL-99–357), United States Department of Labor, Washington, D.C., 2000.

Charney, W., The lift team method for reducing back injuries: a ten hospital study, *AAOHN J.*, 45, 300–304, 1997.

Cohen, A.L., Gjessing, C.G., Fine, L.J., Bernard, B.P., and McGlothlin, J.D., *Elements of Ergonomics Programs: A Primer Based on Workplace Evaluations of Musculoskeletal Disorders*, United States Department of Health and Human Services, Washington, D.C., 1997.

Corlett, E.N., Lloyd, P.V., Tarling, C., Troup, J.D., and Wright, B., *The Guide to the Handling of Patients,* 3rd ed., National Back Pain Association, Middlesex, England, 1992.

Daltroy, L., A controlled trial of an educational program to prevent low back injuries, *New Engl. J. Med*, 337, 322–328, 1997.

Evanoff, B., Bohr, P., and Wolf, L., Effects of a participatory ergonomic team among hospital orderlies, *Am. J. Ind. Med.*, 35, 358–365, 1999.

Feletto, M. and Graze, W., *A Back Injury Prevention Guide for Health Care Providers*, CAL/OSHA Consultation Service, November, CA, 1997.

Fragala, G., *Ergonomics: How to Contain On-the-Job Injuries in Health Care,* Joint Commission on Accreditation of Healthcare Organizations, Oakbrook Terrace, IL, 1996.

Friele R.D. and Knibbe J.J., Monitoring the barriers with the use of patient lifts in home care as perceived by nursing personnel, in *Occupational Health for Health Care Workers*. Hagberg, M., Hofmann, F., Stobel, U., Westlander, G., Eds.. ECOMED, Landsberg, Germany 1995.

Garb, J. and Dockery, C., Reducing employee back injuries in the perioperative setting. *AORN J.*, 61 (6), 1046–1052, 1995.

Garg, A. and Owen, B.D., Reducing back stress to nursing personnel: an ergonomic intervention in a nursing home, *Ergonomics*, 35, 1353–1375, 1992.

Hardicre, J., Put your back out of danger, *Nursing Standard*, 7(5), 54, 1992.

Health Service Executive, *The Manual Handling Operations Legislation*,: HMSO, London, 1992.

Huey, F., Let's get our backs up, editorial, *Am. J. Nursing*, 93, 7, 1993.

Jensen, R., Nestor, D., Myers, A., and Rattiner, J., *Low-Back Injuries Among Nursing Personnel: An Annotated Bibliography*, The Johns Hopkins University, Baltimore, MD, 1989.

Kane M. and Parahoo K., Knowledge and use of lifting techniques among a group of undergraduate student nurses, *J. Clin. Nursing*, 3, 35–42, 1994.

Klaber-Moffett, J.A., Hughes, G.I., and Griffiths, P.Z., A longitudinal study of low back pain in student nurses. *Int. J. Nursing Stud.*, 30(3), 197–212, 1993.

Klein, B., Jensen, R., and Sanderson, L., Assessment of workers' compensation claims for back strains/sprains, *J. Occup. Med.*, 26, 443–448, 1984.

Knibbe, J.J. and Knibbe N.E., The workload on the back during the transfer from the wheelchair to the toilet, in *Locomotion*, Zwaag, The Netherlands, 1990, pp. 1–10.

Knibbe, J.J. and Friele, R.D., Back pain and patient lifting in nurses working in home care, in Hagberg, M., Hofmann, F. Stobel, U., and Westerland, G., Eds., *Occupational Health for Health Care Workers* ECOMED, Landsberg, Germany, 1995, 298–301.

Lagerstrom, M. and Hagberg, M., Evaluation of a 3-year education and training program for nursing personnel at a Swedish hospital, *AAOHN J.*, 45, 83–92, 1997.

Marras, W.S., Davis, K.G., Kirking, B.C., and Bertsche, P.K., A comprehensive analysis of low-back disorder risk and spinal loading during the transferring and repositioning of patients using different techniques. *Ergonomics*, 42(7), 904–926, 1999.

McGuire, T. and Dewar, B.J., An assessment of moving and handling practices among Scottish nurses, *Nursing Standard*, 9(40), 35–39, 1995.

Nelson, M. and Olson, D., Health care worker accidents reported in a rural health care facility, *AAOHN J.*, 44(3), 115–122, 1996.

Owen, B.D., The magnitude of low-back problems in nursing, *West. J. Nursing Res.*, 11(2), 234–242, 1989.

Owen, B.D., Decreasing the back injury problem in nursing personnel, *Surg. Nursing Manage.*, 5(7), 15–16,19–21, 1999.

Owen, B.D. and Garg, A., Patient handling tasks perceived to be most stressful by nursing assistants, in Mital, A., Ed., *Advances in Industrial Ergonomics and Safety I*, Taylor & Francis, Ltd, Philadelphia, 1989, pp. 775–781.

Owen, B.D. and Garg, A., Back stress isn't part of the job, *Am. J. Nursing*, 93, 48–51, 1993.

Owen, B.D., Keene, K., and Olson, S., Patient handling tasks perceived to be most stressful by hospital nursing personnel, *J. Healthcare Saf., Compliance Infect. Control*, 5(1), 19–25, 2000.

Owen, B.D., Keene, K., and Olson, S., An ergonomic approach to reducing back/shoulder stress in hospital nursing personnel: a five year follow up, *Int. J. Nursing Stud.*, 39, 295–302, 2001.

Owen, B.D., Keene, K., Olson, S., and Garg, A., An ergonomic approach to reducing back stress while carrying out patient handling tasks with a hospitalized patient, in Hagberg, M., Hofmann, F., Stobel, U., and Westlander, G., Eds., *Occupational Health for Health Care Workers*, ECOMED, Landsberg, Germany, 1995, 298–301, 1995.

Owen, B.D. and Staehler, K., Approaches to decreasing back stress in home care, *Home Healthcare Nursing Manual*, 21(3), 180–186, 2003.

Owen, B.D., Welden, N., and Kane, J., What are we teaching about lifting and transferring patients? *Res. Nursing Health*, 22, 3–13, 1999.

Personick, M.E., Nursing home aids experience in serious injuries, *Mon. Labor Rev.*, 13, 113–137, 1990.

Pheasant, S., *Ergonomics, Work and Health*, Aspen Publishers, Gaithersburg, MD, 1991, p. 295.

Ronald, L., Yassi, A., Spiegel, J., Tate, R., Tait, D., and Mozel, M., Effectiveness of installing overhead ceiling lifts, *AAOHN J.*, 50(3), 120–127, 2002.

Royal College of Nursing, *RCN Code of Practice for Patient Handling*, London, 1996.

Smedley, J., Egger, P., Cooper, C., and Coggon, D., Manual handling activities and risk of low back pain in nurses, *Occup. Environ. Med.*, 52, 160–163, 1995.

Stubbs, D., Buckle, P., Hudson, M., Rivers, P., and Baty, D., Backing out: nurse wastage associated with back pain, *Int. J. Nursing Stud.*, 23(4), 325–336, 1986.

St. Vincent, M., Lortie, M., and Tellier, C., A new approach for the evaluation of training programs in safe lifting, in Asfour, S., Ed., *Trends in Ergonomics/Human Factors IV*, North-Holland Publishing, Amsterdam, 1987, pp. 847–854.

U.S. Department of Health and Human Services, *Work Practices Guide for Manual Lifting* (DHHS/NIOSH Publication No. 81–122), U.S. Government Printing Office, Washington, D.C., 1981.

U.S. Department of Health and Human Services, *Healthy People 2000: National Health Promotion and Disease Prevention Objectives* (Publication No. PHS 91–50212), Washington, D.C., 1990.

U.S. Department of Health and Human Services, *Healthy People 2000: National Health Promotion and Disease Prevention Objectives: Healthy People 2000 Review.* (Publication No. PHS 98–1256), Washington, D.C., 1997a.

U.S. Department of Health and Human Services, *Ergonomics: Effective Workplace Practice and Programs,* United States Public Health Service, National Institute for Occupational Safety and Health, Washington, D.C., 1997b.

U.S. Department of Health and Human Services, *Healthy People 2010: National Health Promotion and Disease Prevention Objectives,* 2nd ed., U.S. Government Printing Office, (2000). U.S. Department of Health and Human Services, *Ergonomics for the Prevention of Musculoskeletal Disorders: Guidelines for Nursing Homes.* United States Department of Labor, Occupational Safety and Health Administration, Washington, D.C., 2002.

Villeneuve, J., The ceiling lift: an efficient way to prevent injuries to nursing staff, *J. Healthcare Saf. Compliance Inf. Control*, 2(1), 19–23, 1998.

3 A Word about the Nurses' Stories

Anne Hudson

CONTENTS

INTRODUCTION

The stories herein are representative of the thousands of healthcare workers who have suffered irreparable harm from largely preventable injuries while providing care to others, only to find they may not be allowed back to the job they love. Many expressed feeling exposed and vulnerable as they wrote their stories; negative emotions resurfaced as the injured relived their trauma. For they have, indeed, experienced trauma.

Just as repetitively lifting patients can cause cumulative trauma injuries to spinal structures, needless disability, months or years of painful procedures and treatment, stress and uncertainty with the workers' compensation system (which seems to protect the employer and punish the injured worker), and potential rejection as damaged goods may act to inflict cumulative psychological trauma.

The stories included in this book are heart-breaking. It may not be possible to read them dispassionately, realizing that much of the suffering was entirely avoidable. The wish of several story authors to remain anonymous is being respected. It is more important for the injured to tell their stories and to let others know what can and does happen, than it is to identify the individuals.

A recurrent theme is the depression that often accompanies disability and loss of career. It may be difficult to not take personally being needlessly disabled from a preventable injury and then possibly being discarded as if broken and worthless.

Read the stories with sad realization that the disabilities described may have been averted if strict zero-lift policies had mandated the use of lift teams and safe mechanical lift equipment — measures proven effective in preventing most injuries related to lifting and moving patients.

Unless patient handling practices are changed, the stories in this book will continue to repeat themselves, taking a heavy toll in pain and suffering to the injured and loss of valuable healthcare workers to the nation. A preventable disabling back injury could happen to any healthcare worker required to physically lift and move patients. Next time, it may be you, your spouse, parent, child, or friend.

1-56670-631-9/03/$0.00+$1.50

Injured Nurse Story #1: Betrayal in the Temple of Healing

by Litigation Assistant for Injured Workers

I am a litigation assistant on behalf of injured workers. This essay describes the psychological effects of back injury I have observed in nurse clients, and is a call-to-arms for nurses to rise up and claim their rights.

Nurses are naturally nurturing, caring, supportive people. When they hire into a hospital or care facility, they perceive their place of employment as a temple of healing, and their employer as being motivated by a desire for the highest good of all persons within that temple, including the nursing staff.

When a nurse is injured on the job, this perception is shattered when the hospital's workers' compensation defense department denies responsibility for the injury, using the ignoble ruse of "pre-existing injury" or "pre-existing degenerative disc disease" to escape its duty to care for the injured nurse in his or her time of need. The injured nurse experiences a deep and disheartening form of psychological trauma that goes unaddressed in the workers' compensation process or her physical recovery period.

Not only is she physically damaged, she, or he, is financially damaged, emotionally drained, estranged from her employer, her perception of herself as "healer" is challenged by her own slow healing, and worst of all, her perception of her industry as a compassionate healing profession is wrenched through the employer's and insurer's heartless, shameless efforts to avoid responsibility for its own injured employees. The vision of the temple of healing shatters, and the vision of an insurance processing plant appears, where the only things worshipped are profit margins and the bottom line. When the injured nurse becomes aware of the zero lifting techniques and equipment, available world-wide, and yet ignored by her employer, she experiences a rage and frustration that can lead to feelings of despondency and depression. She realizes that the employer chooses to use nurses as throw-away expendables, rather than purchasing zero-lift equipment and training the nurses in its use.

She realizes her back injury was completely preventable. She realizes that it was totally unnecessary for her to lift those hundreds, perhaps thousands, of patients, each one creating micro-trauma in her back, each one contributing to the pain and disability she now experiences, the pain and disability for which the hospital denies responsibility. Often, in the return-to-work/retraining portion of the workers' compensation process, she is offered menial jobs filing chart notes, or offered retraining as a computer keyboarder, when she knows that there are many nursing positions available that do not require lifting, or could be done with the use of zero-lift techniques. She perceives her employer's attempts to avoid creating a position for her dignified return to the nursing profession as abandonment, breach of faith, and betrayal. It would not be healthy or wholesome to describe the horrific financial consequences of long-term denial of financial assistance to back-injured nurses. It is far more important for the back-injured nurse to realize that any depression or despondency she feels sources from refusing to see the truth about her employer.

Rather than feeling like a victim, it is time for back-injured nurses to feel like the angry women who have created beneficial changes in society — the angry suffragettes who got women the vote, the bread and roses marchers who created the concept of weekends for workers, the angry mothers

1-56670-631-9/03/$0.00+$1.50

against drunk drivers who have initiated such beneficial legislation, all the angry groups of workers who rose up and created change in the system that abused them.

It is time for back-injured nurses to allow that depression to become clear-seeing, which then becomes a healthy rage and motivation to change the system.

Injured Nurse Story #2: Preventable

by Anne Hudson

I'm a registered nurse; now, a back-injured registered nurse with a cumulative trauma spinal injury from years of lifting and moving patients. I hope telling the story of how my career as a floor nurse came to a halt will be helpful to others.

I worked at an acute care hospital on medical/surgical, telemetry, and intermediate care units. The patients were generally elderly and acutely ill with a variety of cardiac, medical, and surgical conditions. They were typically heavy-care patients with limited mobility. Many were unable to move themselves up or turn side-to-side in bed, to sit up, stand up, or transfer to the chair or bedside commode without being physically pulled, lifted, or, occasionally, even picked up and carried. Many were both very heavy and too weak to assist, becoming passive dead weight to lift and turn. Confused patients sometimes resisted, increasing the strain. Much heavy lifting was required.

I didn't mind this hard work. I wanted to provide care that would help patients achieve their best possible outcome. This included keeping patients pulled up in bed into the most comfortable and therapeutic position, maintaining mobility by assisting patients up to their walkers and into their bedside chairs for supper, and helping weak patients maintain both their continence and their dignity by getting them up to the bedside commode. Many an evening, as I left work, I felt grateful that I had not been injured.

There was one mechanical lift in the hospital. You never knew which floor or unit it was on and, in order to use it, you had to call and wait for a sling to arrive. Nurses used the lift only under extreme circumstances when they absolutely could not manually move the patient. I saw it used maybe once every 6 or 8 months. We received perhaps a 15-minute in-service on the lift at hiring and were in-serviced when the old lift was replaced with a new one. The lift was used so infrequently that nurses forgot how to use it. I never saw it used for patients who were down on the floor. A

FIGURE 1 Anne Hudson, R.N., one of thousands of nurses disabled and removed from hospital nursing by a preventable spinal injury from lifting and moving patients.

1-56670-631-9/03/$0.00+$1.50

group of staff would gather and do the one-two-three lift from the floor. The mechanical lift wasn't even used when it would appear necessary.

Nurses relied on each other for assistance with lifting and turning patients. Working together, we used the under-axilla method for lifting and transferring, and the draw sheet for turning and pulling patients up in bed. There was no organized method of obtaining assistance. When help was unavailable, we did what we could by ourselves with lifting, turning, and pulling patients up in bed; lifting them from supine to sitting; using the hug lift to raise them to their feet; holding them up while pivoting on their stronger leg for transfer to the chair or commode. There were no friction-reducing products other than a hard plastic sliding board, or plastic garbage bags, for pulling patients bed to stretcher.

In nursing school, we learned to manually lift and move patients, independently and with a partner. During clinicals in the nursing home, we were trained to use the Hoyer for total lifts. After going to work in the hospital, there was no further training on manual patient handling. The annual Safety Fair included a review of body mechanics with the standard instructions: bring the load close to your body, impossible with patients lying in the center of the bed. Bend your knees, lift with your legs, equally impossible with much of the lifting, pushing, and pulling involved in patient care.

I first experienced severe low back and leg pain while walking through my kitchen during a scheduled day off. I had sudden severe, incapacitating pain. I couldn't walk. I couldn't sit. I could hardly move. I could only manage a little shuffle with my feet. I couldn't sit to get into a vehicle and certainly couldn't push the clutch. I couldn't move and couldn't get to the doctor's office. I was in a world of pain. I didn't recognize my pain as severe muscle spasms in response to spinal injury. All I knew was that I had pain like I had never experienced before. A deep severe ache and intense burning settled into my lower back and I had pain and burning into my lower legs and sometimes into my feet.

I thought, "I'll be better tomorrow. This can't be a bad injury. It's just a muscle or something. I'll get better and will be able to go to work on Saturday." I called the doctor's office Friday but the office was closed on Fridays. I kept thinking I would be better and would go to work Saturday. Periodically, the most severe part of the pain partially eased off, giving me hope that I wasn't really injured. But, Saturday, June 3, 2000, I called and reported that I could not come to work because of back pain.

I now had a cumulative trauma injury to lumbar discs and my life as a nurse, as I had known it, was over. Other than a brief unsuccessful attempt a few months later, I have been unable to return to floor nursing.

I went into the hospital on Monday and completed paperwork, including the Workers' Compensation Form 801. Thus began my experience with the workers' compensation system. I also went to my internist that day and left with a diagnosis of lumbar radiculopathy.

Since then, I've been seen by neurologists, orthopaedic surgeons, neurosurgeons, and a chiropractor. I was diagnosed with degenerative disc disease, lumbar strain, and bulging or herniated discs. I've had two MRIs, two discograms, a series of lumbar blocks, and, ultimately, in May 2002, an "ALIF" — anterior lumbar interbody fusion of L4/L5 and L5/S1, with donor bone grafts to replace the discs, and posterior fixation with four titanium screws.

As surgery was being scheduled, workers' compensation closed my claim necessitating filing forms through my attorney to re-open the claim. When the surgery was performed, they called me a "360" because of the anterior and posterior approaches. Fusion of L4/L5 and L5/S1 transfers stress and increases risk to L3/L4 and the sacroiliac joints and can lead to "adjacent segment disease."

Leading up to surgery, I tried a number of conservative measures including over-the-counter and prescription medications, application of heat and cold, analgesic rubs, lumbar supports and cushions, chiropractic, physical therapy, Back School, and an inversion machine — wonderful relief upside down, crushing pain with turning upright.

I have appreciated and benefited much from physical therapy, before and after surgery, with ultrasound, massage, heat, and stretching and strengthening exercises. I've learned the importance

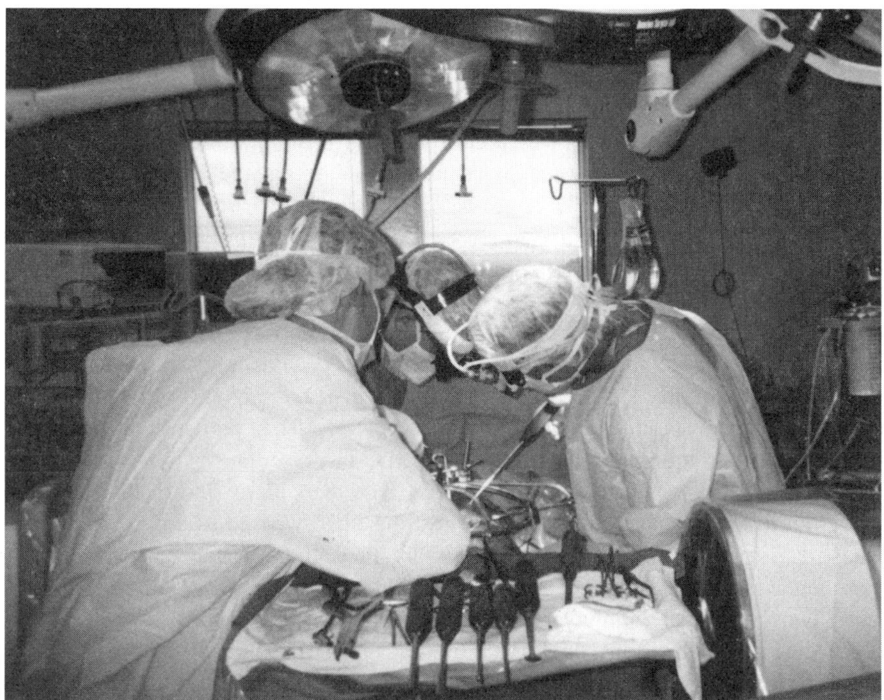

FIGURE 2 The preventable injury led to a 6-hour surgery. Direct medical and compensation costs have exceeded $100,000. Indirect costs may be many times that amount.

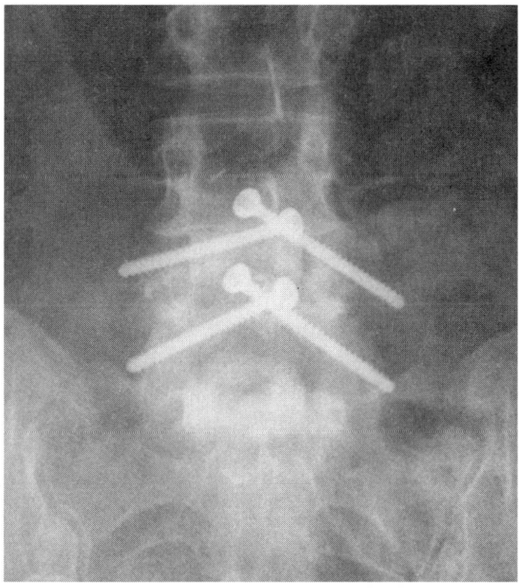

FIGURE 3 Four titanium screws and two cadaver bone grafts cannot replace a healthy spine or repair a nurse's career.

of stretching (I pay if I don't do my stretches!) and I'm stronger, though I still have significant back pain. Since being injured, I recognize that hospital physical therapists may be at increased risk of back injury because nurses frequently summon PT to perform the heaviest lifts. PTs have said they feel like "human cranes" or "a lifting and moving service."

About 4 months into my injury, I saw a chiropractor for the first time in my life. I believe he helped me more single-handedly, by the sheer numbers, than anyone else. He gave me a lift for my right shoe, taught me to avoid sleeping on my stomach by wearing a shirt with large spools rubber-banded onto the front, taught me to sleep side-lying with a pillow between my knees, and gave me ultrasound treatments as well as spinal adjustments. Leaving the office after my first visit, I felt significant pain relief for the first time in 4 months. The chiropractor told me I was having muscle spasms and suggested I get a prescription for a muscle relaxant from my doctor, which I did and that also helped. He suggested I get a discogram because he insisted I had disc pain and not just lumbar strain, and provided the name of an expert in discography. Turns out he was right, too.

About a year into my injury, I had done my homework. I had learned that lift teams, no lift policies, and patient lift equipment had proven effective in preventing nurses' back injuries for at least a decade. I had been a nurse for just 10 years. The technology and methods for preventing nurses' back injuries had been available the entire time I had worked as a nurse.

I requested to speak at the hospital's Back Injury Prevention Task Force meeting. I presented information on the incidence of back injuries to nurses, related medical and compensation costs, ineffectiveness of body mechanics for injury reduction, figures on compressive forces with lifting and moving patients exceeding safe lifting limits established by the National Institute for Occupational Safety and Health (NIOSH), pathophysiology of cumulative trauma to spinal discs, and for reduction of injuries, the necessity of both commitment from the top, for use of safe patient lift equipment backed by a strict no-lift policy, and a leader, such as a Back Safety Nurse. They said, "We've been going for three years. We know all that." What an awful discovery — my injury could have been *prevented.*

"Preventable" became a haunting word. My health and life have been permanently altered, my career has been derailed, and I have not known a pain-free day since suffering a *preventable* injury. On top of that, the hospital offers no permanent light duty for back-injured nurses. And, our nursing contract contains no language for safe patient handling or provision of work for injured nurses.

I've had several types of pain. Rising to my feet in the morning, crushing low back pain would greet me. Deep aching would worsen later in the day. Rather than asking if my back hurt, it would be more accurate to ask which pain was worse at the time and how badly it was hurting — the deep low back ache, burning, stabbing, or shooting pain down one or both legs. I had burning pain in my right lateral lower leg, sometimes in my left. This pain would sometimes go down across the top of my foot, occasionally into my toes. Sometimes I had pain down the backs of my legs into my heels. My legs gave out on me three times, one leg one time, both legs twice. Fortunately, I was in a position where I could grab and hang onto something to keep from collapsing and the weakness in my legs was only momentary. There have only been a few days when I limped badly and I have not needed to use a cane.

Six months post-op, I wish I could say I'm pain free. They say to be patient, that it may be a year, or even two, before I know how I will ultimately feel. I no longer have some of the same types of pain as before surgery. I don't have the same intense burning or crushing low back pain, have very rare stabbing or shooting pain, and have none of a peculiar slicing pain which felt like a horizontal blade across my spine.

I still, however, have almost constant deep aching in my lower back; numbness, pain, and burning in areas of my legs and into my feet; and increased pain in my sacroiliac joints, especially the right side. I still have daily pain to varying degrees. I continue with lifting restrictions and am still limited in many everyday activities. I am glad I had the surgery, though, as I no longer fear losing the use of my legs or becoming incontinent.

Almost everything in my life has been affected by being back-injured. I've been unable to do many things that I formally did with ease. I used to garden, cut grass, do yard work, care for my animals, do housework and laundry with ease. I can no longer flip a mattress or even assist to turn a mattress. I need help changing sheets on the bed. I can't vacuum as before and either let it go, get help, or wait until my analgesic kicks in and then do as much as I can. I'm concerned about potential side effects and don't like taking any kind of medication for pain, over the counter or prescription.

Sitting for any length of time is painful. Driving is painful, because of the vibration and seated position. I've driven over 5000 miles with this injury for multiple out-of-town appointments including one 340-mile round-trip to see two workers' compensation physicians, eleven 240-mile round-trips, and five 420-mile round-trips.

I'm no longer able to pick up small children. As a nurse, I lifted adults. Now, I can't lift a child. After being injured, the first thing that upset me greatly was realizing that I may not be able to pick up a potential grandchild someday.

Sleep has not been the same since I was injured. I only recall one night of what was previously normal sleep for me — one night of going to sleep, being unaware of the passage of time, and waking refreshed in the morning. I'm awake many times every night trying to get comfortable.

At times, I've been unable to put on socks or hose and have put bare feet into slip-on shoes. At times, I've required help from my husband to get dressed. Stairs are painful and I sometimes seek an elevator to avoid taking the stairs. I had to give up my regular-sized purse for a very small one. I can't do laundry as before. I'm unable to pick up and carry a basket with laundry. It's painful to lift wet things out of the washer and to remove things from the dryer. Retrieving items from lower parts of the refrigerator and cabinets causes pain.

I can no longer bend over to wash my hair in the sink, which has always been my preferred method. It has been painful to lift more than one piece at a time of my Buffalo China dinner plates from the cabinet. I can't carry a two-gallon watering can to water plants in the yard. Even picking up one gallon has been painful.

When grocery carts are lodged together in line, the sharp jerk required to separate one has been more than I can handle. I either go to another line of carts or look elsewhere until I find a free one. I feel better in the store pushing a cart and realize that I use it as support like an elderly person might. Dealing with groceries has been a continual distress since being injured. Being on my feet for any length of time increases my pain. I find myself unconsciously saying aloud, "My back hurts" and only realize I've spoken when I notice other shoppers looking at me.

I've gone to the store for certain items and have left without them, like a large bag of cat food or a gallon of milk. I've substituted a smaller size or have just left because it would hurt to put heavy items into the cart and I didn't want to ask for help. It's painful as well to push a shopping cart with any amount of groceries in it. I have sometimes purchased fewer things than I really wanted because of thinking ahead to dealing with the groceries when I got home. If there is the slightest incline exiting the store to the parking lot, I've required help. That is a big distress because I appear to be perfectly fine to any observer. When cashiers ask if I'd like help out, I don't like saying "yes" because it brings home that I don't just want help, I need help. I don't like feeling dependent and needy in public.

It has been wearing to visit multiple physicians and go over and over my injury, my history, and so forth. My history is not all that interesting, but it's mine and I find it most unpleasant to have all these records and forms generated about me. There is now a huge file on my case with multiple strangers reading all about me, having meetings about me, deciding what they are going to do with, or to, me. I feel like a pawn in a game where they know the rules and I don't. But, then, as a pawn, rather than as a player, it is not required to inform or consult me. I feel as though I am there to be moved about by the players. It appears I have been moved from the Asset side to the Liability side of the game board.

I had a ganglion cyst removed from my right wrist a few years ago and have seen other nurses with ganglion cysts on their wrists and hands. I've since discovered that this type of cyst may be the manifestation of overuse injuries and believe it is possible that both ganglion cysts and carpal tunnel syndrome among nurses may be the result of repetitive use, from years of pulling patients up in bed with the rolled-up edge of a draw sheet. This can amount to several hundred or even thousands of pounds every shift.

After being injured, I was permitted modified duty while the decision on my workers' compensation claim was pending. I worked for 60 days in Nursing Education and in the Business Office, receiving RN wages. I managed to work in spite of the injury by changing position frequently, using lumbar support in the chair, and ibuprofen or acetaminophen for pain control during the day. The day that workers' compensation denied my claim, I was dismissed, sent home to use up accumulated leave. There is no permanent light duty.

After being off work for a number of weeks, concerned over the potential loss of my job, I requested a work release from one of the physicians I saw. He released me for full duty wearing a "Sacro-Cinch Support" with metal stays. The brace prevented some bending and twisting but it did nothing to diminish compressive force and pain with lifting. I lasted 3 weeks wearing that brace. A 400+ pound patient I had for 3 or 4 consecutive days finished me off. Job or no job, I could not continue to lift and haul patients.

I went off work again and filed another workers' compensation claim. After a couple of weeks of paper work, figuring out appropriate modified duty, and approval from my physician, I was again permitted modified duty. This time, I worked in Medical Records, again for RN wages. Thirty days later, workers' compensation denied my second claim saying the second injury was related to the first injury, which they had denied as not work-related. I was once more dismissed, sent home to use up accumulated leave.

I received short-term disability insurance payments when my claim was denied by workers' compensation. I was appealing the workers' compensation denial with a court date set a few months ahead. Short-term disability was turned over to long-term disability, which was denied. I then had no income. With exhaustion of accumulated leave, I lost my health insurance. I began making COBRA payments of $332 per month to maintain my health insurance. The cost rose to $502 per month and will increase next month to $938 per month to maintain health insurance under COBRA.

The day of my workers' compensation hearing finally arrived, the day I would appeal denial of my claim. Two nurse witnesses and I testified about the type and amount of heavy lifting required of floor nurses, which I don't believe is very well known. It was ironic that at 7:00 p.m. that evening, the annual hospital Awards Banquet was held where I received my 10-year pin and was one of three to accept, on behalf of the local nurses bargaining unit, the hospital's first ever Team Spirit Award. This was to recognize our union's fund-raising efforts to help a fellow nurse with her young daughter's medical needs. I was off work, without pay, and with a bleak outlook of returning to work at the hospital.

Through negotiations with the hospital, I was permitted 90 more days of modified duty. I worked in the pre-op holding area, where it was a joy to be back working with patients. I, also, helped the department catch up with TB tests, respirator mask fit-testing, and flu shots. I was well-able to perform this work. Studies have shown that injured workers do better when they're kept working. My only restriction all along has been no lifting over 20 pounds. At the conclusion of the 90 days, according to policy, I was again dismissed.

The workers' compensation judge's decision came back in my favor determining that my injury was work-related. I began receiving workers' compensation time loss payments, two thirds of my wages. I did not want time loss payments. I wanted to continue working at the hospital where I loved being a nurse, had good benefits, and a nice start on retirement. This *preventable* back injury has been a huge setback. With our sons through college, this is the time we should be saving for retirement.

The hospital appealed the judge's decision. It took about a year, but the next decision, this time from the workers' compensation board, was also returned in my favor confirming that my injury was work-related. Legal unpleasantries continue as the hospital is appealing one of the disc's involvement with the original injury.

Since being injured, all I have wanted is to keep working as a nurse at the hospital. During the time that I've received time loss payments, my replacement has also been paid. Two of us are being paid but only one is working. The modified duty experiences demonstrated that I could have been working and contributing throughout the time leading up to surgery rather than being a drain on scarce resources.

Workers' compensation insurance carriers might consider reducing time loss and vocational rehabilitation costs by encouraging employers to retain injured nurses — to keep those nurses who are still able to work employed in the facility where they were injured. This may be a logical piece for cost containment.

As I mentioned, workers' compensation closed my claim rather prematurely while surgery was being scheduled. Part of claim closure is evaluation of eligibility for vocational rehabilitation. Workers' compensation offers vocational rehabilitation to those with work-related injuries who, due to their injury, are unable to return to their previous position. In my case, I have been unable to resume floor nursing because of continued lifting restrictions. It was determined that I was eligible for vocational rehabilitation with the following transferable skills:

> Ability to apply nursing and healthcare techniques which include administering injections, medications and treatments, analyzing medical data, patient activity, applying clinical problem solving techniques, applying human anatomy, physiology and biology knowledge. Ability to apply infectious materials procedures, institutional care procedures, interpersonal communication techniques, life support procedures, medical lab techniques, nursing practices and procedures, patient observation and care procedures, personal care procedures, ability to apply sanitation practices to healthcare, apply uniform tests or procedures. Assist in examining and treating patients, ability to chart medical data, collect blood samples, collect clinical data, ability to comprehend body response variations, comprehend composition of drugs, ability to conduct patient assessments, draw blood, maintain medical records, perform routine medical tests, routine medical treatments. Ability to plan and organize work, prepare patient reports, prepare patients for exam and treatments, prepare patients for tests. Ability to read and understand operating and technical manuals, ability to set up patient care equipment, ability to take vital signs, ability to comprehend and use medical terminology, ability to understand drug products, properties and composition of drugs. Ability to utilize pharmacological terminology, ability to apply principles of gerontology, ability to make decisions based on response to care and treatment.

What can I do? All of the hospital-based nursing skills listed above. What can I not do? Lift heavy weights.

Now, if I am ready, willing, and able to perform all of these hospital-based nursing skills right now, without re-training, would it not be more cost-efficient to just put me back to work, performing the skills listed above, in the facility where I was working at injury, rather than spending scarce resources on re-training and continued time loss payments? Just an idea for workers' compensation carriers to consider.

I made numerous requests, verbally and in writing, to work in a variety of capacities that could benefit the hospital and keep me working. Lifting restrictions have been the limiting factor. If I'm unable to return to the rigors of my previous position, it appears I may not be permitted back at all. I have applied for available positions that may have been a good match, and would not routinely require lifting, without success: "We want someone with recent experience." "We hired someone who is working here now on call." "Some patients will need assistance transferring."

Thinking I might need training in an area away from the bedside, I requested to use my earned education funds for a certification course in case management, hoping I might then qualify for a position in the Social Services Department. My unit manager denied the request saying case management was not directly related to my position on the unit and I should be more concerned with keeping up my clinical skills. The director of nurses backed her up.

My experience is not unique. I worked with nurses who complained of chronic back pain and several who had injuries. One nurse asked, "Do you know anyone who's been a nurse for any length of time without a bad back?" Some nursing staff who were injured were able to return, some with, some without surgery. Others disappeared and I never knew what became of them. Before I was injured, it didn't really register with me that what happened to them could happen to me.

Nurses, I hope my story helps you recognize your vulnerability — your health and careers may be on the line. I hope you are inspired to protect yourselves by working toward zero-lift policies, lift teams, and permanent light duty for injured nurses.

Caring for patients and receiving their trust touches nurses deeply. I loved being a hospital floor nurse. Still, my experience as a back-injured nurse has provided the incentive to learn and the motivation to speak out. I'm grateful for the opportunity to be part of the larger effort for reduction of back injuries to healthcare workers.

4 Biodynamics of Back Injury: Manual Lifting and Loads

John D. Lloyd

CONTENTS

INTRODUCTION

Patient handling tasks are often performed under unfavorable conditions and at significant risk to nursing staff. Nursing staff have one of the highest incidences of work related musculoskeletal injuries of all occupations (Magora, 1970; Cust et al., 1972; Leighton and Reilly, 1995; U.S. Department of Labor, 1999, 2001). Back injuries are frequent among patient caregivers, causing personal suffering, lost time, threats to patient safety, workers' compensation expense, and staffing difficulties. Preventive interventions are critically needed to control the hazards and economic burdens associated with performing patient handling and movement tasks (Genaidy et al., 1994).

The four factors that most contribute to musculoskeletal injuries in nursing are: (1) caregiver, (2) load lifted (patient), (3) work environment, and (4) job tasks (Nelson et al., 1997). Efforts to decrease risk of injury related to providing patient care have been largely unsuccessful over the past two decades. Those strategies most frequently cited include (1) education and training in body mechanics, (2) use of standard mechanical lifting aids/equipment, and (3) alterations in lifting techniques. Unfortunately, none of these strategies have been found to be successful in isolation (Daltroy et al., 1997; Girling and Birnbaum, 1988; Harper et al., 1987; Nelson et al., 1997).

SCOPE AND MAGNITUDE OF THE PROBLEM

Musculoskeletal injuries have become so prevalent that they now account for more than 62% of workers compensation claims (Haag, 1992). Across occupations the highest incidence of work related back problems have been found in heavy industry workers and nursing staff (Magora, 1970; Cust et al., 1972). Nursing personnel ranked fifth among all occupations (Klein et al., 1984). Worker injury claim rates from the Bureau of Labor Statistics were 8.6 cases per 100 full-time workers across all occupations, but for nursing facility staff that figure nearly doubled to 15 cases per 100 workers (Gold, 1994). The annual prevalence of all cases of back pain in nursing was reported to be 431 per 1000 at risk and the annual incidence of new cases of back

1-56670-631-9/03/$0.00+$1.50

pain was 77 per 1000 at risk (Stubbs et al., 1983). In 1989, Owen asserted that the back injury problem in nursing may be even greater than published statistics indicate, given that more than 80% of the nurses who had episodes of occupationally related back problems did NOT fill out an incident report. Incidence rates continue to climb: from 1980 to 1990 incidence of back injuries increased over 40% (Fragala, 1992).

The Spine

The human spine is a flexible column of 24 moveable vertebrae (7 cervical, 12 thoracic, 5 lumbar) plus the sacrum and coccyx. Each vertebra has two parts: the vertebral body in the front and the neural arch behind. These form the vertebral column through which runs the spinal cord. Cartilaginous endplates cover the end surfaces of the vertebral bodies and serve as the upper and lower surfaces of the intervertebral discs, which separate the vertebral bodies and act as shock absorbers. An annulus, in the form of concentric layers of fibrous material, surrounds the disc nucleus. The nucleus of each disc absorbs fluids from surrounding tissues to maintain equilibrium.

When tension is placed on the spine, due to posture and/or loading, fluid is slowly forced from the intervertebral discs, the annulus of which may then be more prone to tearing. Eventually, the nuclear material may become extruded, pressing on nerves from the spinal cord, causing intense pain in the distribution of these nerves. This is called a prolapsed intervertebral disc.

Back Injury Mechanics

Biomechanical logic suggests that damage occurs to a structure when the imposed loading exceeds the structure's mechanical tolerance. Loading characteristics and tolerance levels can be influenced by physiological responses. The musculoskeletal structures of the low back may be influenced by either adaptation to or intensification of the load. The tolerance may be mediated by pain responses or discomfort. If the loading of the structure exceeds tolerance limits, then the situation can result in injury. Numerous biomechanical workplace evaluations have demonstrated a positive correlation between increased biomechanical loading and increased risk for work-related back disorder (Chaffin and Park, 1973; Punnet et al., 1991; Marras et al., 1995; Norman et al., 1998).

Damage to the vertebral cartilaginous endplates may lead to back problems in workers. Constriction of the endplate nutrient supply has been found to result in damage to the disc and disruption of the spinal function, which can lead to low back pain (Moore, 2000). Studies have shown that the endplate is the first structure to be injured when the spine is loaded (Brinkmann et al., 1988; Callaghan and McGill, 2000). The tolerance of the endplate decreases 30 to 50% with repetitive loading. Shear loading in the anterior-posterior plane is also responsible for producing endplate damage (Callaghan and McGill, 2000).

The intervertebral disc itself is subject to direct damage with sufficient loading. Herniation may occur under compression or when the spine is flexed excessively. Complex spinal postures can also produce disc herniation (Adams and Hutton, 1982, 1985). Lotz et al. (1998) demonstrated that compressive loading of the intervertebral disc can lead to degeneration and that the pattern of response is consistent with a dose-response relationship that is central to the idea of cumulative trauma. In general, the issue of cumulative trauma is significant for low back pain causality in the workplace.

Injury Risk Characteristics

Numerous studies have explored *characteristics of the nurse* that affect risk of injury. The underlying assumption of this research is that staff could be screened for employment or placed in jobs based on level of risk. This approach, viewed by many as discriminatory, has not been successful. Personal risks identified include level of fitness (Legg, 1987), obesity (Patenaude and Sommer, 1987; Gold, 1994; Lagerstrom et al., 1995), genetics (Gold, 1994), height (Dehlin et al., 1976) muscular strength (Kilbom, 1988), age (Lavsky-Shulan et al., 1985; Kelsey and Golden, 1988; Lagerstrom et al., 1995), and stress (Hawkins, 1987). Nurses with a previous history of back injury are deemed at higher risk for re-injury (Fuortes et al., 1994; Stubbs et al., 1983). Some health-related behaviors and habits might to some extent confound associations between occupation and low back pain, including drug/alcohol consumption (Manning et al., 1984; Bigos et al., 1986) and cigarette smoking (Kelsey, 1975; Frymoyer et al., 1980; Frymoyer et al., 1983; Kelsey et al., 1984; Heliovaara et al., 1987).

Characteristics of the load lifted affect risk of injury. Unlike industry, nursing cannot selectively modify the load size and shape to promote safe manual handling. The problem of lifting a patient extends beyond overcoming a heavy weight (Bell, 1984; Garg et al., 1992). A patient's unique size, shape, physical impairments, and balance deficiencies affect the way that patient handling tasks can be carried out. Furthermore, fluctuations in the patient's physical condition, level of fatigue, and cognitive functioning make it difficult for the nurse to gauge the level of cooperation and assistance a patient will offer at any given point in time. Patients are unpredictable and may suddenly become combative, resist efforts, or go limp during a transfer, causing the nurse to lose balance and/or make sudden, unexpected movements.

Several *characteristics of the work environment* also affect risk of injury, including the nursing practice setting, role, space, lack of appropriate well-maintained equipment, and under staffing. There is greater risk of injury for nurses in nursing homes, geriatric units, and spinal cord injury units than in general hospital units (Valles-Pankratz, 1989; Fragala, 1992; Jensen 1987). Lynch and Freund (2000) found equipment and environmental factors were important contributors to patient transfer behaviors; a reason that training alone is not sufficient for prevention. While some studies identified nursing assistants (NAs) to be at greater risk of injury than registered nurses (RNs) or licensed practical nurses (LPNs) (Personick, 1990; Fuortes, 1994), this finding is likely related to level of exposure, since nursing assistants typically provide much of the manual labor on a unit and therefore have increased exposure to risk. In a study conducted at the James A. Haley Veterans' Hospital in Tampa, FL, no significant differences were found in injury rates between RNs, LPNs, and NAs when controlled for time spent in providing direct patient care (Nelson, 1996). Furthermore, Nelson found that risk increases when optimum body postures cannot be assumed due to space limitations or equipment deficits. These realities interfere with the nurse's ability to adopt optimal body mechanics. Much of the problem of back injuries in nursing facilities has been blamed on a lack of appropriate equipment in working condition and understaffing (Gold, 1994). Larese and Fiorito (1994) found that units with low nurse-to-patient ratios had more back injuries than units with higher ratios. It is generally accepted that some of the lifts and heavy patient care tasks require two to three staff to accomplish safely, staffing deficits make this teamwork difficult and nurses often attempt these tasks alone. A study by Marras et al. (1999) found that even when two persons used a draw sheet to transfer a patient, spinal loads were still relatively high, which necessitates the use of lifting equipment.

The nature of patient handling tasks in nursing practice predisposes nursing staff to risk. *Characteristics of job tasks* that present significant exposure to musculoskeletal injuries include the following: reaching and lifting loads far from the body; lifting heavy loads; twisting while lifting (see Figure 4.1); unexpected changes in load demand during the lift; reaching low or high to begin a lift (see Figure 4.2); and moving a load a significant distance. Lifting patients is the most frequent reason for work related back pain in healthcare (Ferguson, 1970; Cust et al., 1972; Dehlin et al., 1976; Bell et al., 1979; Stubbs et al., 1981; Videman et al., 1984; Harber et al., 1985; Owen 1985; Jensen, 1985; Greenwood 1986; Williamson et al., 1988). A few studies have quantified levels of biomechanical stress induced by patient lifting and transferring tasks (Stubbs et al., 1983; Gagnon et al., 1986; Torma-Krajewski 1987; Garg et al., 1991; Garg et al., 1992). The risk for lifting injury increases for nurses who hold patients away from the body while lifting and when bending and twisting during patient handling tasks (Andersson, 1981; Kelsey et al., 1984). Such awkward postures frequently occur during bathing and feeding and are exacerbated by sustained stretching and reaching (Damkot et al., 1984) or postural stress (Baty and Stubbs 1987, Garg et

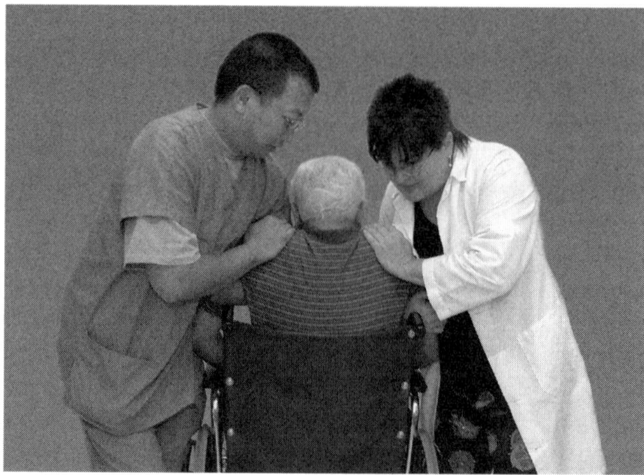

FIGURE 4.1 Example of twisting while lifting: lifting a patient up in a chair.

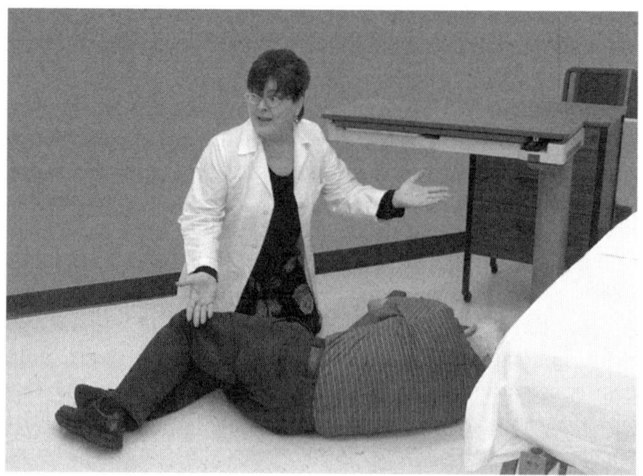

FIGURE 4.2 Example of reaching low to begin a lift: lifting a patient up from the floor.

al., 1991). Marras et al. (1999) found that the greatest risk of injury was associated with one-person transferring techniques. Most evidence indicates that failing to bend the knees while lifting is also harmful (Kelsey et al., 1984), although this does not apply when lifting a patient on a horizontal plane, such as transferring a patient from a bed to a stretcher. Sudden maximal effort from unexpected events, such as preventing a patient from falling, is also associated with high risk for injury (Magora, 1973; Molumphy et al., 1985).

It is generally accepted that the ability to get close to the patient, keep the back in good alignment, flex the knees, and keep the feet apart with one foot in the direction of the move so as not to rotate the spine are all considered important. Transfers are safer when a gentle rocking motion is used to provide the kinetic energy needed so that a pulling rather than lifting action can be used to transfer (Owen and Garg, 1990). Additional biomechanical evaluations are needed to address optimal lifting and patient handling techniques for caregivers and nursing staff.

The etiology of musculoskeletal discomfort and injury in nursing personnel is multifactorial, with host, agent, and environmental factors all playing a part. Host factors include personal characteristics of the caregiver, while agents include characteristics of the load lifted (for example, level of cooperation), as well as the task performed (for example, frequency and duration).

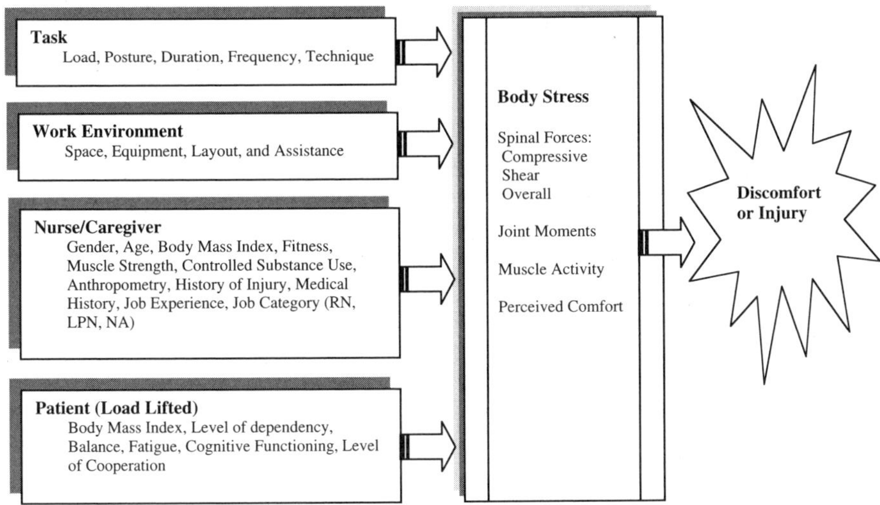

FIGURE 4.3 Injury risk characteristics: conceptual model.

Environmental factors include not only the physical layout of the patient care space, but also the amount of assistance available to the caregiver.

The following conceptual model was derived from the scientific literature and illustrates the effect of the above-discussed characteristics on risk of injury (Institute of Medicine, 2001; National Institute for Occupational Safety and Health, 1997; U.S. Department of Labor, 1999; U.S. Department of Labor, 2000) (see Figure 4.3).

Biomechanical Evaluations

The biomechanics of lifting and handling loads provides a scientific basis as to the development of musculoskeletal injuries in high-risk nursing tasks. Two different biomechanical considerations must be met to achieve safe levels for patient handling: stress to the low back should not exceed "safe limits," and physical demands of the tasks should not exceed workers' strength, range-of-motion and reach capabilities. Workplace assessments should not be limited to spinal force measurement. Gross (1988) declared that the following measurement variables should be addressed during an ergonomic work place assessment: dynamic working postures, balance, forces applied to specific body parts, duration, task frequency, loads handled, cardiovascular stressors, and production demands. "Dynamic" implies a continuous measurement of motion of the body through space. A task-by-task assessment allows for a greater level of sensitivity and precision.

Biomechanical Evaluations in Nursing

Few studies have evaluated the biomechanics of nursing practice, most within the past 7 to 8 years. During that time, methods for data collection have improved substantially, progressing from observation and still photography (Dehlin and Lindberg, 1975; St. Vincent et al., 1987) to two-dimensional videotaping (Takala and Kukkonen, 1987; Owen and Garg, 1991; Garg and Owen, 1992; Garg et al., 1992). A recent study concluded by Nelson et al. (in press) used three-dimensional electromagnetic tracking technology (HumanTRAC) to evaluate the biomechanics of high-risk nursing tasks in a laboratory. As the complexity of data capture systems improves, we are able to more accurately calculate the true biomechanical impact on the musculoskeletal system.

In an earlier study (Dehlin and Lindberg, 1975), the lifting burden of a nursing aide in a geriatric ward was determined using a force plate. Three lifting operations were performed for patients needing lifting assistance. The tasks were photographed at the same time as the forces were recorded. The lifting burden during nursing tasks often equaled or exceeded the recommendations for maximum weight loads during different types of lifts. The lifts were often performed under unfavorable conditions and seldom with an ideal lifting technique.

In a study conducted in 1987, nursing staff were biomechanically analyzed during performance of three bed-related and two transfer activities of geriatric patients (St. Vincent et al., 1987). Across 1400 observations, a stressful bent back posture was identified 90% of the time. Further analysis revealed two commonalties of inappropriate use: (1) If proper posture causes sacrifice of balance during a transfer, the correct technique was discarded, and (2) If the handling distance (body-to-body center of gravity) is shorter despite a bent back, the wrong body mechanics were used.

Data collection methods progressed from observation and still photography to two-dimensional videotapes. In a study conducted by Takala and Kukkonen (1987) the association between different lifting practices and low back pain were investigated in five hospitals. A total of 143 RNs, LPNs, and NAs participated. Video tapes were analyzed using the following criteria: type of lift, total time, loading time, bending time, load reduction, space, type of lifting aids used, and the number and compatibility of lifters. Mechanical lifts were regularly used on only one of the seven wards. Lifting with a mechanical lift was slower than without an aid, but the total extra time needed for their use accounted for only 3 to 6% of an eight-hour work shift. Stooped and twisted trunk positions occurred less often when lifting aids were used than without an aid, however, some nurses adopted poor spine loading postures even when using lifting aids. The issue of time for task completion with various techniques emerged as a critical finding in this study.

Garg and Owen have conducted several studies evaluating the biomechanics of lifting techniques in nursing practice. A prospective epidemiological study was conducted in two units with 57 nursing assistants (NAs) in a nursing home. This study demonstrates the efficacy of an ergonomic intervention strategy to reduce back stress in nursing personnel (Garg and Owen, 1992). The study was conducted over 4 years and was divided into the following phases: (1) the identification of 16 patient-handling tasks perceived to be most stressful by the NAs (Owen and Garg, 1989); (2) an ergonomic evaluation of the work performed by NA prior to the introduction of change (Garg et al., 1992); (3) a pilot study to identify and locate assistive devices, to establish criteria for their selection, and to perform preliminary trials of these devices (Owen and Garg, 1990); (4) a laboratory study to select patient-handling devices that were less stressful than existing methods in the nursing home (Garg et al., 1991); (5) the introduction of selected devices in the nursing home and training of NAs in their use with patients (intervention); (6) postintervention measurement of back injury incidence and severity rates, acceptability rates, biomechanical task demands, and perceived level of physical stress.

Owen and Garg (1990) identified 16 patient handling tasks perceived by nursing staff to be most stressful. In rank order these included: (1) transferring client from toilet to chair, (2) transferring client from chair to toilet, (3) transferring client from chair to bed, (4) transferring client from bed to chair, (5) transferring client from bathtub to chair, (6) transferring client from chair lift to chair, (7) weighing a client, (8) lifting a client up in bed, (9) repositioning a client in bed side to side, (10) repositioning a client in a chair, (11) changing an absorbent pad, (12) making a bed with a client in it, (13) undressing a client, (14) tying supports, (15) feeding a bed ridden client, and (16) making a bed while the patient is not in it.

An ergonomic evaluation of 7 of the above 16 high-risk tasks was conducted. The tasks selected were all lifting tasks. In the laboratory study conducted by Owen and Garg (1991), six NAs performed eight different methods for performing each of the selected high-risk tasks to determine which methods were most effective and least stressful. A two-dimensional biomechanical analysis of the different methods was conducted. No significant differences between the methods in relation to degrees of trunk flexion, trunk lateral bending, or trunk rotation were identified. However the four methods involving pulling actions required lower forces than the methods involving lifting. Based on the static strength simulation, 81% of female workers would be capable of transferring patients from the toilet to wheelchair using the walking belt two person method; 74% would be capable of using the gait belt, two person method; but only 43% would be capable of transferring with the traditional lifting method. In their subjective data analysis, significant differences were noted between techniques and between the three lift transfers, in terms of patient comfort and security ratings. The amount of time required for each patient handling task was important.

A pilot study was then conducted to investigate the feasibility of using various patient handling methods and assistive devices. Each subject practiced the technique until they were comfortable. Following task performance the participants were asked to rate their perceived exertion for the shoulder, upper back, lower back, and whole body. Mechanical data were collected, including body angles. These angles were verified from videotapes. Pulling forces were measured by attaching a hand forced dynamometer to the transfer belt or sling. A static biomechanical model was used to estimate erector spinal force, compressive and shear forces at L5/S1 disc, moments on the L5/S1 disk in three dimensions, and the percentage of the female population that would be capable of performing each method based on static strength. The estimates of compressive force to the L5/S1 disk according to percentile of client weight were all above the action limit permitted as safe (NIOSH, 1981), but below the maximum permissible limit.

Nelson et al. recently completed a definitive biomechanical study of nine "high-risk" patient handling tasks that involve vertical lifting, lateral transfers, frequent bending, or sustained awkward positions (in press). Building upon preceding work (Dehlin and Lindberg, 1975; St. Vincent et al., 1987; Takala and Kukkonen, 1987; Owen and Garg, 1991; Garg and Owen, 1992; Garg et al., 1992; Nelson, 1996), their objective was to scientifically identify patient handling and movement strategies that would reduce the incidence and severity of occupational musculoskeletal injuries in nursing.

Participants included RNs, LPNs, and NAs, each with a minimum of 6 months experience and a job description that included a minimum of 80% direct patient care responsibilities. A total of 134 participants, comprised of 71 subjects in the control group and 63 in the intervention group, completed this study.

Interventions, developed by an expert panel, included engineering solutions (e.g., equipment) or administrative solutions (e.g., changes in work method to reduce the level of exposure to a stressful task).

Data were collected using a three-dimensional electromagnetic tracking system (HumanTRAC) (see Figure 4.4), surface electromyography (EMG), and questionnaires that were used to capture demographic data, anthropometric measurements, and perceived comfort.

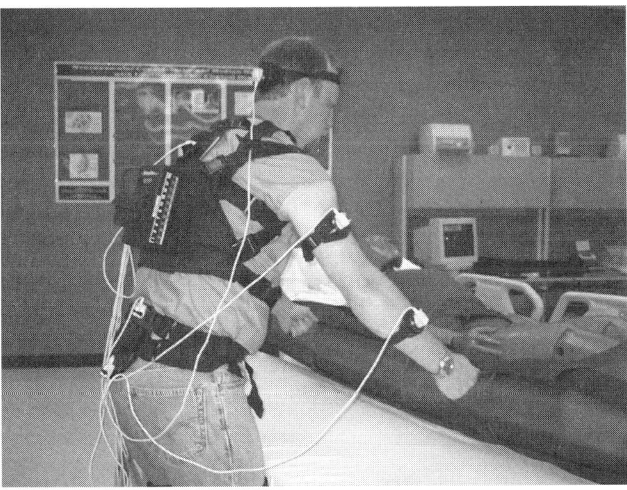

FIGURE 4.4 Subject performing a patient handling task as measured by HumanTRAC. Participant is performing a bed-to-stretcher lateral transfer task. Throughout task execution, postural demands were acquired through real-time measures of human motion using HumanTRAC. This system measured the position of sensors placed on body segments, which are then mapped onto a computer-generated dynamic model of the human form.

Based upon the findings of their study, the researchers concluded that:

Ceiling-mounted patient lifts reduce the risk for musculoskeletal injuries in nursing staff. Mechanical lifting devices of any type have been shown to be far safer for both nurses and patients. However, several limitations interfere with their use in practice, including difficulty using in confined spaces, the extra time required, lack of accessibility or availability, difficulty using and storing, and poor maintenance (Fragala, 1993). Many of these barriers can be eliminated through the use of ceiling-mounted lifts over each patient bed. Nelson et al. (in press) discovered that using a ceiling-mounted lift significantly decreased spinal moments by 79% when compared to operation of a floor-based lift. The main disadvantage of the ceiling mounted lifts is cost for purchase and installation.

To be accepted into clinical practice, interventions must be perceived by staff to be advantageous. Participants rated tasks using the ceiling-mounted lift, friction-reducing device, and stretcher that converts to a chair as more comfortable to perform than the associated control tasks. It is therefore likely that these engineering interventions will have high acceptance and adoption.

National Institute for Occupational Safety and Health (NIOSH) guidelines for safe lifting should not be applied to nursing tasks. NIOSH (NIOSH, 1981; Waters et al., 1993) provided the scientific basis for safe practices for lifting and handling in the U.S. An equation was developed for calculating appropriate lifting loads under varying conditions. Under ideal conditions this equation sets the maximum recommended weight limit at 51 pounds. Based on population strength characteristics, the equation is applicable to more than 90% of men and at least 75% of women.

For nursing care, a number of the standard conditions of the NIOSH formula do not apply. Firstly, a patient who may oppose or make unexpected movements is considered an unstable load and therefore specifically disqualifies utilization of the NIOSH revised lofting equation (Waters et al., 1994). Furthermore, the work areas in which patient handling tasks are often performed are quite restricted due to bathroom fixtures and hospital furniture (Waters et al., 1994).

Nevertheless, several biomechanical studies have evaluated nursing tasks against NIOSH lifting guidelines. Garg and Owen (1992) computed spinal compression forces of 3600 to 4715 N during the execution of typical patient transfer tasks (bed to chair/bed to commode/chair to commode), which is considerably higher than recommended spinal tolerances. Various two-person lifting techniques evaluated by Wilkelmolen et al. (1994) also had substantial static compressive loads, each greater than 3315 N.

Unlike manufacturing operations, nursing tasks involve a predominantly horizontal component to load transfer, which exert shear stresses on the spine in the anterior-posterior and lateral planes. An analogy can be drawn between the spine and a column of bricks; the column is able to withstand extraordinary forces directed along the length of the column, however, a minor force applied perpendicularly could cause the column to fail. This proposes that shear forces in the anterior-posterior and lateral forces are a major contributor to back injuries among nursing professionals.

Marras et al. (1999) conducted a laboratory evaluation of nursing lifting and transferring tasks. It was determined that neither one-person nor two-person transfer techniques satisfy reasonable tolerance limits. Results of maximum lateral shear, anterior-posterior shear and compressive forces for various one-person patient-handling tasks are presented in Figure 4.5. Compressive and shear tolerance limits are represented for clarity.

In all cases, an appropriate administrative or engineering ergonomic intervention strategy must be devised to substantially reduce spinal forces, thereby reducing the risk of injury. Solutions that minimize anterior-posterior and lateral shear forces acting on the spine, without translating those forces to other planes or joints, are urgently required.

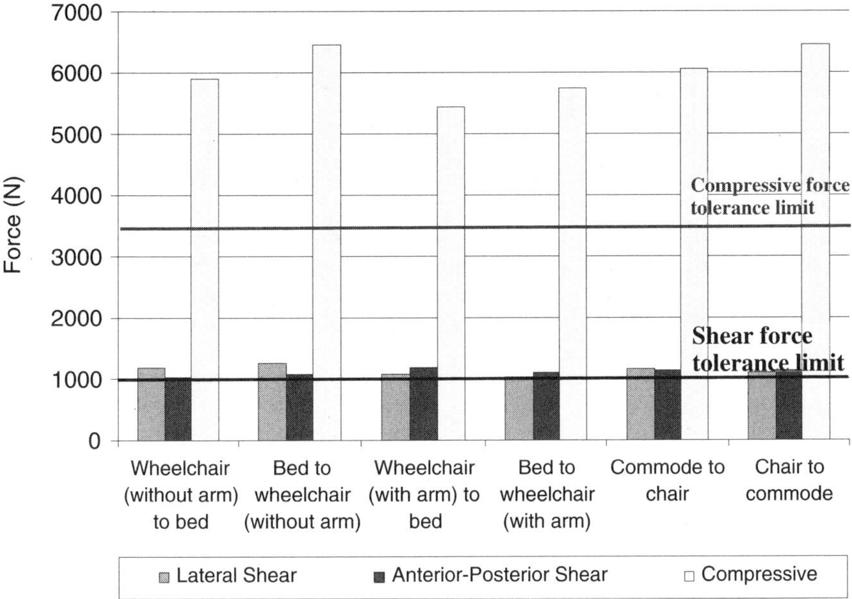

FIGURE 4.5 Maximum lateral shear, anterior-posterior shear and compressive forces presented for one-person patient handling tasks. (From Marras, W.S., Davis, K.G., Kirking, B.C., and Bertsche, P.K., A comprehensive analysis of low back disorder risk and spinal loading during the transferring and repositioning of patients using different techniques, *Ergonomics*, 42(7), 904–926, 1999. With permission.)

The effective utilization of friction reducing devices for the lateral transfer of patients is an excellent example of a recommendation that minimizes shear forces acting on the lumbar spine without translating those forces to other planes or joints. Using the traditional draw-sheet technique, caregivers experience considerable compressive and anterior/posterior forces at the L5/S1 disk. Selection and effective utilization of a quality friction reducing technology can substantially reduce these forces without translation to other planes or joints (Lloyd and Baptiste, in press).

REFERENCES

Adams, M.A. and Hutton, W.C., Prolapsed intervertebral disc: a hyperflexion injury, *Spine,* 7(3), 184–191, 1982.

Adams, M.A. and Hutton, W.C., Gradual disc prolapse, *Spine,* 10(6), 524–531, 1985.

Andersson, G., Epidemiologic aspects on low-back pain in industry, *Spine*, 6, 53–60, 1981.

Baty, D. and Stubbs, D., Postural stress in geriatric nursing, *Int. J. Nursing Stud.*, 24(4), 339–44, 1987.

Bell, F., Dalgity, M.E., Fennell, M.J., and Aitken, R.C., Hospital ward patient-lifting tasks, *Ergonomics*, 22(11), 1257–1273, 1979.

Bell, F., *Patient Lifting Devices in Hospitals*, Croom Helm, London, 1984.

Bigos, S., Spengler, D., Martin, N., Zeh, J., Fisher, L., Nachemson, A., and Wang, M.H., Back injuries in industry: a retrospective study. II, Injury factors, *Spine*, 11(3), 246–241, 1986.

Brinkmann, P, Biggemann, M., and Hilweg, D., Fatigue fracture of human lumbar vertebrae, *Clin. Biomechanics*. 3(Suppl 1), S1–S23, 1988.

Callaghan, J., McGill, S.M., Intervertebral disc herniation: studies on a porcine model exposed to highly repetitive flexion/extension motion with compressive force, *Clin. Biomechanics*, 16(1), 28–37, 2000.

Chaffin, D.B. and Park, K.S., A longitudinal study of low-back pain associated with occupational weight lifting factors. *Am. Ind. Hyg. Assoc. J.,* 34(12), 513–525, 1973.

Cust, G., Pearson, J., and Mair, A., The prevalence of low back pain in nurses, *Int. Nursing Rev.,* 19(2), 169–179, 1972.

Damkot, D., Pope, M., Lord, J., and Frymoyer, J., The relationship between work history, work environment and low-back pain in men, *Spine,* 9, 395–399, 1984.

Daltroy, L.H., Iversen, M.D., Larson, M.G., Lew, R., Wright, E., Ryan, J., A controlled trial of an educational program to prevent low back injuries, *New Engl. J. Med.*, 337, 322–328, 1997.

Dehlin, O. and Lindberg, B., Lifting burden for a nursing aide during patient care in a geriatric ward, *Scand. J. Rehabil. Med.,* 7, 65–72, 1975.

Dehlin, O., Hedenrud, B., and Horal, J., Back symptoms in nursing assistants in a geriatric hospital, *Scand. J. Rehabil. Med.*, 8(2), 47–53, 1976.

Ferguson, D., Strain injuries in hospital employees, *Med. J. Aust.,* I, 376–379, 1970.

Fragala, G., Implementing ergonomic approach promotes workplace safety, *Provider,* 18(12), 31–34, 1992.

Fragala, G., Injuries cut with lift use in ergonomics demonstration project, *Provider,* 19(10), 39–40, 1993.

Frymoyer, J.W., Pope, M.H., Costanza M.C., Rosen, J.C., Goggin J.E., and Wilder, D.G., Epidemiologic studies of low-back pain. *Spine*, 5(5), 419–423, 1980.

Frymoyer, J., Pope, M., Clement, J., Wilder, D., MacPherson, B., and Ashikaga, T., Risk factors in low back pain: an epidemiologic survey, *J. Bone Joint Surg. Am.,* 65, 213–218, 1983.

Fuortes, LJ., Shi, Y., Zhang, M., Zwerling, C., and Schootman, M., Epidemiology of back injury in university hospital nurses from review of workers' compensation records and a case-control survey, *J. Occup. Med.*, 36(9), 1022–1026, 1994.

Gagnon, M., Sicard, C., and Sorois, J., Evaluation of forces on the lumbro-sacral joint and assessment of work and energy transfers in nursing aides lifting patients, *Ergonomics*, 29, 407, 1986.

Garg, A., Owen, B., Beller, D., and Banaag, J., A biomechanical and ergonomic evaluation of patient transferring tasks: bed to wheelchair and wheelchair to bed, *Ergonomics*, 34(3), 289–312, 1991.

Garg, A. and Owen, B., Reducing back stress to nursing personnel: an ergonomic intervention in a nursing home, *Ergonomics*, 35(11), 1353–1375, 1992.

Garg, A., Owen, B., and Carlson, B., Ergonomic evaluation of nursing assistants' jobs in a nursing home, *Ergonomics*, 35(9), 979–995, 1992.

Genaidy, A., Davis, N., Delgado, E., Garcia, S., and Al-Herzalla, E., Effects of a job-simulated exercise programme on employees performing manual handling operations, *Ergonomics*, 37(1), 95–106, 1994.

Girling, B. and Birnbaum, R., An ergonomic approach to training for prevention of musculoskeletal stress at work, *Physiotherapy,* 74(9), 479–483, 1988.

Gold, M., The ergonomic workplace: charting a course for long-term care, *Provider,* 20 (2), 20–22, 23, 26, 1994.

Greenwood, J., Back injuries can be reduced with worker training, reinforcement. *Occup. Health Saf.*, 26–29, 1986.

Gross, C., Ergonomic workplace assessments are first step in injury treatment, *Occup. Health Saf.,* 57(5), 16, 18–19, 84,1988.

Haag, A., Ergonomic standards, guidelines, and strategies for prevention of back injury, *Occup. Med.: State of the Art Reviews*, 7(1), 155–165, 1992.

Harber, P., Billet, E., Gutowski, M., SooHoo, K., Lew, M., and Roman, A., Occupational low-back pain in hospital nurses, *J. Occup. Med.*, 27(7), 518–524, 1985.

Harper, P., Shimozaki, S., Gardner, G., Billet, E., Vojtechy, M., and Kanim, L., Importance of non-patient transfer activities in nursing-related back pain: II. Observational study and implications, *J. Occup. Med.*, 29, 971–974, 1987.

Hawkins, L., An ergonomic approach to stress, *Int. J. Nursing Stud.,* 24(4), 307–318, 1987.

Heliovaara, M., Knekt, P., and Aromaa, A., Incidence and risk factors of herniated lumbar intervertebral disc or sciatica leading to hospitalization, *J. Chronic Dis.*, 40(3), 251–258, 1987.

Institute of Medicine, *Musculoskeletal Disorders and the Workplace: Low Back and Upper Extremities*, National Academy Press, Washington, D.C., 2001.

Jensen, R., Events that trigger disabling back pain among nurses, Proceedings of the 29th Annual Meeting of the Human Factors Society, Human Factors Society, Santa Monica, CA., 1985.

Jensen, R., Disabling back injuries among personnel: research needs and justification, *Res. Nursing & Health*, 10, 29–38, 1987.

Kelsey, J., An epidemiological study of acute herniated lumbar intervertebral discs, *Rheumatol. Rehabil.,* 14, 144–159, 1975.

Kelsey, J., Githens, P., White, A., Holford, T., Walter, S., and O'Connor, T., An epidemiological study of lifting and twisting on the job and the risk for acute prolapsed lumbar intervertebral disk, *J. Orthopedic Res.*, 2 (1), 61–66, 1984.

Kelsey, J., and Golden, A., Occupational and workplace factors associated with low back pain, *Occup. Med.: State of the Art Reviews*, 3(1), 7–16, 1988.

Kilbom, A., Isometric strength and occupational muscle disorders, *Eur. J. Appl. Physiol. Occup. Physiol.*, 57(3), 322–326, 1988.

Klein, B.P., Jensen R.C., and Sanderson, L.M., Assessment of workers' compensation claims for back strains/sprains. *J. Occup. Med.*, 26(6), 443–448, 1984.

Lagerstrom, M., Wenemark, M., Hagberg, M., and Hjelm, E.W., Occupational and industrial factors related to musculoskeletal symptoms in five body regions among Swedish nursing personnel, *Int. Arch. Occup. Environ. Health*, 68, 27–35, 1995.

Larese, F. and Fiorito, A., Musculoskeletal disorders in hospital nurses: a comparison between two hospitals, *Ergonomics*. 37(7), 1205–1211, 1994.

Lavsky-Shulan, M., Wallace, R.B., Kohout, F.J., Lemke, J.H., Morris, M.C., and Smith, I.M., Prevalence and functional correlates of low back pain in the elderly: the Iowa 65+ Rural Health Study, *J. Am. Geriatrics Soc.*, 33(1), 23–28, 1985.

Legg, S., Physiological ergonomics in nursing, *Int. J. Nursing Stud.*, 24(4), 299–305, 1987.

Leighton, D.M. and Reilly, T., Epidemiological aspects of back pain: The incidence and prevalence of back pain in nurses compared to the general population, *Occup. Med. (London)*, 45(5), 263–2678, 1995.

Lloyd, J.D. and Baptiste, A., Biomechanical evaluation of friction reducing devices for lateral patient transfers. *J. Rehabil. Res. Dev.*, in press.

Lotz, J.C., Colliou, O.K., Chin, J.R., Duncan, N.A., and Liebenberg, E., Compression-induced degeneration of the intervertebral disc: an in vivo mouse model and finite-element study, *Spine*, 23(23), 2493–2506, 1998.

Lynch, R.M. and Freund, A., Short-term efficacy of back injury intervention project for patient care providers at one hospital, *AIHAJ*, 61(2), 290–294, 2000.

Magora, A., Investigation of the relation between low back pain and occupation, *Ind. Med.*, 39(11), 31–37, 1973.

Manning, W.G., Leibowitz, A., Goldberg, G.A., Rogers, W.H., and Newhouse, J.P., A controlled trial of the effect of a prepaid group practice on use of services, *New Engl. J. Med.*, 310(23), 1505–1510, 1984.

Marras, W.S., Lavender, S.A. Biomechanical risk factors for occupationally related low back disorders, *Ergonomics* 38(2), 377–410, 1995.

Marras, W.S., Davis, K.G., Kirking, B.C., and Bertsche, P.K., A comprehensive analysis of low back disorder risk and spinal loading during the transferring and repositioning of patients using different techniques, *Ergonomics*, 42(7), 904–926, 1999.

Molumphy, M., Unger, B., Jensen, G.M., and Lopopolo, R.B., Incidence of work-related low back pain in physical therapists, *Phys. Ther.*, 65(4), 482–486, 1985.

Moore, R.J., The vertebral end-plate: what do we know? *Eur. Spine J.*, 9, 92–96, 2000.

National Institute for Occupational Safety and Health (NIOSH), *Work Practice Guide for Manual Lifting*, DHHS (Publication No. 81–122), Washington, D.C., 1981.

National Institute for Occupational Safety and Health (NIOSH), Musculoskeletal disorders and workplace factors: A critical review of epidemiologic evidence for work-related musculoskeletal disorders of the neck, upper extremity, and low back, Cincinnati, OH, 1997.

Nelson, A., Identification of patient handling tasks that contribute to musculoskeletal injuries in SCI nursing practice, James A. Haley VA Medical Center Study, 1996.

Nelson, A., Gross, C., and Lloyd, J., Preventing musculoskeletal injuries in nurses: directions for future research, *Spinal Cord Injury Nursing*, 14(2), 45–52, 1997.

Nelson, A., Lloyd, J., Gross, C., and Menzel, N., Redesigning Patient Handling Tasks to Prevent Nursing Back Injuries, *AAOHN*. In press.

Norman, R., Wells, R., et al. A comparison of peak vs. cumulative physical work exposure risk factors for the reporting of low back pain in the automotive industry, *Clin. Biomechanics*. 13, 561–573, 1998.

Owen, B., The lifting process and back injury in hospital nursing personnel, *West. J. Nursing Res.*, 7(4), 445–459, 1985.

Owen, B. and Garg, A., The magnitude of low-back problem in nursing, *West. J. Nursing Res.,* 11(2), 234–242, 1989.

Owen, B. and Garg, A., Assistive devices for use with patient handling tasks, in *Advances in Industrial Ergonomics and Safety,* Das, B. Ed, Taylor & Frances, Philadelphia, PA, 1990.

Owen, B. and Garg, A., Reducing risk for back pain in nursing personnel, *AAOHN J.*, 39(1), 24–33, 1991.

Patenaude, S. and Sommer, M., Low pain: etiology and prevention, *AORN J.*, 46(3), 472–475, 477, 479, 1987.

Personick, M., Nursing home aides experience increase in serious injuries, *Monthly Labor Rev.*, 113(2), 30–37, 1990.

Punnett, L., Fine, L.J., Keyserling, W.M., Herrin, G.D., and Chaffin, D.B., Back disorders and nonneutral trunk postures of automobile assembly workers, *Scand. J. Work, Environ. Health*, 17(5), 337–346, 1991.

St. Vincent, M., Lortie, M., and Tellier, C., A new approach for the evaluation of training in safe lifting in *Trends in Ergonomics/Human Factors VI,* Asfour, S.S., Ed., Elsevier Science Publishers, North Holland, 1987.

Stubbs, D.A., Rivers, P.M., Hudson, M.P., and Worringham, C.J., Back pain research, *Nursing Times*, 77(20), 857–858, 1981.

Stubbs, D., Buckle, P., Hudson, M., Rivers, P., and Worringham, C.J., Back pain in the nursing profession: I. Epidemiology and pilot methodology, *Ergonomics*, 26(8), 755–756, 1983.

Takala, E.P. and Kukkonen, R., The handling of patients on geriatric wards, *Ergonomics*, 18, 17, 1987.

Torma-Krajewski, J., Analysis of lifting tasks in the health care industry, in *Occupational Hazards to Health Care Workers*, Pub. No. 0170, American Conference of Governmental Industrial Hygienist, Cincinnati, Ohio, 1987, pp. 52–68.

U.S. Department of Labor, Occupational Safety and Health Administration, Ergonomics program; Proposed rule, Federal Register, November 23, 1999.

U.S. Department of Labor, Bureau of Labor Statistics, News release: Workplace injuries and illnesses in 2000. http://www.bls.gov/

Valles-Pankratz, S., What's in back of nursing home injuries? *Ohio Monitor*, 62(2), 4–8, 1989.

Videman, T., Nurminen, T., Tolas, S., Kuorinka, I., Vanharanta, H., and Troup, J., Low back pain in nurses and some loading factors of work, *Spine*, 9(4), 400–404, 1984.

Waters, T.R., Putz-Anderson, V., Garg, A., and Fine, L.J., Revised NIOSH equation for the design and evaluation of manual lifting tasks, *Ergonomics,* 36, 749–776, 1993.

Waters, T.R., Putz-Anderson, V., and Garg, A., *Applications Manual for the Revised NIOSH Lifting Equation*, National Institute for Occupational Safety and Health (NIOSH), Washington, D.C., DHHS Publication No. 94–110, 1994.

Wilkelmolen, G.H.M., Landeweerd, J.A., and Drost, M.R., An evaluation of patient lifting techniques, *Ergonomics,* 37, 921–932, 1994.

Williamson, K., Turner, J., Brown, K., Newman, K., Sirles, A., and Selleck, C., Occupational health hazards for nurses (part 2), *Image — The Journal of Nursing Scholarship*, 20(3), 162–168, 1988.

Injured Nurse Story #3: Who Will Care for the Nurses?

by Gave My All

This is my story. I've been a registered nurse for over 25 years holding certification in psychiatric and mental health nursing. For the past several years, I worked night shift in an acute care facility with psychiatric and older adult patients. There was usually one technician on each unit to assist with the patients. We were extremely busy, often with crisis admissions at night, as this was the way to best access managed care by being an emergency admission at night.

My injury occurred while assisting the technician pull up an obese, agitated, Alzheimer bed patient, in order to give her medications. The technician did not pull up the patient on the count of "three." I had reviewed with the technician the procedure of lifting on the count of three, but it appears there was some type of language miscommunication.

During this time, for cost containment, there was no orientation for nurses or technicians other than on the unit by the staff. The old "see one, do one, teach one" served as orientation. There were no lifts for patients and the bed was in one position because it was broken.

Initially, I was diagnosed with a thoracic-lumbar sprain/strain by the house physician and kept working while taking NSAIDs and using heat. My prognosis was good. After working for a week, I leaned over to change my infant's diaper and "locked up." I went to the emergency room and received muscle relaxers. I was then out of work for 6 months, during which I saw workers' compensation doctors and attended physical therapy. I improved, but never completely.

I returned to work part-time when workers' compensation cut me off after an independent medical exam said I was cured. I quickly decompensated and had to return to physical therapy and later switch to a nursing supervisor position as I could no longer lift, carry, or restrain patients. I was now having nerve involvement. This began the long battle with workers' compensation.

After 8 years of part-time work, and struggles with physical therapy, I became totally disabled in March of 2002. During these years I have had recurrent exacerbations, with diminished abilities after each. My employer was willing to allow me to stay on as a part-time night supervisor during this time. But, as the administration changed, and changes for the worse were made, I was unable to leave and seek employment elsewhere.

No one can use a nurse unable to meet the job requirements of lifting, restraining, etc. Supervisory positions without lifting were scare as supervisors now had to work on the nursing units as well as act as supervisors. Other typical positions for nurses with back injuries were also hard to come by. Basically, any nurse position requires the ability to sit, stand, bend, lift, etc., none of which I can do without pain.

Presently, I am unable to lift 5 pounds and have daily pain. I am limited in my life with my husband and children. I have pain everyday, though it varies in intensity. Such common things as brushing my teeth or my child's hair can be extremely painful. I have become dependent on others to tie my shoes, my 9-year-old shaves my legs, and my 15-year-old son has become my back. He has become very independent in shopping, etc. This also puts a strain on my marriage as even "relations" with my husband can be extremely painful afterwards and wish to be avoided.

1-56670-631-9/03/$0.00+$1.50

My family and I have been videotaped and accused of every trick in the book by workers' compensation lawyers. Sometimes I feel useless. But I keep going with the bare necessities as a type of physical therapy and I refuse to just lie around.

This also puts a terrible strain on my finances. It is amazing how all the insurance companies can't wait to sign me up for coverage. But, when the crisis hits, they try to reject me. I keep being told that perseverance is the only way to have success. But as I go bankrupt and fear losing my house, husband, and family, and see people on websites being sent money to get them out of bankruptcy, I wonder if there isn't some sort of help available.

I was a nurse and gave my all for my patients. As nurses, we give up meals and bathroom breaks, do mandated overtime, and work conditions no other professional or union would tolerate — all to give the best nursing care possible, no matter what the cutbacks, because we care as nurses. But when you are injured, no one seems to care. It is a system based on money, bottom line. I guess, as a nurse who writes CAREplans, it is hard to accept when no one seems to care.

Now, I have to pay a lawyer to win my workers' compensation case. I've applied for social security disability and was, as predicted, denied my first attempt. This is because "nothing is seen on the MRI/x-ray" but this is muscle with nerve impingement. And, "everyone has degeneration of their spine with aging," but mine is decades ahead of my chronological age in my injured areas. So, now a lawyer must file an appeal and wait another year. My employer's long-term disability insurance company is dragging its feet, trying to deny me, or wait for financial ruin to set in and try to force me back to work.

I have learned the hard way that nurses have to be prepared. Don't depend on your employer for your disability insurance. Nurses must compare policies against each other and look for loopholes or reasons that the insurance will not pay.

There is also a charitable organization in need of donations to assist nurses (short term) in time of severe financial crisis. Nurses House, a national fund for nurses in need since 1922, is located at www.nurseshouse.org.

The Nurses House budget is small and you must be living practically on the streets to be considered. I hope I don't have to go that far. Please help publicize Nurses House as a worthwhile charity for nurses to donate money to help other nurses. If, and when, I get back on my feet, I will make them my charity of choice.

Nurses must also work together to identify and develop a nursing careplan, so to speak, to address these issues. These are not isolated incidents. Nurses need access to disability insurance, lifting equipment, adequate training, adequate trained staffing, nurse-oriented lawyers, and referral sources. I hope my experience can help someone else, and, if anyone can help me, before I lose everything, that would be nice, too.

5 How to Accomplish a Responsible Cost-Benefit Back Injury Analysis in the Health Care Industry

William Charney

CONTENTS

INTRODUCTION

Workers' on-the-job back injury protection is seldom discussed as an ethical issue. Rather, the real discussions are about money. If the prevention programs are expensive and budgets are not easily achievable, the program is seldom, if ever, implemented and worker health and safety is sacrificed. Despite this logic, there have been rigorous efforts by the federal government and some state governments to enact ergonomic legislation (Washington State and California). The common industry responses in the public hearings have been less than supportive, mostly due to the fear of the economic impact the legislation would have on the business bottom line. It is my theory that if a rigorous science existed on cost-benefit analysis of injury, industry would be much more cooperative in initiating and implementing prevention, as it would understand the obvious monetary rewards at the front end. Prevention is always cheaper than paying for the injury. This truism becomes lost in an administrative inability to understand and incorporate the real cost of the injury into a cost-benefit analysis. This is happening for several reasons:

1. Lack of study and understanding of the science of cost benefit
2. Lack of study and an inability to translate the dollar costs due to a decrease in productivity because of injury
3. An inability to translate low morale issues due to injury into dollar values
4. A lack of understanding of the monetary impact of indirect costs for the injury, which can often be considered four times the direct costs
5. A lack of scientific understanding at the CEO and CFO level of the real cost of worker injury

While researching this article, it became apparent that there was a body of already existing research on both the cost benefit of the "direct cost" analysis and, even more importantly and less

1-56670-631-9/03/$0.00+$1.50

understood, "indirect cost" analysis. So, then, it becomes less a case of not having the body of science in the literature on calculating cost-benefit, but rather the chief executive officers (CEOs) and chief financial officers (CFOs) not having studied this all important data.

DISCUSSION

Back injuries in healthcare due to moving patients account for between 33% and 65% of a system's workers' compensation cost in dollars.[1] This can amount to between thousands of dollars and millions of dollars, depending on the size of the facility. The average cost per injury varies from state to state and whether the injury requires surgery. In California, the average back injury cost is $8250 and with surgery it is $25,090. In other states, costs can run as high as $13,000 without surgery to $75,000 with surgery. One permanent disabling nursing back injury in Florida has already cost $450,000 and is ongoing. Now, there are many different studies that show excellent results in solving the problem of back injury while moving patients,[2–6] yet cost-benefit must still be proved beyond any reasonable doubt. At times, just comparing direct cost dollars does not equal out. That is why the literature on indirect cost is so important.

This article has been written more in the spirit of a literature review than creating new science. The author has cited much within this article from *Cost-Benefit Analysis of the Ergonomic Standard* (May 2000).[7] It is hoped that the more these data are collated and published, especially in one coordinated source, the more it will be accessed and cited. The reason there is so much need for an emphasis on cost-benefit is that, despite the epidemic of healthcare worker injury the administrations of healthcare programs insist on seeing that the program will pay for itself. There is rarely an implementation due to ethical considerations alone.

It is striking that, despite the hundreds of millions of dollars of incurred costs in the hospital industry due to occupational back injury, hospital officials rarely apply a cost-benefit analysis. The Bureau of Labor Statistics (BLS) indicates that the healthcare industry is severely affected by back injury.[9] Nursing and personal care facilities continue to report especially high rates of injury. In 1995, the lost workday rate was 600 injuries and illnesses per 10,000 workers, more than double the national rate of 250 per 10,000.[9] Overexertion injuries, primarily while moving and lifting patients, accounted for over half of the lost workday rates in nursing homes.[9] In 1998, the BLS reports 261,000 occupational injuries suffered by hospital workers and 199,000 occupational injuries suffered by nursing home workers, among which one-half to one-third were overexertion injuries or ergonomically related.[9] The real loss — costs in dollars and productivity — is not well understood by the healthcare sector. Workers' compensation fund and associated fund administrative costs are the direct costs that employers bear for work-related musculoskeletal disorders (WMSDs). Workers' compensation costs and the medical costs associated with these injuries are the direct costs, and are somewhat easily calculated if the hospital has a decent retrieval system for these data. Not all hospitals have this capability, and, if they do not, they are entirely in the dark in terms of financial liability for these injuries. There is no law that requires hospitals to have reliable cost data. However, these direct costs are more easily understood as they appear in hard dollars. What is less understood or accepted in the industry when a cost-benefit analysis is done are the indirect costs. Research has shown there are sizable injury-related costs that are not covered by insurance, which nonetheless are borne by the employer. These costs are commonly referred to as the uninsured or indirect costs. These include:

1. Productivity losses due to an injured worker's absence and reduction in coworkers' productivity following the injured worker's return to work
2. Lost workdays: (wages × hours lost)
3. Lost time for managers
4. 21% lost productivity of individual worker injured
5. Overtime paid to others during lost workdays

6. Personnel and training time to hire replacements
7. Cost of emergency treatment
8. Cost of light duty
9. Claims processing costs
10. Additional recruitment costs for dealing with the injury episode.
11. Legal expenditures required to defend an employer from litigation.
12. A hospital with a high level of WMSD injuries may have to resort to over-employment in order to keep the business viable.

A number of researchers have attempted to estimate the magnitude of indirect costs to employers. Estimates of the size of indirect costs range from 10 to 2000% of direct costs. The review includes many studies on the issue of indirect costs, and it has been done to lend convincing references from the literature to the argument that indirect costs must be added to the formula of calculation for total back injury costs. Without this addition, the total cost of the injuries remains obfuscated and prevention budgets are harder to rationalize.

Heinrich was probably the first investigator to evaluate systematically indirect costs and direct costs of workplace injuries.[10] He evaluated a very large number of industrial accidents from the 1920s through the 1940s and determined an indirect-to-direct cost ratio of 4:1. According to the Washington State Department of Labor and Industries, which compiled the literature review, adjusting for Heinrich's inclusion of insurance administrative costs in the indirect-cost category and the lower wage-replacement rate for injured workers 60 years ago, roughly estimated current indirect costs were 200% for this data set. The more recent literature that was reviewed by the Washington State Department of Labor and Industries provides strong support for indirect costs being at least 100% of direct costs.[7] In Table 5.1, we see that one major study found a 1 to 3 dollar variation and a 5 to 50 dollar variation in material damages. Another study found a 1:4 ratio from the 0 to 3000 dollar range and then a descending scale the higher the incurred cost was. Andreoni[11] found a median ratio of 1:4. Leopold and Leonard[12] found 1:4.5, Klen[13] found 1:5, and Oxenburgh[14] found a 1:2 ratio.

As a conservative assumption, the Washington State Department of Labor and Industries has decided upon indirect costs of 75% of direct workers' compensation incurred costs.[7]

Lost Productivity as a Factor in Cost-Benefit

The issue of reduced worker productivity due to chronic musculoskeletal pain while the worker is back on the job is not well understood or economically calculated. It has been addressed in one study,[15] which combines direct measures of worker productivity while on the job with measures of absenteeism and disability. The study found that workers with musculoskeletal injuries lost approximately 21% of their potential productivity due to a combination of absenteeism, disability leave, and lower productivity while working.

A study based on a cohort of claimants found that administrative measures commonly used to quantify injury duration, such as days to first return to work and time on temporary disability, substantially underestimate the length of work disability following low back injury.[16] Dassinger found that the number of lost workdays as measured by self-reported experience of claimants far exceeded the duration of wage-replacement benefit payments. Workers' compensation administrative data underestimated total cumulative lost work days by more than 50% over a 1 to 3.5 year period following the injury.

The Politics of Cost-Benefit Analysis

The U.S. is twenty-seventh on the list for providing per capita occupational safety for its workers, at $3.97 per worker. In this age of nursing shortages (124,000 unfilled positions) it is of the utmost importance to integrate the concept of a safe working environment into the equation of the nursing shortage. In an OSHA testimony hearing on the Federal Ergonomics Bill it was stated that the U.S.

TABLE 5.1
Summary of studies on indirect costs of injuries

Authors	Terms	Results
Heinrich (1931, 1959) - U.S.	Direct costs: Victim's indemnity costs, medication, and hospitalization. Indirect costs: 1. Time remunerated (paid back) but no work by the accident victim the day of the accident 2. Time remunerated (paid back) but no work by the other employees, which are stopped because of the accident 3. Time lost by the supervisor and other managers 4. Time lost for security, medicine, or infirmary 5. Damages to machines, tools, materials 6. Production interruption, delayed delivery, lost goods/sales 7. Social benefits paid without production 8. Loss of profits caused by productivity loss from lost employee 9. Wage paid to accident victim with reduced production 10. Reduced employee morale and heightened fear of accidents 11. Electricity, heat, rent.	4:1 indirect to direct costs; Note: Insurance administrative cost included in the indirect costs category.
Simonds and Grimaldi (1956, 1984) - U.S.	Insurance costs: Assessment of the insurance system, includes administrative costs, plausible funds, and prevention costs of the insurance system for the noninsured. Heinrich's definitions: less recruiting of replacements, productivity loss of other workers, machine stoppage. Expenses: heating, electricity, and rent.	Lost time case $465 (1982 dollars); medical intervention case $115; first aid case $25; no lost time case $850; 1:1 uninsured to insured cost.
Bird (1974) - U.S.	Insured and noninsured.	$1 to $3 various expenses (hiring, training, replacement, inquiry [interview/evaluation], wages, and $5 to $50 material damages for each dollar of insured costs.
Levitt, Parker, Samuelson (1981) - U.S./Construction sector	Direct costs: Re: Heinrich's definitions — they add administrative costs without more precise indirect costs; Re: Simonds and Grimaldi definition — they add the cost of productivity loss of other workers.	Lost time direct costs — Ratio direct:indirect 0–$3000 — 1:4.1 $3000–$4999 — 1:1.6 $5000–$9999 — 1:1.2 + 10,000 — 1:1.1 No lost time — Ratio direct:indirect 0–$200 — 4/2 $200–$399 — 5.1 + $400 — 9.2
Andreoni (1986)	Review of methodologies for estimating the financial costs of occupational injuries	Direct:indirect Lower limit 1:1.58; median 1:4.10; upper limit 1:20
Leopold and Leonard (1987) - U.K./Construction sector	Direct costs: Additional accident payments; added premiums, material damages, indirect legal costs; only wage costs.	Indirect to direct costs 1:4.5 Note: Some direct costs would be considered indirect (i.e., variable portion of insurance premium) by other authors.

TABLE 5.1 (CONTINUED)
Summary of studies on indirect costs of injuries

Authors	Terms	Results
Laufer (1987) - Israel/ Construction sector	Nonverifiable costs: Fixed insurance costs not variant in terms of verifiable accident costs.	Conventional method (insured/noninsured costs); costs of noninsured represents 1.59% of profits before taxes: the method of verifiable/nonverifiable costs increases the incentive for prevention (the author does not cite figures); estimated uninsured to insured ratio 0.2:1.0.
Klen (1989) - Finland	Direct costs: Compensation paid to the victim, transport to the hospital; difference between wage and compensation, social benefits, wage paid the day of the accident. Indirect costs: Investigation, administrative costs, lost time from the other workers, productivity loss, material damages, interest in the insurance premiums.	Indirect to direct costs 1 to 5: 60% of accident costs are absorbed by the employer, 30% by public administration and 10% by the worker.
Rinefort (1977)	Survey of 140 Texas chemical, paper, and wood product firms. Followed the analysis recommendations of Simonds.	Found that firms with low workers' compensation costs spent more on safety than firms with high workers' compensation costs; Uninsured to insured ratio approx. 1:1.
Hinze and Applegate (1991)	Survey of 103 construction industry injuries.	Indirect to direct ratios: Medical only 1:1 (1.6:1.0 with claims) LWD 0.35:1 (4.4:1.0 with claims).
Brody, Le'Tournequ and Poirier (1990)	Review article and report on authors' 13-industry analysis.	Indirect costs: Quebec industry $1100 (Canadian) Canadian road transportation $2900 Indirect to direct ratio approx. 1:1
Oxenburgh (1991)	Estimation of indirect costs for occupational injuries at a Swedish automotive industry.	Direct:indirect 1:2.1.
Oxenburgh (1993)	Results from a survey of manufacturing and manual handling jobs in Australia; direct and indirect costs reported by employers.	Direct:indirect 1:0.75.

looses 5% of its nurses annually because of disabling injury. In a survey conducted by the Federation of Nurses and Health Professionals (FNHP), 56% of nurses surveyed said they would leave the profession due to the excess of stress and physical demand. In a survey conducted by 1199 in New Jersey, 40% of CNAs surveyed said they were leaving the profession due to the danger of back injury. Turnover rates in some nursing homes are 100% due to different factors among which is a dangerous workplace. Because there is really no federal regulation on "timely reporting" of compensation claims, injury costs increase by 18% if reported late by 1 week and an increase in injury cost by 43% if reported late by 1 month. In Washington Hospital Services data bank it is calculated to cost two to three times more to replace an injured nurse's salary.

Some data suggests that workers compensation costs in most healthcare systems is less than 1% of the total operating cost budget. This small percentage does not raise administrative "red flags." But this data point only tells a very small part of the story. First, 1% of a 50 million dollar operating budget is still $500,000 hemorrhaging out of the system due to injury costs. Second, the 1% is calculated on "direct-costs" and is not inclusive of the "indirect cost," which for the purposes of this chapter is a factor of three times the direct costs, which is a compromise factor between what is found in the latest peer review[17,18] (see reference 17, which gives a factor of 4× and reference 18, which gives a factor of 2×).

Some formulas in use at present give compelling arguments as to what a competent safety program can save. A formula using the profit ratio of the hospital is:[19]

1. Estimated cost savings of the program divided by the profit ratio = dollars not having to be billed. Example: Hospital A spends $170,000 on a back prevention program and calculates the 1 year saving of workers' compensation costs to be % 54,000. The $54,000 divided by the profit ratio of 2.8% equals $2 million dollars the system does not have to bill out to cover the cost of the injuries.
2. Using a formula for calculating cost rate, which is: cost rate = dollars actually spent/100 FTE to calculate cost rate numbers use: cost rate = (dollars spent/area/yr) × 200,000 hours worked/100 FTE divided by numbers of hours worked/area/year. This formula lets you actually compare dollars saved after implementation of a program.

The positive cost-benefit pictures are presented in several chapters in this book (see Chapters 6, 8, and 9). These programs have shown in peer review science a heavily weighted positive cost-benefit. The Occupational Health and Safety Administration (OSHA) has cited in its Federal Ergonomics Proposal over 50 success stories for zero lift. Chapter 6, also published in peer review,[20] shows positive cost-benefit in all 12 hospitals studied. In fact, all lift team studies published to date in peer review have shown positive cost-benefit.[21–27] Two ceiling lift studies[28,29] have been published that clearly show the positive cost-benefit when ceiling lifts were installed in two nursing homes, one in the U.S. and the other in British Columbia. In the U.S. study, lost days were reduced to 0. Both Garg and Charney have shown in published studies that the equipment purchased for a back prevention program pays for itself within 12 to 15 months. In the British Columbia study on ceiling lifts it was shown that benefits exceeded costs by a ratio of 6:1 or a rate of return of 17.9%.[28]

CONCLUSION

The addition of indirect costs to the direct costs must become an axiomatic part of the formula for calculating workers' compensation costs. It is no longer scientific just to assume the direct costs as the final calculation. Too many "real" dollars are missed. Hospital administrators and CFOs must study the referenced citations, especially on indirect costs, and be able to calculate injury costs professionally.

A conservative formula that is supported by the peer-reviewed literature would be:

Direct costs (compensation + medical) × 4 (indirect costs) = dollars spent on injury

Everywhere peer review science looks there is a positive (+) cost-benefit analysis to implementing back injury prevention programs. This benefit of prevention is not just for the hospital systems involved but also for society as some injured workers end up on SSI. Saving nurses backs so they can remain on the job is essential in combating one of the important causes of the national nursing shortage. Many countries have implemented, either by regulation or policy, a "no manual lift" philosophy including Great Britain, Australia, New Zealand, Ireland, South Africa, Sweden, and Denmark (see Chapter 7). Ethics should become a greater part of the decision making process for implementing these programs, as the case for economics has been proved. Protecting the backs of healthcare workers should be part of the social contract.

The state of Washington, in calculating the cost-benefit of its ergonomic rule, stipulated that "comparing the costs and benefits of the ergonomics act demonstrates the benefits to society greatly exceed the costs of compliance: a benefit ratio of 4.24 to 1.0, and this was true in nearly every single digit SIC [standard industrial classification] code, including healthcare."[7]

There should be an act or regulation requiring employers to implement safety programs to reduce injuries, even when the cost-benefit applies on a 1:1 basis. The analysis by the state of Washington, quoted above, only reiterates the obvious, that prevention is much cheaper than paying for the accident. Adding the direct to the indirect costs will almost always apply the 1:1 ratio, and more often will create a savings ratio much greater than 1:1. This analysis should be required reading for CFOs.

REFERENCES

1. Charney, W. Ed., *Handbook of Modern Hospital Safety*, CRC Press/Lewis Publishers, Boca Raton, FL, 1999, pp. 701–757.
2. Donaldson, A., Lift team intervention: a six-year picture, *J. Healthcare Saf. Compliance Infection Control*, 4(2), 65–68, 2000.
3. Meittunen, E.J., The effect of focusing ergonomic risk factors on a patient transfer team to reduce incidents among nurses associated with patient care, *J. Healthcare Saf. Compliance Infection Control*, 3(7), 306–312, 1999.
4. Charney, W., Lift team method for reducing back injuries: a 10 hospital study, *AAOHN J.*, 45(6), 300–304, 1997.
5. Charney, W., Reducing back injury in nursing: a case study using mechanical equipment and a hospital transport team as a lift team, *J. Healthcare Saf. Compliance Infection Control*, 4(3), 117–120, 2000.
6. Villeneuve, J., The ceiling lift: an efficient way to prevent injuries to nursing staff, *J. Healthcare Saf. Compliance Infection Control*, 2 (1), 19–25, 1998.
7. State of Washington, *Cost-Benefit Analysis of the Ergonomic Standard*, Department of Labor and Industries, Olympia, WA, 2000.
8. Charney, W., *Epidemic of Healthcare Worker Injury: An Epidemiology*, CRC Press, Boca Raton, FL, 1999.
9. National Safety Council, *Accident Facts*, National Safety Council, ltasca, IL, pp. 48–70, 1998.
10. Heinrich, H.W., *Industrial Accident Prevention: A Scientific Approach*, 4th ed., John Wiley & Sons, New York, 1959.
11. Andreoni, D., *The Costs of Occupational Accidents and Diseases*, International Labor office, Geneva, 1986.
12. Leopold, E. and Leonard S, Costs of construction accidents to employers, *J. Occup. Accidents*, 8, 273–294, 1987.
13. Klen, T., Costs of occupational accidents in forestry, *J. Saf. Res.*, 20(31), 31–40, 1989.
14. Oxenburgh, M., *Increasing Productivity and Profit Through Health and Safety: Case Studies*, CCH International, Melbourne, Australia, 1991.
15. Burton, W., The role of health risk factors and disease in worker productivity, *J. Occup. Environmental Med.*, 41, 863–877, 1999.
16. Dassinger, L.K., Duration of work disability after low back injury: a comparison of administrative and self-reported outcomes, *Am. J. Industrial Med.*, 35, 619–631, 1999.
17. Fragala, G., How to contain injury in healthcare, *Ergonomics,* 4, 68–70, 1996.
18. Johnson, A., Occupational injury in US, *Arch. Int. Med.*, 3, 1557–1568, 1997.
19. Liberty Mutual Cost Benefit Formula.
20. Charney, W., Lift teams, an 18 hospital study, *J. Healthcare Saf.*, 1(1), 2003.
21. Charney, W., The lifting team, *AAOHN,* 39(5), 1991.

22. Charney, W., The lifting team, a follow-up study, *AAOHN*, 41(8), 1992.
23. Charney, W., Lift team method for reducing back injury: a ten hospital study, *AAOHN,* 45(6), 1997.
24. Donaldson, A., Lift team intervention; a six year picture, *J. Healthcare Saf. Compliance Infection Control*, 4(2), 2000.
25. Davis, A., Birth of a lift team, *J. Healthcare Saf. Compliance Infection* Control, 3(8), 76–80.
26. Caska, B., Implementing and using a nurse staffed lift team, *J. AOHP,* 2000.
27. Meittunen, E. et al., The effect of focusing risk factors on a patient transfer team, *J. Healthcare Saf. Compliance Infection Control*, 3(2), 1999.
28. Speigel et al., Implementing a resident lifting system and cost benefit, *AAOHN,* 50(3), 2002.
29. Teisman, H., Charney, W. et al., Ceiling lifts in residential care, *J. Healthcare Saf.*, 1(1), 2003.

Injured Nurse Story #4: My Last Day as a CNA

by Susie Lyons Toftum

I was 24 years old and had been a nurse's aide for 3 years. In the nursing home, we had two Hoyer lifts for four wings. I didn't have a lot of training on the Hoyer — maybe 2, less than 3 hours. Almost all of the training was on manual lifting. We didn't have gait belts. We used sheets to lift or body-lifted the patients.

I was on the heavy-care wing with 15 patients that day. There were only two of us. Normally, we have four people on the wing but two were out sick. We were getting the patients up for supper. I went into "Mr. L.'s" room. He was belligerent and didn't want to get up. He was a major stroke victim and paralyzed on the left side. He didn't want to get out of bed. I talked with the nurse who said it was on his chart and he had to get up. I had lifted Mr. L. several times the same way, so didn't think there was a need to get the Hoyer.

I finally talked Mr. L. into getting up and scooted him to the edge of the bed. On the count of three, he was going to stand up. I put my arms around him to body lift him to a standing position. When I got to the count of two, he decided to stand straight up and caught me totally off guard. I heard a loud pop in my back and felt excruciating pain but, because I already had him mid-turn in the standing position to turn to the right and set him into his wheelchair, I barely got him into his chair because of so much pain in my back.

I screamed for help. I was bent over, couldn't stand up, couldn't move basically. That was about four-thirty or five o'clock. Because we were so short-handed, I stayed until they got somebody to replace me and went home about eight o'clock. That was the last day I ever worked there.

The next day I was hospitalized. I had crushed discs L3, L4, and L5. They put me into traction for 10 days with major morphine and Demerol. I had physical therapy for about 3 months, three times a week. Then, when therapy failed, the doctor opted to see if I was a candidate for surgery. He put me into a body cast for 2 months. The cast was all the way down my torso and half way down one leg to my knee, on the left.

My husband had to help me get dressed in the mornings. I could only put on one shoe and one sock. I lived in sweatsuits — very attractive. I was a real babe. I went two months without a shower and I'm a shower buff. It was horrible. I even tried soaking the cast off one night. I said this is enough, it's coming off, but it was fiber glass and wouldn't soak off.

While in the body cast, I had no pain and didn't have to take any muscle relaxers. It was basically just packing an extra 60 pounds. The cast kept my back completely straight. As long as my back was straight, I had no pain. The cast was just cumbersome.

I had to keep going back to the doctor every two weeks and had x-rays. I begged the doctor to take the cast off. To me, it was damaging because I was losing all my muscle tone. When the cast came off, I had to work even harder in physical therapy, had to go back on pain medicine and muscle relaxers.

The doctor decided I was a good candidate for surgery. He told me I would be an excellent candidate. The only reason I didn't, and I'm glad that I didn't do it, was that he wanted to put 4 steel bars around my spinal column to fuse my back with these metal bars. I had talked with too many people who had the same surgery I was going to have. They had more pain after surgery and there was no guarantee the surgery would fix my back.

1-56670-631-9/03/$0.00+$1.50

My husband had to do everything for me — tie my shoes, help me get dressed. Being injured affected everything, including my sex life. I couldn't drive or anything. I hated it. My husband did the floors because there was no way I could scrub them. As far as the tub and shower, forget it. There was no way I could do it. My husband worked an 8- or 10-hour job and then came home and helped with the housework and laundry. I had to have him do everything for me and I hated that. I'm very independent.

I had to give up so much. I was unable to go water skiing, horse back riding, rock climbing. I was 126 lbs. and did all that but still can't to this day. I had four horses I had to give up. My back still won't tolerate riding. I have to be real careful when I ride my ATV now. I've given up a lot of stuff because of my back. I have degenerative disc disease because of the injury. My back is so fragile that if my husband were to hug me real hard, it would break my back.

My employer basically did nothing. They kept in touch with me for about 3 weeks but they knew I wouldn't be coming back. They said if I couldn't get better and come back to work, there was nothing they could do. The aides did all the heavy lifting. They offered me nothing else. I was a nurse's aide, plain and simple.

There was another aide who was hurt. She got hurt on the same wing. She was off on workers' compensation and was never able to return. She got lucky and got another job, a desk job.

I ended up on workers' compensation and it took me 4 years to get off of that. The workers' compensation was only about half of what my wages were. I fought workers' compensation for 4 years. I had to go through their doctors, my doctors, their psychiatrist, and the pain clinic where they said it was all in my head. They had my records, my MRIs, x-rays, everything from my doctor, but they still don't believe me. I had to go to their psychiatrist twice.

I had to keep going back at least once a month to their doctors in order to keep my workers' compensation. They measured me with instruments and a tape measure every time, bent me over in front of three or four other male doctors every time I went. It made me feel horrible. It was degrading and they don't take your word for it.

That was a 400-mile round trip each time. I had to stop about five times and stretch on the way. I finally quit because I couldn't afford going any more. I finally had enough. My husband was taking me to an appointment. I was thinking about it and started crying real hard. I was shaking and hyperventilating with uncontrollable crying from frustration at trying to deal with the system that wasn't going to do anything for me. Enough was enough. I said I wasn't going back again. My husband asked if I was sure that was what I wanted. I said enough was enough and I wasn't going back. He said okay if that was what I wanted.

At first my attorney said we were looking at $150,000. Then, he said $90,000, then $30,000. I told him just to settle. It didn't matter if I got a penny. I just couldn't do it anymore and wanted to get on with my life. I wanted to go back to work. Workers' compensation vocational rehabilitation had told me I could be re-trained as a med aide. I would need to take classes at the college where I would have had to walk to the classes. Even walking was real hard for me to do.

I had to go through their attorneys and all that. They settled my claim for $3,777.75, for the rest of my life. Because I was tired of jumping their hoops and hated being on workers' compensation, I said that was enough. I was just tired of it all. The last time I drove up for an appointment, my tire blew out and I couldn't even change my tire.

Depression was terrible because I was used to working all the time. The first 2 weeks were like vacation but after that it was horrible. All I had to look forward to was doctors' appointments and attorneys' appointments. I wanted to go back to work.

I was done with workers' compensation and the doctors and wanted to go back to work. I was turned down for, like, three jobs that involved lifting. For two jobs, I had to go for a back x-ray. I had one job until they got the x-ray back and said I was too much of a risk. I was crushed.

A friend of mine worked in a grocery store and helped me get a job there where I didn't have to do too much heavy lifting. Some men at the store did most of the lifting. I've been in retail ever since. That was in 1987.

The doctor said I was a good candidate for surgery but couldn't guarantee that I'd be fine. I wasn't going to have the surgery and I was able to find the job with the grocery store. I'm glad now that I didn't have the surgery.

The doctor said to be very careful what I do and to know my limitations. I know my limitations. I've given up horseback riding, rock climbing, hiking. Anything on uneven ground just kills me. I ride my 4-wheeler but know what I can and can't do. I have a ramp that I use to load and unload it. I ride it right up into the truck. I never try to push or pull it. When riding on the sand dunes, you just sit on it and the machine does all the work.

I used to love climbing rocks on the jetty and down at the beach. I used to be pretty athletic. Not any more. I ended up with $3000. I guess they figured I'd lost that in wages. I lost a whole lot more than that.

6 Striving for Zero-Lift in Healthcare Facilities

Guy Fragala

CONTENTS

INTRODUCTION

Musculoskeletal disorders (MSDs), specifically back injuries, have been an ongoing problem within the healthcare industry. Direct patient-care staff that must lift and assist dependent patients with mobility needs are constantly placed at risk. Recently, more attention has been directed at the fact that healthcare workers are one of the most at-risk groups for occupational injuries based on these back-injury exposures. Many organizations are also recognizing that the traditional approaches, which have been utilized over the years and are based on teaching workers proper body mechanics to conduct manual lifts, have not yielded wide-spread success in preventing injuries and reducing injury rates. Many organizations now recognize that the best approach to back-injury prevention programs for healthcare workers involved in patient care is to identify the high-risk job tasks that place workers at risk and make physical changes to redesign these high-risk activities.

It is not a simple process to know what needs to be done for improvement and to actually achieve the desired goal. How does an organization get from knowing what needs to be done for improvement to implementing actions that will move the organization to this desired improvement, then actually achieve some level of significant improvement? In the case of ergonomics and patient handling activities, where new work practices must be developed and changed, appropriate programs, policies, and procedures need to be developed and implemented.

TRADITIONAL PREVENTION EFFORTS

The traditional approach to reducing back injuries in the healthcare industry was to develop and implement a training program in body mechanics and how to conduct lifts properly.

Numerous studies have been conducted over the past 30 years regarding the effectiveness of such training in reducing the impact of occupational back injuries as a whole and, specifically, to the healthcare industry. These studies concluded that lifting instruction had little or no effect in injury prevention.

1-56670-631-9/03/$0.00+$1.50

Work done in the 1970s and Brown's findings, with respect to occupational back injuries related to patient handling tasks, show little evidence to suggest that intensive training schedules had decreased back injuries over a 35-year period (Brown, 1972). In another study conducted by Dehlin in a geriatric hospital, results demonstrated that lifting instruction given to nursing aides had little or no effect upon the occurrence of back pain symptoms (Dehlin, 1976).

Additional work was done in the 1980s. Anderson states, "Although instruction on manual handling and lifting is widely believed to have prophylactic value, no scientific evidence was found that it is in fact effective in reducing the frequency or severity of back pain" (Anderson, 1980). Dawes' findings indicate that following the introduction of a training program for nursing within a health authority in southeast England, there was a pronounced decrease in the number of reported back injuries. However, after this impressive start, the results turned out to be disappointing. He states that these studies raised the question as to whether existing training for the nursing profession with respect to manual handling is appropriate (Daws, 1981). Buckle, who in his thesis work published a case study of patients with severe low back pain attending a rehabilitation unit, found that there was no difference related to history of previous training in lifting techniques, indicating that amount of training received was not a factor separating those who had experienced low back pain problems. Work done by Stubbs indicated that further evidence shows no relationship between the time spent training in such lifting techniques and subsequent prevalence of back pain. The current emphasis on training was questioned and the need for controlled perspective trials stressed. An approach requiring the development of intrinsically safe systems at work, with particular emphasis on the contribution of ergonomics, was recommended (Stubbs, 1983). In a study conducted by St. Vincent, results show that the handling principles taught — that is working with a straight back and using legs — were not frequently used in the healthcare workplace. These results further suggest that actual training is not well adapted to patient handling. The first problem was that, what was taught in the actual training programs could not always be applied and the methods taught could be questioned, particularly to the emphasis given to the use of legs (St. Vincent, 1989).

We must understand the difficult lifting tasks that are required within our healthcare facilities. Access to patients can be very difficult because of small spaces, such as bathrooms. It can be very difficult for healthcare workers to position themselves properly when trying to assist a dependent patient with toileting activities. Healthcare rooms are often very crowded and awkward postures are often required when trying to gain access to a patient in a bed. We must also remember that lifting people is not like lifting boxes. People are live, dynamic loads and can do unexpected things. With a well-designed piece of lifting aid equipment, we can reduce many of the variables related to unexpected behavior and create a safer situation for the healthcare worker and the patient. If we watch patients being manually lifted within a healthcare facility, we very soon come to the conclusion that there must be better ways to conduct this activity.

Reviewing work done in the 1990s, there is continued support for the premise that training alone in lifting techniques is not the answer in developing effective prevention strategies. Owen reports that although the most common approach to the prevention of back injuries has been education and training in the biomechanics of lifting techniques, little evidence was found to support this approach (Owen, 1991). Harber's work indicates that training at nursing school or on-the-job did not have a protective effect in preventing back injuries among new nursing graduates. Further implementation of engineering job redesign was suggested (Harber, 1994). Investigations done by Larese indicate that training courses are often useless when work organization and the number of nurses involved in patient care do not change (Larese, 1994). Lagerstrom reported on the effectiveness of a 3-year education and training program where 90% of the subjects questioned were positive about participating in the program and 88% expected that participation would lead to decrease in musculoskeletal disorders; however, no decrease in the prevalence of musculoskeletal symptoms was reported during the study time period (Lagerstrom, 1997).

As a final reference, I would like to report results from a large study conducted involving 4000 postal workers. I am using this reference to support the previous studies cited, which were specific to the healthcare industry. This study confirms the concept that traditional back injury prevention programs involving teaching proper lifting techniques is not the answer for improved performance. Daltroy's findings indicate that a comparison of the intervention and control groups found that the education program did not reduce the rate of low back injury, the median cost per injury, the time off from work per injury, the rate of related musculoskeletal injuries, or the rate of repeated injury after return to work. It seems only the subjects' knowledge of safe behavior was increased by the training (Daltroy, 1997).

The one dozen references cited above indicate the lack of effectiveness of traditional training programs. Researchers are continually telling us that traditional back injury prevention training programs have not yielded beneficial results when considering prevention programs aimed at reducing back injuries and back pain. Why have these traditional approaches been ineffective? A key reason is that in a classroom setting, participants are taught theoretical principles under optimum conditions. However, when they get out into the real work environment, because of the design of patient care areas and the equipment in use, it is often difficult to apply these optimum theoretical principles to real life situations. The environment in which we care for patients can be very unpredictable and is constantly changing. If the patient becomes weak and his legs buckle, when a healthcare worker is assisting a patient from a bed to chair, the worker has neither the time nor opportunity to consider theoretical principles of lifting. A device that can assist in this transfer process can remove the opportunity for this unexpected event to occur.

Consider lifting and transferring heavy patients as a final point of consideration regarding the value of traditional training program. The heavy weights involved and the posture that must be assumed while conducting these lifting tasks are an unsafe situation. Even if there are optimum lifting principles that might be applied, does the use of proper technique remove the risk for the healthcare worker? No, because of the heavy loads that must be lifted. Many organizations are spending much time and effort on their back injury prevention programs with little resulting improvement. With some guidance, injury prevention efforts could be directed to where existing resources are best utilized. I am not advocating that we abandon training, rather that we redirect our training efforts to teach all within an organization how to identify which tasks are dangerous and unacceptable and with this knowledge, work together to select the best options for improvement and then train workers on how to best utilize these new tools, devices, and techniques.

HIGH-RISK ACTIVITY

Think of some high-risk patient handling activities with the idea of changing the high-risk components of the job. Tasks involving a bed to chair or chair to bed transfer can be very difficult. Consider moving someone out of a bed and into a chair, the difficulty of the task will vary depending upon the dependency level of the person to be moved. With a totally dependent person, staff members must reach across an obstacle (the bed) to have access to the person they need to assist. This involves reaching and it is usually not possible to position oneself with bent knees since the worker is usually leaning up against a bed. The patient needs to be physically lifted and, considering weight, the loads involved in the lift are unacceptable. Movement into a chair involves moving the person being assisted to a different height level and there is usually some carrying involved. The unacceptable risk factors of this job involve reaching, lifting a heavy load, suboptimal lifting postures, and carrying a load a significant distance. In order to redesign this task effectively, the optimum situation would be to eliminate these high-risk activities. Lifting aid devices are applicable to this situation. These lifting aid devices include full body slings, which are very useful for the totally dependent patient or resident. In addition, the bed to chair transfer can be converted into a bed to stretcher transfer. Through the use of convertible wheelchairs, which bend back and convert into stretchers, and with height adjustment capabilities, a slide transfer rather than a lift may result.

If the patient is not totally dependent, a transfer such as bed to chair may be done by first getting the patient to a sitting posture. Again, the amount of assistance required will depend upon the patient's status. Once in a sitting posture, a stand and pivot transfer can be conducted. Some healthcare workers are highly skilled in this transfer technique and have done it many times without suffering any occupational injuries. However, loads involved are heavy and if the patient does something unexpected, such as collapses from a weakness in the legs, the healthcare worker must react and often times these unexpected occurrences result in occupational injuries. Again, through application of some lifting aid devices the risk associated with this type of transfer can be minimized. A device that could be considered in this situation would be a standing and repositioning lift, which is a lifting device with a simpler sling for patients or residents with weight bearing capabilities.

Another difficult task for the caregiver to perform, which places him at risk for occupational injuries, is the constant need to move patients up in bed. Patients will tend to migrate toward the foot end of the bed requiring ongoing repositioning. Because of the horizontal reach involved in gaining proper access to the patient and the weight of the patient, this task is very difficult and has been found to be a major source for occupational injuries. Using new design concepts, which improves the bed head articulation function, the surface and the patient can be combined in the fashion that minimizes the patient's migration toward the foot end of the bed. As the head portion of the bed is raised, the patient will stay properly positioned, providing better patient comfort and reducing the need for repositioning. Again, this technology currently exists and demonstrates how effective equipment design can eliminate the need to conduct some high-risk activities.

EFFECTIVE ERGONOMIC INTERVENTIONS FOR PATIENT HANDLING TASKS

A number of research and case studies show a large number of feasible controls that are effective in reducing occupational injuries. Many of these effective interventions are affordable engineering controls, which are the preferred methods of controlling workplace hazards. A number of these engineering controls have been able to completely eliminate certain manual handling tasks. For example, there are beds on the market that convert into a chair position with the touch of a control, thus completely eliminating the need for some lifting and transferring tasks. These beds are particularly helpful in hospital surgical or intensive care units where lifting and transferring can be further complicated due to presence of ventilators and other medical equipment. The care plan for many hospital patients includes "up in chair" orders as written by the physician. The frequency recommended by the physician will vary depending upon the patient's condition, however, close to 70% of the "up in chair" orders written require a patient to be put in a seated position at least two times per day. In many situations, there is low compliance for these "up in chair" orders and often caregivers are unable to comply at all. With this new technology and using the concepts of ergonomics, the high-risk activity is eliminated. Innovations in bed design can also reduce other risks (Fragala and Shelton, 1998).

A variety of mechanical lifts (lifting aid equipment) on the market have eliminated lifts involved in a number of patient handling tasks, such as bathing and helping a patient who has fallen on the floor. The use of patient lifting aid equipment is growing as the equipment coming on the market has provided more options. In the past several years a number of organizations and facilities have begun selecting lifting aid equipment that will best meet their needs. These organizations identify the tasks that require changes and then investigate the equipment to see what will meet the needs of both the workers and patients. One of the best ways to investigate such equipment is to have a variety of models brought into the facility for hands-on evaluation by the workers who would be using them. Vendors are very helpful and agreeable in allowing facilities to try various models on a trial basis. Some vendors now also provide guarantees on their equipment — if injuries are not reduced as a result of using the equipment and an ergonomics management program, the vendor will refund the purchase price.

Tasks for which mechanical lifts provide the optimum solution include the following:

- Assisting patients with some weight-bearing capability when transfering from bed to chair or from bed to standing position.
- Assisting patients with some weight-bearing capability with toileting activities.
- Helping totally dependent patients with transfer from bed to chair using a full-body sling. This eliminates lifting, carrying, and lowering the patient.
- Assisting totally dependent patients with bathing activities using a full-body sling.

Another group of engineering controls has successfully eliminated manual lifts by replacing them with lateral slide transfers, a task that does not require as much physical force. Height adjustable beds have made it possible to replace stretcher-to-bed lifting tasks with slide transfers. Transfer chairs that convert into stretchers have also replaced bed-to-chair lifting with lateral slide transfers.

Other engineering controls have reduced the force required in certain manual handling tasks. As mentioned above, friction reducing sheets have reduced the amount of force workers must exert when repositioning patients, for instance. Grasp-assist mechanisms attached on the sides of beds allow patients with some weight-bearing capacity to assist in the transfer and standing process, thus reducing the amount of weight the worker must handle. Appendix 6A includes a list and description, including average cost, of some of the engineering controls that healthcare facilities have found to be effective in reducing back injury rates.

It should be pointed out that some changes in manual handling tasks can be accomplished using current equipment and utilizing basic ergonomic concepts. For example, the height adjustment on hospital/nursing beds can reduce the force required to do patient transfer, especially where the patient has some weight-bearing capability. Those patients, by sitting on the edge of the bed, can be raised to almost a standing position without lifting them. In addition, rails and other ambulatory assist devices can be installed to provide added support for standing or transfer. Adjusting the bed height in accordance with ergonomic principles is also a way that a healthcare worker can reduce awkward postures when repositioning a patient in bed or performing other bedside procedures. These results are achievable at no cost through understanding ergonomic concepts and exploring their application to redesign manual handling tasks.

Further benefits of patient handling devices can be demonstrated with biomechanical modeling. Through this approach, various healthcare lifting tasks can be quantitatively analyzed to provide some basis for judging the degree of risk presented by a particular task. Through modeling, a system can be represented in an understandable presentation, then quantitatively demonstrate improvement achieved through a task redesign. In a small laboratory study conducted by the author, disc compression for transfer of patients between beds and chairs was evaluated. Results were consistent with other, more intensive studies conducted (Fragala and Shelton, 1998; Zhuang, 1999; Marras et al., 1999). For a redesign of the task utilizing a chair converting to a stretcher and a friction-reducing device, compressive forces experienced during the transfer were significantly reduced. A small field study utilizing the same job task redesign was conducted. A study instrument that had been used in previous studies to gather subjective ratings from both workers and residents was used. Each time workers conducted a transfer they were asked to rank what level of physical exertion they felt on the body parts under consideration on scale from 0 to 10 — 0 meaning no exertion and 10 meaning high exertion. In previous studies, this approach of exertion scale skill rating had been found to be just as valuable and accurate as more labor intensive biomechanical modeling evaluation techniques (Owen, 1993). Exertion felt by workers was much less utilizing the new chair design and friction-reducing device (Owen and Fragala, 1999).

ERGONOMIC INTERVENTION CASE STUDIES

Facilities that have developed and implemented ergonomic-based injury prevention programs utilizing effective engineering controls have demonstrated much success. Below, examples are provided of some facilities where the author participated and is familiar with their experience, and in addition, other success stories cited in the literature are presented. These facilities have seen improvement within a very short time period once engineering controls have been implemented and the facilities are very pleased with the new lifting and transferring patient-handling techniques. Experience indicates that once a facility has made the commitment to implement engineering controls in some areas, they will usually continue the process and continue to add new equipment and devices.

In 1991, the Commonwealth of Virginia implemented a program at its Northern Virginia Training Center where mechanical lifts were supplied to four high-risk units. The average reduction in injuries over the four units after the intervention was 73.75% (Werner and Scott, 1992). A similar facility in the state of Wyoming reduced injuries related to lifting residents by 60% after lifting aid devices were introduced (Stensaas and Leonard, 1992).

A pilot study in 1992 in the state of Connecticut implemented an injury prevention program using new technologies in lifting aid devices. The facility had some experience with the use of older lifting devices and was eager to try new technology. Prior to the intervention related to the study, the facility had experienced 12 reportable strains and sprains in nursing departments in 1991 and, in the first 9 months of 1992, had eight workers compensation cases related to strains and sprains caused by moving residents. Two new types of lifts were introduced — a sling lift used for a total transfer and a standing assist lift used to replace the manual stand and pivot transfer. After the new lifts were introduced to the facility, only two strains or sprains were reported over a 6-month period, neither of which involved lost time or workers' compensation. One injury occurred on the third day of the study period before lifting aid equipment was completely installed. No occupational injuries involving resident transfers were experienced during the study period when the lifting aid equipment was used (Fragala, 1993).

As part of the study, employees were asked which lift technique is the easiest to use when transferring a totally dependant resident and 16 out of 17 respondents selected the new sling lift. When asked which lift was the easiest to use when transferring a moderately dependant resident, 10 of 15 respondents selected the new stand-assist lift provided. Six employees said that they felt the sling lift saved their back from pain and injury, and nine responded that this technique was more secure and safer for the resident. When asked reasons why workers liked and used the stand-assist lift, five felt it was easier on their back and six thought it was easier for the resident (Fragala, 1996).

Another study conducted at a 568 bed long-term care facility in the state of Connecticut found that back injuries were reduced approximately 74% over a 3-year period due to a hospital-wide total quality management initiative that included an ergonomics-based back injury prevention program involving mechanical lifting devices. Preintervention workers' compensation assessment for back injuries was $174,412. Following program implementation, workers' compensation costs were $4500 for an equivalent time period. Annual lost workdays were reduced from 1025 preintervention to 81 postintervention (Fragala et al., 1995).

A case study conducted at a Texas hospital tells the story of a quality improvement team directed at costs associated with occupational injuries within their facility. After much consideration, it was obvious to team members that effective patient handling devices were needed as a part of their intervention strategy. Appropriate lifting equipment was purchased in February of 1994 and a facility-wide orientation program was implemented. The average annual direct cost associated with back injuries resulting from patient handling based on an average of 3 years' experience was $111,159. In the year following implementation of an ergonomics-based back injury prevention program using effective mechanical lifting aids, the direct cost was reduced to $743 (Fragala et al., 1995).

Lawrence and Memorial Hospital of New London, Connecticut, established an ergonomics implementation plan using a five-step process for program development. As part of the process, problems were identified and priorities set. Work began on redesigning high-risk activities. By categorizing patient populations in two high risk units within the facility, two types of lifting devices were selected to reduce many of the injuries. The devices identified were a standing and repositioning lifting aid with a toilet attachment and a full body sling, with a bed scale attachment. On one high risk unit, there was an 83% improvement in the number of occupational injuries. The number of lost work days and restricted work days decreased dramatically with lost work days dropping from 69 to 0 and restricted work days from 122 to 2. On the second high-risk unit, there was a 75% improvement in the number of occupational injuries and lost work days were reduced from 48 to 0 and restricted days dropped from 11 to 4 (Fragala and Santamaria, 1997).

The above examples cited are those in which the author had some direct experience. There are other success stories appearing in the literature. In January of 1993, England introduced the EC legislation on lifting and manual handling, which states all hazardous manual handling tasks are to be avoided wherever possible. If hazardous manual handling tasks are unavoidable, they must be assessed in advance. Once they are assessed, action should be taken to remove or reduce the risk of injury. A safe manual handling policy, incorporating training and assessment, must take place. Dangers and hazards must be identified and equipment provided for safer working practice for staff and caregivers. Before any moving and handing procedure can be performed, the nurse should undertake a full risk assessment, completing the appropriate documentation. As a result of these regulations, facilities in England have experienced a reduction in injuries among caregivers (Logan, 1996).

To measure the impact of the EC Legislation, a study was conducted by the National Audit Office where 30 acute trusts, a grouping of healthcare facilities, were surveyed. It was found that with the implementation of an ergonomic program similar to the OSHA elements, in 1994 through 1995, lost work hours from patient moving and handing injuries dropped by 84%. Preintervention, more than 11,635 hours of work were lost at the Wigan and Leigh NHS Trust from 1993 to 1994 because of occupational injuries. Over 6720 of these hours were due to injuries caused by moving and handling of patients. Under the direction of the health and safety advisor at the Trust, the group took advantage of a free equipment assessment offered by a lift manufacturer. A moving and handling coordinator was appointed to undertake a detailed audit of manual handling activity and related issues. An action plan was developed and equipment purchased. The results were dramatic yielding an 84% decrease in lost work hours, thus saving 5638 hours of work. Once the program was implemented, good injury experience was maintained and the yearly costs in absenteeism resulting from lifting and handling injuries has been reduced by 97.5%. The moving and handling coordinator appointed was a nurse with over 30 years experience who played a key role in persuading staff to change their practice and use new equipment. Initially, there was a common misconception among staff that using equipment to move patients took longer. Staff soon began to realize that once they were skilled and confident with using the equipment, it actually was a lot quicker. The ergonomic approach meant that hospitals had to improve all aspects of their work systems. This involved trusts implementing a policy with appropriate management support, equipment, and training, and sustaining intervention over a long period of time (Waters, 1997; Gaze, 1997).

In a study conducted at Surrey Memorial Hospital in Surrey, British Columbia, where more than 5800 work days had been lost in a year at the hospital costing an estimated $950,000, an ergonomics-based program was implemented. The program included the development of a no manual lift policy with standardized lift and transfer procedures. At the end of the pilot project, injuries had been reduced by 95% (Bruening, 1996; Perrault, 1995). Through the engineering controls implemented, the intent was to eliminate the need to conduct a manual lift, with the objective to eventually eliminate all manual lifting within the facility.

A health sciences center in Winnipeg, Manitoba, Canada, implemented an ergonomics program in 1990. It was stated that the most important facet of the ergonomics program was getting employees to recognize the benefits of interventions. The interventions primarily being lifting aid devices. Since the program began, back injury incidence rates have fallen 23% and lost time hours have dropped 43%. In selected nursing units, there was a drop of 39% in injuries and 83% in lost time hours. Interventions included worker assisting devices, equipment to help employees move patients from one bed to another and from one sitting position to another. Each nurse completed a 45-minute session on injury prevention and a 45-minute hands-on training session with equipment (Health Science Center).

In a study done in Quebec, ceiling mounted lifts were installed in a 200-bed facility. Preintervention, the facility was experiencing approximately 26 lost time injuries per year from patient care activities. This is based on 4 years of data with an average of 983 lost days per year. In the two years following installation of the lifting devices, the average number of injuries dropped to 6.5 per year and the annual average days lost was 67 (Villeneuve, 1998).

The Kennebec Health System, a nonprofit corporation in the state of Maine experienced 1097 lost work days in 1990. In 1995, after an ergonomic management program and engineering control interventions were implemented, only 48 work days were lost. The experience modification factor for the entire health system dropped from 1.8 to 0.69 and insurance premiums fell from $1.6 million to $770,293. A major part of the intervention program used to achieve these results was the purchase of mechanical lifting aid devices for facilities (BNA Workers' Compensation, 1996).

Another study in the state of Maine done by Maine Employers Mutual Insurance Company demonstrated a drop in medical and indemnity costs from lifting injuries from $75,000 in 1993 to less than $5600 in 1997, after an ergonomic management program had been put in place with a no manual lift policy as the program's cornerstone (Hospital Employee Health, 1999).

The concept of lifting teams has been employed within some facilities. Lifting teams designate workers who will conduct lifting and provide teams with mechanical lifting devices to conduct lifts. With lifting teams, not all healthcare workers conduct patient transfers and lifts. Designated groups are considered to be lifting teams and they are trained in the use of new lifting aid devices. Some facilities like this approach since they can focus training activities regarding new equipment on a smaller group of workers. The number of lifting teams within a facility can vary depending on facility needs. A statistically significant decrease in lost time injuries was achieved by using these teams provided with equipment at a San Francisco hospital (Charney et al., 1991). As a follow-up to this work, a ten-hospital study was pursued where significant reduction in back injuries was achieved through the application of lifting teams. In this study, lifting teams helped to control injuries by successfully reducing injury rates by almost 70% and lost work days by 90% (Charney, 1997). A Toledo hospital tried the lifting team concept and a 6-month lifting team pilot program on one floor reduced nursing injuries by 100% (Hospital Employee Health, 1994).

The National Institute for Occupational Safety and Health (NIOSH) has funded some large studies to investigate the effectiveness of ergonomic-based injury prevention programs. In a study done by Garg (1999), results demonstrated a 62% decrease in the number of injuries suffered as determined on an annual basis considering experience for 3 years both preintervention and postintervention. For the same time period, annual lost work days were decreased an average of 86% and restricted work days decreased 64%. Calculations were also done on workers' compensation costs and the average decrease was 84%. In another large NIOSH study done by Collins et al. (2002) involving 6 nursing homes, injury rates due to resident lifting/transferring were reduced dramatically when considering experience for 3 years both preintervention and postintervention. The resident lift injury rates for the 3 years preintervention were as follows: 17.7, 18.8, and 10.8. For the 3 years postintervention these dropped to: 9.0, 8.4, and 6.4. Workers' compensation costs for the same time periods also improved. For the 3 years preintervention costs were as follows: $176,848, $161,377, and $138,688. For the 3 years postintervention costs were as follows: $63,408, $15,765, and $105,913.

The experience of these facilities demonstrates how an ergonomic management program using effective lifting aid devices and engineering controls can successfully reduce occupational injuries associated with patient and resident handling. Many healthcare organizations are now forming ergonomic committees to address occupational injury problems. These committees are looking for guidance and direction as to how to proceed.

SUMMARY

In order to achieve improvement related to reducing musculoskeletal disorders among healthcare workers involved in direct patient care, many are now reaching the conclusion that difficult and demanding jobs must be redesigned by applying the principles of ergonomics. High-risk jobs must be changed and modified and a strategy based on answers to the following questions is suggested:

1. Can the need to do the high-risk activity be eliminated, such as by eliminating a bed to chair transfer using a bed which converts into a chair configuration?
2. Can the high-risk activity be redesigned using devices such as mechanical lifts?
3. Can the high-risk activity be improved through risk reduction using some type of lifting aid device, such as a gait belt with handles?

At this time, everyone recognizes the need for ergonomic design improvements including healthcare workers, management and administration, and equipment manufacturers. The future should bring new concepts and innovations, which can provide many benefits. Beyond the potential reduction in caregiver injuries, there are many possibilities to improve patient outcomes through better equipment design. Applying the principles of ergonomics enhances and increases caregiver productivity through a reduction of patient transfers required and minimizes staff required to ambulate patients. Another added value to applying ergonomics to equipment design might be the development of convertible furnishing, which might create a reduction in operating and capital expenses by reducing the need for some furnishings. Equipment such as cardiac chairs, sling scales, patient chairs, special rental surfaces and other features might be incorporated into bed design. Ergonomics makes sense and provides opportunities to create win/win situations throughout the healthcare industry. Current and future innovations will provide improvements resulting in improvements where everyone will benefit. These benefits will include a higher quality of work life for healthcare workers and an improved quality of care for patients.

REFERENCES

Anderson, J., Back pain and occupation, in Jayson, M.I.V., Ed., *The Lumbar Spine and Back Pain*, 2nd ed., Pitman Medical Ltd., London, 1980, pp. 57–82.

BNA Workers' Compensation Report, 7, Empowering Workers Helps Nursing Home Find Answers to Injury Problem, Cut Costs, 1966, p. 483.

Brown, J., Manual Lifting and Related Fields. An Annotated Bibliography, Labor Safety Council of Ontario, 1972.

Bruening, J., Keeping Healthcare Workers Healthy, *Ergonomics News*, March/April, 1996, pp. 20–21.

Charney, W. et al. The lifting team: a design method to reduce lost time back injury in nursing, *AAOHN J.*, 39(5), 231–234, 1991.

Charney, W., The lift team method for reducing back injuries: a 10 hospital study, *AAOHN J.*, 45(6), 300–304, 1997.

Collins, J.W., Wolf, L., and Hsiao, H. Intervention Program for Transferring Residents in Nursing Homes. Presentation at: Safe Patient Handling & Movement Conference, sponsored by VISN 8, Patient Safety Center of Inquiry, Clearwater, FL, January 16–18, 2002.

Daltroy, L. et al. A controlled trial of an educational program to prevent low back injuries. *New Engl. J. Med.*, 337, 5–322–328, 1997.

Daws, J., Lifting and moving patients, a revised training programme. *Nursing Times*, 2067–2069, 1981.

Dehlin, O. et al., Back symptoms in nursing aides in a geriatric hospital, *Scand. J. Rehab. Med.*, 8, 47–53, 1976.

Fragala, G., Injuries cut with lift use in ergonomics demonstration project, *Provider*, 39–40, October 1993.

Fragala, G., *Ergonomics*: *How to Contain On-the-Job Injuries in Health Care*, Joint Commission on Accreditation of Healthcare Organizations, Oakbrook Terrace, NY, 1996, Chap. 6, pp. 55–57.

Fragala, G., Ergonomics: the essential element for effective back injury prevention for healthcare workers, *Am. Soc. Saf. Eng.*, 23–25, March 1995.

Fragala, G. and Shelton, F., Applying the concepts of ergonomics to improve healthcare bed design, *Rehab Ther. Prod. Rev.,* May/June, 34–38, 1998.

Fragala, G. and Shelton, F., Assessing risks of healthcare lifting tasks and making improvements, *Rehab Ther. Prod. Rev.,* Jan/Feb, 52–59, 1998.

Fragala, G. et al., Backs to the drawing board, *Health Facilities Management,* 24–28, May 1995.

Fragala, G. and Santamaria, D., Heavy duties? *Health Facilities Management,* 22–27, November 1997.

Garg, A., Long-Term Effectiveness of Zero-Lift Program in Seven Nursing Homes and One Hospital, *NIOSH,* August 1999.

Gaze, H., Occupational health: get back on track, *Nursing Times,* 93(40)), 40–41, 1–7, 1997.

Harber, P. et al., Personal history, training and worksite as predictors of back pain of nurses, *Am. J. Ind. Med.,* 25, 519–526, 1994.

Health Science Center at Winnipeg, Manitoba Canada, Giving health-care workers a helping, mechanical hand, *CTD News,* 73–77.

Hospital Employee Health, Lifting teams can help hospitals eliminate costly back injuries to nurses, 13(7), 81–87, July 1994.

Hospital Employee Health, Sacrificial lamb stance is killing healthy backs, 29–33, March 1999.

Lagerstrom, M. and Hagberg, M., Evaluation of a 3 year education and training program, *AAOHN J.,* 45, 2–83–92, 1997.

Larese, F. and Fiorito, A., Musculoskeletal disorders in hospital nurses: a comparison between two hospitals, *Ergonomics,* 37(7), 1205–1211, 1994.

Logan, P., Moving and handling, *Community Nurse,* April, 22 –24, 1996.

Marras, W.S. et al., A comprehensive analysis of low-back disorder risk and spinal loading during the transferring and repositioning of patients using different techniques, *Ergonomics,* 42(7), 904–926, 1999.

Owen, B. and Garg, A., Reducing risk for back pain in nursing personnel, *AAOHN J.,* 39(1), 24–33, 1991.

Owen, B., Garg, A., and Jensen, R., Four methods for identification of most back-stressing tasks performed by nursing assistants in nursing homes, *Int. J. Ind. Ergonomics,* 9, 213–220.

Owen, B. and Fragala, G., Reducing perceived physical stress while transferring residents. *AAOHN J.,* 47(7) 316–323, 1999.

Perrault, M., Investing in ergonomics, *OH&S Canada,* Sep/Oct., 39–45, 1995.

St. Vincent M. and Tellier, C., Training in handling: an evaluative study, *Ergonomics,* 32(2), 191–210, 1989.

Stensaas, L., Wyoming State Training School, November 10, 1992.

Stubbs D. et al., Back pain in the nursing profession II. The effectiveness of training, *Ergonomics,* 26(8), 767–779, 1983.

Villeneuve, J., The ceiling lift: an efficient way to prevent injuries to nursing staff, *J. Healthcare Saf., Compliance Inf. Control,* Jan. 1998: 19–23.

Waters, J., Reducing the risks from lifting, *Nursing Times,* 93(50), 52–54, 10–16, 1997.

Werner, S., Department of Mental Health, Mental Retardation and Substance Abuse Services, Internal Report, Fairfax, VA, October 22, 1992.

Zhuang, Z., Biomechanical evaluation of assistive devices for transferring residents, *Appl. Ergonomics,* 30, 285–294, 1999.

Appendix 6A: Examples of Engineering Controls for Patient Handling Tasks

Sliding boards. For bed to stretcher type transfers, low cost sliding boards are available. Sliding boards are usually made of a smooth rigid material with a low coefficient to friction. The lower coefficient of friction allows for an easier sliding process. These boards assist when lying slide transfers are done. Some, but substantially reduced, force is still required to move the patient; however, sliding boards do offer some improvement at a minimal cost. Sliding boards offer a starting point, with a low initial investment to begin to improve the way patient transfers are conducted. (Approximate cost $80 to $120.)

Air-assisted lateral sliding aids. These are devices where a flexible mattress is placed under a patient in the same manner as a sliding board. There is a portable air supply attached to the mattress, which inflates the mattress. Air flows through perforations in the mattress and the patient is moved on a cushioned film of air allowing staff members to perform the task with much less effort. (Approximate cost $1200 to $1600.)

Friction reducing lateral sliding aids. These devices are positioned under the patient or resident similar to a sliding board, but rather than moving with the patient they provide a surface for the patient to be slid over more easily due to the friction reducing properties of the device. These are simple low cost devices, usually made of a smooth fabric, which is foldable and very easy to store. Properly designed handles can reduce horizontal reach, such as experienced with traditional draw sheets. (Approximate cost $100 to $120.)

Flat stretchers with transfer aids. Stretchers are now available which are height adjustable and have a mechanical means of transferring a patient on and off the stretcher. Some are motorized and some use a hand crank mechanical device. These mechanical means of mechanizing the lateral transfer are also available as independent options able to be used with most beds and stretchers. These devices eliminate the need to manually slide the patient further reducing risk to the worker. (Approximate cost $1800 to $8000.)

Convertible wheelchairs. Since bed to chair transfers are difficult because lifts are involved, some new wheelchairs can convert into stretchers where the back of the wheelchair pulls down and the leg supports come up to form a flat stretcher. These devices eliminate the need to do transfer in and out of wheelchairs. There are wheelchair devices that convert to stretchers, which also have a mechanical transfer aid built in for a bed to stretcher or stretcher to bed type transfer. (Approximate cost $1000 to $3500.)

Gait belts. An object with handles improves the grasp opportunity for the worker and reduces the risk. Gait belts are belts that go on patients or residents, usually around the area of the waist, that provide handles for a worker to grasp when assisting or transferring a dependent patient or resident. Small hand held slings, which go around the patient, can also facilitate a transfer by providing handles. These options are available for patients with weight bearing capability who need only a small assist. (Approximate cost $8 to $20.)

Full body sling lifts. Probably the most common lifting aid device in use is a full body sling lift. There are a number of models and configurations available. The majority of sling lifts are mounted on a portable base; however, use of ceiling mounted sling lifts is

1-56670-631-9/03/$0.00+$1.50

growing. Portable base and the ceiling mounted devices have their advantages. With a ceiling mounted device, there is no need to maneuver over floors and around furniture with bases. These units are quite easy to use; however, transfers are limited to where overhead tracks have been installed. Where overhead tracks are not available or practical, portable bases can be used to suspend full body sling lifts. At this time, portable base units are much more prevalent than overhead track mounted lifts. Sling lifts are usually used for highly dependent patients or residents. They can be used to move patients out of beds, into and out of chairs, for toileting tasks, bathing tasks, and for any type of lift where a patient or resident is highly dependent. These lifts are available with all types of features and, there is a wide variation in the types of slings available. Newer sling design is much easier to put on the patient or resident and in some cases, does not even require the patient or resident to be lifted or rolled to place the sling under them. (Approximate cost $1200 to $6000.)

Standing assist and repositioning lift. These lifts provide an alternative to full body sling lifts where a simple sling is used, which passes under the arms of the patient or resident, similar to the type of sling used for helicopter rescues at sea. The sling is of simple design and very easy to place on the patient. These types of lifts are very useful where patients are somewhat dependent and have some weight-bearing capabilities. They are excellent for moving patients in and out of chairs and for toileting tasks. They approach toilets from the front and can maneuver in small bathrooms with restricted areas. There are some variations in the sling design, but the basic concept is to put it around the patients back and under their arms. (Approximate cost $1500 to $5000.)

Standing assist and repositioning aids. Some patients or residents may only need a little support to stand. In this case they can lift themselves if they have a support to grasp on to. Various types of devices can be provided to assist a patient from a seated to standing position by allowing them to hold on to a secure device and pull themselves up. These devices may be free-standing or attached to beds. (Approximate cost $80 to $300.)

Bathing lifts. There are a wide variety of bathing lifts available, some are integrated with the tub unit or may operate independently. A lift that can be used to lift patients or residents from beds in residential areas, then used as a transport device and further used also as a bathing lift will minimize the number of transfers required. Bathing lifts can be used in conjunction with ergonomically designed bathing systems, such as height adjustable tubs which allow for easy transfer of the patient or resident from the tub then bringing the patient or resident up to appropriate height to reduce static bent over postures by the worker while providing bathing care. (Approximate cost $1200 to $6000.)

Other ergonomic bathing devices. New and innovative bathing devices are constantly being developed. Some new devices now can be brought right to the patient or residents bed minimizing the transfers and transports required. These units have lifting and turning capabilities. Other new bathing and showing units might include an integrated system where a shower chair or wheelchair can roll right into a shower cabinet eliminating the need for a transfer.

Bed improvements. Current bed technology has incorporated many ergonomic improvements. Some examples include beds that eliminate the need for bed to chair transfers by easily converting to a chair configuration. Another innovation in bed design referred to as shearless pivot reduces the need to constantly reposition a patient in the bed by minimizing the amount of slippage down to the foot of the bed experienced by the patient through old methods of raising the head of the bed. Further innovations with bed mattress surfaces can rotate and move a patient as needed in many intensive care units using air bladders incorporated into the mattress surface. This concept can also be used to more easily position friction-reducing devices under a patient prior to a lateral transfer.

Injured Nurse Story #5: The First to Go

by Kmak, Palmdale, CA

When I look back on how my back injury happened, I feel that nurses have been used as pieces of equipment to lift patients. Then, when nurses are disabled from this work, they're cast aside, treated as expendable broken equipment.

Speaking in terms of being a man in nursing, we are used much more often to lift patients in the units. We're called upon to position patients, lift patients, and restrain patients, as well as for "Code Strong" when someone goes out of control. Men are usually mandated to go to such codes, making male nurses more prone to injuries. We're compassionate people and, without realizing, can suffer cumulative injuries from so much heavy lifting.

I have 12 years of experience through medical/surgical, telemetry, DOU (Definite Observation Unit), a step-down unit from ICU (Intensive Care Unit), and in the coronary care unit. We took care of CABG (Coronary Artery Bypass Graft) patients following open heart surgery in an acute care hospital. We had the general type of patient with heart conditions, normally big and heavy. Most, coming from surgery, had a lot of tubing, wires, and other medical devices attached, to give a picture of what the patients looked like.

In the hospital where I worked, there wasn't much training on transferring patients. We had a safety class for 1 hour every year. On lifting and transferring, there were maybe 5 or 10 minutes on a video. So, that's all the training we had. We didn't have training on the equipment on the floor and the equipment was usually broken down or hidden somewhere. We had maybe one Hoyer lift for the whole floor with almost 60 beds on the floor.

I sprained my back three times. The first time was in 1990 when I first started my job. Then, I sprained it again in 1995 and again in 2000. The last time was while lifting up a patient in bed.

Our hospital system has a workers' compensation company that was subcontracted by the hospital. The workers' compensation company doctors are supposed to treat us for 180 days before we had the option of being seen by someone else. The workers' compensation doctor, under contract with the hospital, said it was a sprained back, gave x-rays, ordered hot pads, and put me back to work with modifications. But, the modified duty is normally heavier than you would think — pushing medication carts, admitting patients, pushing beds, putting patients into bed, assisting onto weight scales, starting IVs, and so forth. They called it modified work but it was more of a joke to say it was modified.

My pain got so bad, I couldn't do it anymore. I went back to the doctor and said something's not right. They said everything was okay and you should go back to more of your usual routine and the pain will go away. They have studies to show that people get back to their normal life faster if they increase activity. Being a nurse, I know that x-ray would only show so much and they should do a CT scan, but the contracted doctor would not approve it.

I sought legal counsel, who sent me to an independent orthopedic doctor. This was a real orthopedic doctor. The first doctor was a cardiologist trying to treat my back. The independent orthopedic doctor sent me for an MRI, which showed that I have a slipped disc at the L4/L5 region, 8 to 10 millimeters in size. I'm on the borderline of going for surgery or not. So, I chose not to

1-56670-631-9/03/$0.00+$1.50

have surgery. I went through a whole 6 months of physical rehabilitation and sitting around at home when I couldn't do much of anything.

The injury basically almost ruined my life. I can't lift anything heavy. I can't play with my child. I have physical restrictions for not lifting anything over 25 pounds. And, my final restriction will be to never lift more than 50 pounds. After all the rehabilitation, I guess my disc has "frozen" now. Normally, after a few years, it should "freeze up" or calcify. Right now, I'm considered disabled but stationary with advice not to do what I've done in the past for work. I've been advised to leave nursing.

The hospital did not assist me to return to work or to receive appropriate medical care. I feel they have shrugged their responsibility by having the contract with the workers' compensation company that restricted proper medical care. So, in a sense, they are not supporting their employees when they are hurt.

I went back and said, "Why don't you give me a job that does not require any lifting, like case manager?" They said, "There is no such opening." I realized they were paying case managers less than nurses, so, they wanted to hire people off the street instead of keeping me at nurses' wages. They do not support the nurse. Once injured, you are much more of a liability to them than an asset.

When I sought legal counsel, they still didn't trust the employee who had been with them for 12 years. They were sending out investigators to take pictures of me, in front of my home, doing my daily chores. They wanted to see if I was lying. So much for being supportive.

I couldn't work for almost a year and a half and only received workers' compensation or state disability. My life was almost ruined. I was $30,000 in credit card debt because of house payments and responsibilities as the sole bread winner to my family with a wife, two daughters, and two dogs.

They paid me workers' compensation for a while. When those payments stopped, I had to file for state disability by myself, explaining that workers' compensation had stopped. After the attorney filed against the hospital, workers' compensation started paying again. They will try in any way to cheat the injured workers. So, legal counsel is encouraged at this point.

I think the emotional aspect is very important. I have the feeling that nobody cares. The employer doesn't care. The state system doesn't care. How can they be paying you $490 per week to sustain your life when you were making two or three times that amount? Workers' compensation paid $490 per week until I went into rehabilitation and then paid another rate. My family life is affected. Even sexual activity is affected because of back pain. And, you can't even play with your loved ones, your daughters. So, emotionally, it's a wreck, in terms of being injured.

Even if the lift equipment worked, it was hidden. It was never found. And, that equipment was just to weigh patients. It was never to lift patients up into the chair or lifting them up in the bed, to keep them from sliding down.

Our hospital has a nonlifting policy of not more than 30 pounds and I think this is a joke. I think this policy was written so the hospital could cover themselves when somebody gets injured. Then, they say, "Why didn't you follow the policy?"

And, when you are injured, they would send you to see a one-hour video- class to teach you all about body mechanics. This puts you in the situation of being at fault, as if you didn't follow the policy, or you didn't watch the video, or you didn't have the knowledge of how to prevent back injuries.

All of the fault points to the person with his injury and never points to the hospital to say they should have policies in place and equipment to prevent work injuries.

Ask any nurse on duty right now. They would all know at least a few nurses with back injuries. That's how rampant back injuries are among nurses.

I worked in a CCU unit with 16 beds. Being a man, I helped out every female nurse there with getting patients up to the chair or lifting them up in the bed — at least a few times a day for each patient. That's all cumulative.

"Watch your back!" is what I would say to anyone considering nursing today. Every nurse should be on a mission to educate himself as well as coworkers that back injury does not pay.

With the current nursing crisis, and the best staffing pattern, they are increasing the number of patients to each nurse, which will add more stress to the nurses. This, in turn, will keep RNs further away from this beautiful profession.

We passed safe staffing ratios legislation in California and, hopefully, it will be implemented as soon as possible. This would relieve some of the back injuries to current nurses with the heavy patient loads that they have.

I don't know what they're going to do about injuries, to decrease injuries to men, seeing as they've been used the most for lifting, and whether they are going to do studies about back injuries to male nurses.

In all the hospitals I've worked, there have been no lift teams. So, if the studies point to decreasing back injuries, why isn't the hospital industry following the recommendations for reducing back injuries? I believe the hospital industry does not care. If there are more nurses out with back injuries, they have more reason to not staff the hospitals with adequate RNs, with the excuse that there are no available RNs or they're out with disability.

You would be a liability if they allowed you to return. Once you have a history of back injury, you may be out again in a few months or a year. They know that and they don't want you. You are no more than an old IV pump that is not running. That is how they look at you.

I want people to see the perspective that the men are there to help out, that the men are the muscle. The female nurses say, "I need some muscle" and they don't mean anything wrong by that. Men, of course, being obligated to help, should know that cumulative stress does a lot of harm and, then, you just lose another nurse. Now, I can't use the education I went to school for and can't be at the bedside as a nurse. And that's the end of the nurse's career, right? I'm 42 years old, 5 foot 10 inches, and about 175 pounds — a lot of muscle, but a bad back now.

They expect you to be 100% functioning. They don't want to hire someone with an injured back. If you don't fill in the truth, they can terminate you. It's a disadvantage to someone with a back injury in seeking work. I don't know what else I can do besides educate all my fellow nurses and friends that they have to watch their backs.

Now, I work with the California Nurses Association as a labor representative, spreading the word about nursing, about the power of nurses, to improve the image of nursing, pushing for legislation, enforcing contracts, and so on. It's pretty exciting to be involved with this work, this aspect of nursing.

The effect of back pain still lingers on and is a bad thing. It keeps reminding me and I have to pass along to others to watch their backs. As nurses, we say you lose your hearing first, then your sight, then your brain and your mind. But, for a male nurse, the first to go would be your back.

Injured Nurse Story #6: My Heart is Still There

by Beth DeWees Piknick

While working as a registered nurse in the Intensive Care Unit (ICU), I suffered a career-ending back injury that was devastating, both personally and professionally.

The hospital where I worked was a 240-bed nonprofit community hospital with approximately 400 registered nurses at the time of my injury. The ICU was a 13-bed unit. Our patients were everything from trauma, surgical, medical, neurosurgical, and even cardiac, although, there was a 14-bed coronary care unit across the hall. I considered this unit to be progressive in regard to many things, including our equipment. Our nurse manager was always supportive of the nurses in the unit, always looked out for our best interest. I felt that the hospital, as well as the ICU, was up-to-date with all equipment.

In our annual mandatory safety sessions, which all nurses attended, lifting patients was demonstrated via videos. I always thought the films and advice were rather inappropriate and not applicable to real practice. Most of our patients were not cooperative, were not elderly, frail women sitting at the edge of the bed who needed gentle guidance back to bed. This was our video! There was never any mention of the use of any equipment and the only equipment we had was a sliding board.

Moving patients from their beds to chairs and vice versa were some of the many lifting maneuvers I performed. I won't even get into the occasional fist you had to move away from quickly or the unexpected family member who you were suddenly catching as they fainted, never mind the patients I was trying to prevent from hurting themselves. These are the lifting activities I performed throughout my 21-year career in the ICU. I would twist, bend, pull, and push. It was part of the job. I never had any back problems.

But, on February 17, 1992, while leaning over a patient to assist him to bed, I severely injured my back. I was 42 years old. I went to chiropractors, physicians, surgeons, and physical therapists — determined to get back to work, but I remained in pain. I underwent an MRI, a mylogram, a discogram, and wore a Boston Brace, a brace made of a hard plastic material, worn from your chest to your hips. I was willing to go through whatever I had to, but my major goal was to get back to the job I loved — taking care of critically ill patients and their families.

The discogram showed my injury was cumulative. I had degenerative disc disease. The disc between L2 and L3 was completely gone and the disc between L3 and L4 was all but gone. I basically had bone on bone. Sixteen months after my injury, I had a spinal fusion, L2–L4 was fused with iliac bone and titanium wire. After 5 days in a hospital, which was 2 hours from my home, family, and friends, I was discharged. It was summer and it was certainly hot in the brace! I wore that thing for almost 4 months. As much as I hated it, it was my best friend as it made me feel secure. I was able to remove it while I was bathing, however, I could not bend or move without it. So, in the heat of the summer, my poor husband had to shower, bathe, wash my hair, and shave my legs. I was always cold; therefore, the water was always warm. My husband does not like the heat at all but endured these torturous showers without ever complaining or even making a face. Okay, he did perspire a little! After 6 weeks, I was able to go swimming, or should I say wading, with my brace. This was wonderful!!

1-56670-631-9/03/$0.00+$1.50

I have considered myself quite fortunate in dealing with my hospital and workers' compensation during the two and a half years I was out of work. I returned for a short time before surgery and wasn't able to continue. There was never a refusal or problem with my treatments or payments. The biggest problem I had was when it was decided I should have surgery. My surgeon and I were prepared in early March and could have had it done then. March would have been good for me personally as my husband was home, his work is seasonal, and it was cooler, plus I wouldn't have had so much time to think of all the things that could go wrong. I could have saved myself many tearful phone conversations with the claims person. I had to wait almost 5 months for their approval. This was the only time I had a problem with workers' compensation.

As I stated earlier, my goal was to get back to the job I loved, but after surgery and major rehabilitation that remains impossible. The reality is that, if I want to be pain free, I need to be in control of my environment. You can't be in control of your environment when you take care of patients. I must constantly be aware of my surroundings. I cannot stand for long periods of time without support. I'm usually looking for a wall. I can only sit in certain chairs. I am usually looking for ones with cushions. I cannot lift or do any repetitive bending or twisting motion. My official restriction, per my physician, is no lifting more than 40 pounds (I can really feel problems at 25 pounds) and no repetitive forward bending. Before my injury, I was an active person who enjoyed bicycling, competitive racquetball, water-skiing, and yearly white water rafting trips with my family, all of which I can do no longer.

I was able to return to the hospital as a registered nurse, which many of my back-injured colleagues are unable to do. The hospital was reasonable and my timing was good, as a permanent "light duty" position was created at the time of my return. At the time, I performed TB surveillance activities tracking, monitoring, testing, and counseling hospital employees for the Employee Health Department. This was not why I became a nurse. I became a nurse to care for patients. My identity has been based mostly on being an ICU nurse. The inability to care for patients was devastating and triggered four and a half years of severe clinical depression. I truly realized how fortunate I was to be able to return in any capacity but it didn't seem to matter. The costs of the treatment for clinical depression, which workers' compensation did not cover, mostly came out of my pocket. My family has had to cope with both my physical and psychological problems. The true cost to them can never be quantified, especially for my husband. I was thankful that my children were grown. It was horrendous and 10 years later, I am still somewhat adjusting.

Two years ago, my position in Employee Health was eliminated. Thanks to my union (the MNA), the hospital, and the nurse manager, who gave me accommodations, I was able to get another job. After MNA negotiated with the hospital, the requirements of the Employee Health position were carried over. One of the qualifications for the position was to have a permanent disability from a work-related injury. I'm now working in an endoscopy unit. I only work in the preadmission area doing pre-admission assessments, vitals, and IV starts. I do not work in the recovery area or in the procedure rooms. I don't work where the patients are medicated. I work only three days a week and can't really work 2 days together. It certainly does not have the challenge of an ICU but I am with patients again. However, it is bitter sweet, because I am with patients, even though they are supposedly healthy and not medicated as yet, things can happen and I'm not in control of my environment, which leads me into back pain.

I am often bored and miss the challenge of the ICU. It still hurts to walk through the unit doors. My heart is still there and probably always will be; but, I am grateful that my body can still be with patients. And, once in a while, a patient will come in for a procedure and I feel like I've helped them; I feel like a nurse.

Throughout my career, I have been actively involved in my state nurses association, the Massachusetts Nurses Association (MNA) and, through them, involved with the American Nurses Association (ANA). In 1999, I was asked by ANA and MNA to speak in Washington regarding proposed OSHA ergonomic standards. This was held in a press conference in the Capitol Building.

This was the first time I spoke openly about what had happened to me. I went to Washington again in 2000 at the Department of Labor.

After speaking in Washington, I was given an opportunity, since I was working for Employee Health, now Occupational Health, and involved with MNA's Congress of Health and Safety, to trial mechanical lifting devices. After quite a while, the hospital purchased five lifting devices and I was given the responsibility of providing in-service to the nursing staff throughout the hospital. I was also involved with trialing a lifting device that helps move a patient from bed to chair without lifting. This, I feel, is one of the most dangerous insults to our bodies. Unfortunately, the hospital did not purchase that particular one but we do have the others. These new lifting devices can pick patients up off the floor!

One of the most surprising and disturbing moments in this whole experience was when I learned that my injury could have been prevented. I, like most nurses, assumed that lifting was part of the job. It isn't. I also thought that my hospital was up-to-date with equipment. It wasn't. At the time of my injury, I did not know that there were lifting devices commercially available and that my hospital simply chose not to have them. While I was trialing the lifting devices, I met a nursing assistant who was familiar with the device. She had worked with something similar *15 years* earlier in Florida. I was flabbergasted. I had no idea these existed for so long. I had always prided myself as being up-to-date and aware of new and changing things within my profession. I was stunned that I was unaware of such things. Engineering controls like lifting devices must be the primary means of prevention now and should have always been.

Perhaps, if I had been using such a lifting device for the last 15 years, I would still be living an active life. I would still be able to stand and have a conversation with someone without pain. I would still be doing what I love, caring for critically ill patients and their families.

I've become involved in contract language negotiations at my hospital this year, which I haven't done for 15 years. Some of the language that we were negotiating was "industrial accident leave" language. Also, after the hospital purchased the lifting devices, my position, as I said, was eliminated and the devices sat in the corner. The staff I had in-serviced were aware but quickly forgetful. There are many new people who not only didn't know how to use them but some didn't even know they existed. The committee was able to settle on individuals to be responsible for them. I was one of them. Also, my awareness of depression and its effects are tenfold. I find myself explaining and helping others with themselves or their loved ones regarding this very silent illness which, in today's society, carries so much misunderstanding and shame.

So, I have been able to do some positive things after this injury and I'm more or less settled into my life now. But, the bottom line is I sure wish I could be in the ICU and doing some white water rafting with my children and grandchildren!

7 Introducing a Safer Patient Handling Policy

William Charney

CONTENTS

ARE YOU STILL LIFTING?

No one working in a hospital, nursing home, or community setting should need to lift patients manually any more. Hoists, sliding aids, and other specialized equipment mean staff should no longer have to risk injury while doing their job. Yet manual lifting continues, taking its toll on nurses' health. One in four qualified nurses has taken time off with a back injury sustained at work (Disabled Living Foundation, 1994) and for some it has meant the end of their nursing career. But there is no need for injury to be an occupational hazard, and this chapter offers advice on implementing a safer handling policy. It also highlights some examples of good practice.

WHAT IS A SAFER PATIENT HANDLING POLICY (A BRITISH APPROACH)?

Such a policy might state: "...the manual lifting of patients is eliminated in all but exceptional or life threatening situations... patients are encouraged to assist in their own transfers and handling aids must be used whenever they can help to reduce risk if this is not contrary to a patient's needs." It is important to avoid any handling that involves manually lifting the whole, or a large part of, a patient's weight (this eliminates, for instance, the shoulder/Australian lift).

1-56670-631-9/03/$0.00+$1.50

Manual handling may only continue if it does not involve lifting most or all of a patient's weight. It is acceptable to give a patient some support, or to perform horizontal moves with a sliding aid, if this is done according to agreed safer handling principles. But the policy must include a commitment to use those safe principles and only patients who need light assistance should be handled in this way. The message to get across is that most of the time manual lifting is to be avoided.

IS IT FEASIBLE?

Patients can do a lot for themselves if shown or encouraged, and this will benefit them too. *Beware of pajama-induced paralysis!*

Hoists can transfer nonweight-bearing patients on and off a chair, bed, bath, toilet, or floor. A variety of rigid or fabric sliding devices deal with transfers onto bed or trolley and with moves up, down and around the bed. There are hundreds of handling aids for all needs (many are listed in the Disabled Living Foundation Guide (1994)). Handling techniques have also improved, for instance, patients can often be rolled instead of straight-lifted. Preadmission information to patients can also be useful.

With a safer handling policy, staff must continue to assess the capabilities and rehabilitation needs of patients in order to determine which methods and handling aids should be used. In all cases, the handlers' physical effort should be minimal and well within their skills and capacity.

WHAT IF A PATIENT REFUSES TO BE MOVED WITH EQUIPMENT?

This is unusual, but it can happen. If it does, then someone skilled needs to be brought in, for instance, a handling trainer or coordinator. Sometimes staff have transmitted their own insecurity over the use of equipment, or the patient has been hurt in the past by the clumsy application of slings. In this case, perhaps another method or piece of equipment can be used. The benefits to the patient and relatives should be pointed out. If all attempts at persuasion fail, then a manager must decide whether to refuse the patient an element of care. They will have to weigh the risk of injury to staff against the risk to the patient if a particular procedure is not carried out. In practice this difficult situation is rare. In cases that have cropped up in the community, it has usually been feasible to nurse the patient in bed while steps are taken to persuade them to be moved by a hoist.

IS IT EXPENSIVE?

Hospitals have traditionally assumed that it would be far too expensive to provide enough handling equipment to even consider a radical change in lifting policy. In fact, costs are not that great and the cost of not tackling the issue is likely to be much greater. Modification of care areas, such as widening toilet cubicles, can often be done with a modest outlay. Organizations should estimate costs and allocate budgets, if necessary over several years, so that a safer handling policy can be fully implemented. Overall the total spending on equipment (hoists, sliding aids, special baths, etc.) is likely to be around 0.3% of the organization's annual budget, and most places will have much of the necessary equipment already.

BECAUSE IT BENEFITS STAFF

A safer handling policy means nursing no longer has to be a heavy job. Sickness and injury should drop to levels enjoyed by other workers in similar physically active work. Trusts that have already introduced safer handling policies say their nurses report feeling less tired by the end of their shift. Introducing safer handling policies across the healthcare sector could have a dramatic effect. One review (Pheasant, 1991) of the research in industry showed that people in heavy occupations are 10 times more likely to have back pain than people in jobs involving medium workloads.

BECAUSE IT BENEFITS PATIENTS

Safer handling contributes to quality of care. Encouraging self-help will stimulate patients both physically and mentally, reduce side effects associated with immobility and contribute towards physical rehabilitation programs. Most dependent patients appreciate being lifted more safely and comfortably in a hoist, and obese patients may feel less reticent about asking caregivers to move them.

THE EMPLOYER'S DUTY

> Each employer shall, so far as is reasonably practicable*, avoid the need for his employees to undertake any manual handling operations at work which involve a risk of their being injured. Where that is not reasonably practicable they must make a suitable and sufficient assessment and take appropriate steps to reduce the risk of injury to those employees arising out of their undertaking any such manual handling operations to the lowest level reasonably practicable. From the *Manual Handling Operations Regulations*, 1992.

THE EMPLOYEE'S DUTY

> "Each employee shall make full and proper use of any system of work provided." From the *Manual Handling Operations Regulations*, 1992.
>
> Staff should "take reasonable care for the health and safety of himself and of other persons who may be affected by his acts or omissions" and "co-operate with his employer." From the Health and Safety at Work Act, 1974.

PLANNING A POLICY

A full safer handling policy cannot be implemented until:

- Sufficient and suitable handling equipment and furniture is in place.
- The environment is suitable, for instance, cubicles made bigger, extra hand-rails installed, floor quality improved, etc.
- Staff have received adequate training (including patient assessment and the use of equipment).

A first step is to conduct a risk assessment (as required by the *Manual Handling Operations Regulations),* cost the needs, find acceptable compromises where necessary (e.g., if a bathroom cannot be made bigger), and set aside a budget, possibly spanning several years. Meanwhile, an interim policy should be set up stating that all staff are expected to follow safer handling principles whenever possible but where it is recognized that in practice this will not happen everywhere and all the time.

THE ROLE OF ASSESSMENTS

Staff should continue to assess the needs and capabilities of their patients and devise the best handling methods accordingly. A safer handling policy restricts their choice of methods to those that the organization considers acceptable. Although a member of staff may feel competent to select

* "Reasonably practicable" means weighing up the risk of injury against the cost or effort to introduce changes. Employers could only justify doing nothing if the cost of measures greatly outweighs the risk. The burden of proof rests on the employer. Note that in legal terms, if measures are too expensive for an organization, that does not prevent them from being "reasonably practicable."

another method, the policy should not allow them to do so except in an emergency. Any nonurgent exceptions to the policy needed by a ward or group of staff should be agreed with management, in consultation with a trainer/coordinator.

MONITORING

The policy is worthless if it is not enforced. This is true even in claims for negligence. Before taking disciplinary action against staff who do not comply with safer handling policies it is worth looking for reasons behind it. For example there may not be enough readily available handling aids. One simple way of monitoring compliance is to check the patient care plans. These should clearly indicate the capabilities of the patient, along with the equipment and handling methods used for various moves. Traditional patient care plans do not make it easy for staff to write enough detail, so the format may need to be changed. This is not just for monitoring purposes but to ensure that the results of a patient assessment are clearly communicated to all professionals caring for a patient.

The success of a safer handling policy will be measured through incident or accident reports, sickness absence levels, reports to Occupational Health, and the numbers of civil claims.

ACKNOWLEDGMENTS

The authors are grateful for information supplied by Margaret Milne and Sister Cartlin, who were both dedicated to the no-lifting policy let in Broomhill Hospital Glasgow.

SUGGESTED READING

U.K. Government Health Services Advisory Committee, Health and Safety Commission, *Manual Handling in the Health Services* (ref C00398), HSE Books, Suffolk, U.K., 1998.

H.M. Government Health and Safety Executive, *Manual Handling Operations Regulations 1992: Guidance on Regulations*, HSE Books, Suffolk, U.K., 1992.

Royal College of Nursing, *Code of Practice for Patient Handling*, Royal College of Nursing (Publication code 000 604), London, 1999.

Royal College of Nursing, *Manual Handling Assessments in Hospitals and the Community: An RCN Guide,:* Royal College of Nursing (Publication code 000 605), London, 1999.

Royal College of Nursing, *Taking a Uniform Approach: An RCN Guide to Selecting the Right Clothing for Nurses,* Royal College of Nursing (Publication code 000 993), London, 1999.

Pheasant, S., *Ergonomics, Work and Health*, Macmillan Press, London, 1991.

Seccombe, I. and Smith, G., In the balance registered nurse supply and demand, IES Report 315, 1996.

OTHER READING

U.K. Government HSC/HSAC, *Manual Handling in the Health Services* (ref C100398), HSE Books, Suffolk, U.K., 1998.

Royal College of Nursing *The WING Guide for Injured, Ill or Disabled Nurses*, Royal College of Nursing (Publication code 001006), London, 1999. (Available only to members of the RCN Work Injured Nurses Group - ring 0345 726100 for more details).

Appendix 7A: Memorandum of Understanding (New in its entirety, between Association of Unions and Health Employers Association of British Columbia)*

Re: Manual Lifting

The parties agree to establish a goal of eliminating all unsafe manual lifts of patients/residents through the use of mechanical equipment, except where the use of mechanical lifting equipment would be a risk to the well-being of the patients/residents.

The Employer shall make every reasonable effort to ensure the provision of sufficient trained staff and appropriate equipment to handle patients/residents safely at all times and specifically to avoid the need to manually lift patients/residents when unsafe to do. If the use of mechanical equipment would be a risk to the well-being of the patients/residents, sufficient staff must be made available, to lift patient/residents safely.

The parties agree to take the following immediate steps through the Occupational Health and Safety Agency for Healthcare to achieve this goal through out the sub-sector.

a) Work in partnership with the Workers' Compensation Board, the Ministry of Health and others to establish a financing framework to make funds available to purchase the necessary mechanical equipment.
b) Finalize and distribute clear industry guidelines for safe patients/residents handling.
c) Encourage the full participation of the local joint Occupational Health and Safety Committee in the development, implementation and on-going monitoring of this goal.
d) Recommend to the Ministry of Health that all new healthcare facilities be equipped with appropriate lifting equipment.
e) Produce an annual report card on the progress to date including specific recommendations for the coming year.

* From the Comprehensive Report Facilities Subsector Tentative Agreement (Page 35).

1-56670-631-9/03/$0.00+$1.50

Appendix 7B: International No Lift

Manual lifting of patients has created an epidemic of back injury and associated injuries to healthcare workers. In the U.S. these injuries represent almost a third of ergonomic injuries to healthcare workers and in the United Kingdom it was estimated that one in four of qualified nurses has taken time off work due to back injury and for some it has meant the end of their career. Studies have shown that manual lifting of patients exceed NIOSH recommendations even to the upper limit of 6400 N and many lifts are in the micro-fracture range. Research (Marras, 1999) has quantified compressive forces during many different forms of one and more person transferring of patients and concludes that manual lifting of patients is an extremely risky event for all involved in the lift.

A "no manual lift" philosophy is an approach that involves:

1. Identifying risk
2. Auditing existing equipment
3. Purchasing equipment based on assessed need
4. Training staff on the equipment
5. Mandating the use of mechanical equipment
6. Prohibiting the manual lifting of patient except in emergencies

The equipment costs are normally no more than 0.3% of the facilities overall operations budget and pays for itself according to peer review studies within 12 to 15 months of purchase.

There are many countries that have enacted either zero-lift legislation or policies, preventing their healthcare workers from doing manual lifting, transferring or re-positioning of patients, including England, Australia, New Zealand, South Africa, Ireland, Sweden, and Denmark. Many provinces in Canada as well are beginning to convert over to a zero-lift philosophy. However, the United States is far behind these countries in approaching a zero-lift regulatory attitude, despite the fact that there are 124,000 unfilled nursing positions in the U.S. and that healthcare back injury is playing a major role in this national shortage. In a recent American Nurses Association survey (2001) 39% of respondents said they consider leaving the profession due to fear of a disabling back injury. SEIU 1199 in New Jersey, also in a recent survey of their certified nurses assistants found that 40% leave that job classification due to back injury.

ZERO-LIFT LEGISLATION OR MEMORANDA OF UNDERSTANDINGS OR POLICIES

The province of British Columbia has a memorandum of understanding between the associations of unions and the Health Employers Association that "all parties agree to establish a goal of eliminating all unsafe manual lifts of patients/residents through the use of mechanical equipment…."

In the United Kingdom, the *Manual Handling Operations Regulations* of 1992 came into force in January, 1993. The regulations require employees to make use of equipment provided for them in accordance with training and instruction by the employer and follow appropriate systems of work to promote safety during the handling of loads.

The Australian Nursing Federation adopted a no lift policy for all its nurses in March, 1998. It states in part, "…the manual lifting of patients is to eliminated in all but exceptional or life

1-56670-631-9/03/$0.00+$1.50

threatening situations." This policy was adopted because nurses in Australia had the highest injury rate in the female workforce and they accounted for over half the number of claims in the health industry in 1995/96.

WHAT IS TO BE DONE

It should be axiomatic that states adopt a "no lift" policy to protect against this type of injury in the healthcare industry. Labor needs to put as much muscle and zealousness into passing this legislation as they did for the needlestick legislation. It needs to be a industry specific "no lift" legislation campaign that is introduced state by state until the federal government enacts such regulation. Some hospitals around the country and healthcare delivery systems are voluntarily implementing programs such as zero lift or lift teams. But the numbers of hospitals doing programs remains critically small (probably less than 5%). This is disconcerting when one looks at the rates at which nursing staffs are getting injured compared with the speed at which healthcare systems are implementing programs. The bottom line is that the safety programs are not keeping up with the injury rates, which means devastating injuries, millions of days of lost time, nurses leaving the profession, patients getting injured, increased nursing shortages, and millions of wasted dollars.

REFERENCE

Marras, W.S. et al., A comprehensive analysis of low-back disorder risk and spinal loading during the transferring and repositioning of patients using different techniques, *Ergonomics*, 42(7), 904–926, 1999.

Injured Nurse Story #7: I Won't be There

by Sad L&D Nurse

I've been an RN in obstetrical nursing for more than 20 years. It's all I ever wanted to do since entering nursing school. I loved my work as a labor and delivery (L&D) nurse, helping laboring women and new mothers. Now, all I want is for people to know what happens to injured nurses.

When I was an 18-year-old nurses' aide, I worked in a nursing home where they had a Hoyer lift and we didn't manually lift anybody. It was a small nursing home but they had a Hoyer on every floor. They tried to make it so the staff didn't hurt themselves.

It never crossed my mind when I went to work in the hospital that they weren't protecting us from injuries. I loved my job but, in the end, it made me a physical wreck. When I went to nursing school in 1974, I was never told that nursing involved heavy lifting. At the time, I had a choice of becoming an RN or an iron worker. In the 1970s they were trying to get women integrated into male-dominated workplaces because of women's minority status. I actually thought nursing was going to be the lighter job. But, if I'd hurt my back as an iron worker, or a firefighter, instead of as a nurse, I wouldn't be having all my financial problems – I'd be out on a pension with 50% of my wages plus medical care. I used to be on the nurses' union negotiating team. Most of the nurses felt we didn't need a raise, just a little time off, and would settle for meager changes in the contract. Nurses are co-dependent.

If a man went for a job interview and was told, "We're hiring you full-time, but you could work any shift, nights, days, or evenings, and when the patient census is low, we're going to send you home and not pay you. You'll be exposed to heavy lifting and, when you're injured, you'll have no assistance and no monetary support. You'll be terminated if you can't keep lifting. You'll be exposed to viruses and bacteria. You might have needlesticks and you might get AIDS. You'll be harassed by the physicians and might be assaulted by the patients. Because you're a college graduate, we're going to start you at $32,000 a year, but, you may not get all of that, because, remember you'll be sent home during low census." Now, how many men are going to take this job? I don't believe there are very many men who would do that, but women do because we are caring individuals and we think we can make a difference.

We had Employee Day once a year with some safety videos about 20 minutes long. Nine times out of ten during Employee Day, I would get called away from watching videos to L&D because they were short-handed. The hospital had plastic slider boards for transferring patients from the stretcher to the bed. But in L&D, we worked one-on-one with patients. So, if you needed to get a patient off the toilet or out of the bathtub, you did it yourself. And, many times we worked 12- to 14-hour shifts without a break: no bathroom, no water, no food.

We often placed a laboring woman's foot on our hip for them to push against. Sometimes the woman would forget and kick and we'd get kicked over. For the delivery, we removed the bottom part of the birthing beds which are heavy metal, stainless steel covered with enamel. We had to bend over and pick up the removable part of this ill-designed equipment. Infant warmers are high at the top and heavy at the bottom for stability, but there's drag with pushing them from room to room – more ill-designed equipment to push and pull, as well as pushing, pulling, and lifting the patients.

1-56670-631-9/03/$0.00+$1.50

There were two kinds of bathtubs used for women in labor. We had to bend over and help women up out of the tubs. And, we had to bend over the tubs in awkward positions for a minute or two to listen to fetal heart tones. It's a wonder that my cumulative injury didn't happen sooner. Maybe it's because I was so strong.

Nowadays, we live on a small farm but can't have animals because I can't tend them. I was an avid gardener and landscaper but can no longer do that. I was also a world traveler. I was a part of the women's health delegation to China which looked into women in labor, breastfeeding, and other issues related to women's health. I've been to Africa and Mexico. My husband and I are kind of amateur archeologists. We like bones and civilizations. It's so much fun to go where another language is spoken and try to figure out how to talk to each other. It's fascinating. Now, we can't do any of that because of my injury.

My injury has changed our lifestyle. I made good money at the hospital. We should be at our greatest earnings. My husband and I had goals for retirement that we're not going to achieve. Every day I think, "Should we sell the car?" I was saving for retirement. I'm not saving anything anymore. I think, "What will happen to us in ten years? What is our future?"

MRIs of my lumbar and thoracic spine show herniated discs at T7/8, T8/9, T9/10, and L2/3, L3/4, L4/5, and L5/S1. They didn't MRI my cervical spine but my neck kills me sometimes. What you don't realize is the "domino effect." Your whole spine can be affected. I have a new x-ray report about the amount of disc degeneration, some moderate, some severe. The evidence of extensive spinal injury is there but workers' compensation wants to go on just one incident of back sprain/strain with moving a bed.

I was leaning over, pushing a birthing bed. I was preparing to pull it out of the room. It was our policy to move patients who had delivered from the more narrow birthing beds into regular hospital beds. I had unlocked the bed wheels and thought it was just a little stuck but the electrical cord was wrapped around the front wheel where I didn't see it. I bent over and gave the bed a little push. It was when I pushed the bed that I had sharp, searing pain across the middle of my back. Now, I was injured but I still needed to take care of my patient. She was sitting on the toilet calling me and I still had to pull the bed out. I had to finish my charting and take care of my patients while icing my back.

That day, I filed a report and drove to my nurse practitioner. I could barely get out of the car. My NP gave me injections, Flexeril, ibuprofen, and a note to be off work for a week. Then, she sent me to physical therapy for six weeks. Eventually, the PT said, "You're not getting better. You're getting stronger, and more flexible, but something's wrong here." That's when the MRIs were ordered. The doctor told me not to go back to work. He said I needed fusion surgery but wanted to wait six months to see if I improved. I have not had surgery and am still undecided about it. I've had Prednisone orally and by injection. Prednisone hops me up so that I move around a lot and it doesn't help the pain. I took so many anti-inflammatories that my stomach was an absolute mess. I've been on steroids and ibuprofen for two years. Now, I get rebound migraines from taking ibuprofen. Even the prescription pain medicine, Vicodin, has acetaminophen in it and I worry about the effect on my liver.

I went to a large pain clinic that was so far from home that I stayed at the pain clinic housing five days a week for about a month, then for two days a week. My husband went with me part of the time. They had me lifting progressively heavier weights, like picking things up off the floor and putting things over my head. This was combined with PT and stretching exercises, too. I've been told all along that staying stretched out will help with a quicker recovery if you have surgery.

During my interview at the pain clinic, I was chastised by the doctor when I told him my daily routine with housework, etc. He said I was not an active participant and not a good partner because my husband had to do all the vacuuming. Before I was through at the pain clinic, they decided I was able to go back to work because I was vacuuming more. Injured nurses are so vulnerable in the position we're in – not knowing if they're going to continue paying us, if we even have a job, if we'll ever be out of pain – and, then, to be attacked on that level, accusing me of not being a good marriage partner. It took me a long time to overcome the feeling that I was a bad partner to my husband.

When I walk up hill, I drag my right leg. At the pain clinic, they put me on an incline machine and I flunked it every time. The theory of the pain clinic is to strengthen other muscles to support those that are weak. They told us about the theory of pain, about the "Gate Theory," and how some nerves can become irritated and stay irritated. But the reality is that pain is not measurable. You can't put a gauge on pain.

As nurses, we're taught to believe what patients say about their pain, that pain is whatever the patient says it is. I now have total disregard for those in the medical profession with workers' compensation. I believe some of them categorize all injured people as trying to rip off the system. It seems insurance companies figure nurses don't get hurt. Workers' compensation even has private detectives taking pictures of people who are receiving pensions, to make sure they aren't out doing things like water skiing.

My attorney told me, "If you feel paranoid, it's probably valid." And, there were people at the pain clinic checking up on us, watching us in the halls and between the buildings. One time, one of the PTs parroted what I had said at the lunch table when none of the PT people were present. I had been talking with other pain clinic patients at the table.

The pain clinic tries to identify those who are faking. I looked into this and found that less than one percent is not reporting real pain. Nurses with injuries don't go around moaning and groaning unless we get bumped or do something that pinches a nerve or something. Then, we let out a little yelp, and, we go on. I try to only go out when I feel well and look perfectly normal.

I called the hospital with updates every week. One of the hospital managers tried to say that I was faking my back injury. This is one of the worst parts of being injured – the perception people have that we're faking. I take my cane when I have to go out on days that my pain is bad and have had people make fun of me for using my cane. And, these are people that I know, my acquaintances. That aspect is frustrating. The whole thing has just been a nightmare. It never ends.

Part of the pain clinic was like a work-hardening program, going from nine to three, like a day at work might be. I was able to spend the hours there, so it looked like I could work and they wrote that I was ready to go to work. The pain clinic was very detrimental to me. I was told all kinds of things I was supposed to be able to do for a job, like I could go out next week and do all different things. The psychologist at the pain clinic kept telling me I was fine, that there was nothing psychologically to keep me from a job. I've said all along that activity makes my back worse. They set me up for failure.

Six months into my injury, my employer called to tell me my job was gone. By contract, you can only be out six months. My seniority goes for two years, though. I still believe hospitals could show a little more concern. They could have said, "We're sorry we can't keep your job open for you any longer" but they don't seem to even care about us.

The other time my employer called and left a message for me at the pain clinic saying they had a job for me – full benefits, same rate of pay. They would put me in a new workstation, and provide me with a new chair. They told me I had a job! Then, when I went back to the pain clinic to see my counselor, I was told, "Oh, by the way, they gave your job away." Those were the only times my employer called to say I had a job and then to say that I didn't. It absolutely broke my heart. I thought I would finally be able to work with my friends again. I had been on cloud nine. I called the hospital to find out who had done it, who had given my job away, but no one would say. It still hurts. I thought they were finally going to help me get back to work.

I gave the Federal Register on the Final Rule of the Ergonomics Standard to my attorney and attending physician because of the research on nurses' back injuries. I hope providing this information helps with my case. It's kind of frustrating. The doctor wrote that my injury was severe back strain and sprain from moving a bed and degenerative disc disease. I might have believed that myself until I read the research about cumulative trauma to nurses' spines from lifting patients. It's in black and while. I kept reading and thought, "This is amazing!" If I hadn't found the material, I might have believed that my years of nursing had nothing to do with the damage throughout my spine. I have one very messed up back and what the answer is for taking care of it, I don't know. My point is that this could happen to any nurse and nurses have to protect themselves because no one else is going to.

I had to go three times to workers' compensation IME's, independent medical examiners, their orthopedists and neurologists. Now, workers' compensation is close to closing my claim. They're saying I'm "medically stationary" and won't pay for me to see any more physicians. If I have further problems, I'll have to re-open my claim, which may not be easy to do. I'm stuck with a lifetime problem and other insurance companies don't have to cover it. I only have one more month of health insurance under COBRA. You can only carry COBRA for 18 months but I can't afford it anymore anyway. Before I was medically stationary, I was getting 60% of my wages at the time of injury from workers' compensation. I don't get that anymore. Since the union declared that I'm permanently disabled, I receive a $325 monthly pension from the union, and COBRA is $458. Fortunately, my husband has insurance. We were double-covered, but his insurance isn't as good as what I had with the union.

My application for Social Security Disability was denied. I'm ready to write my letter of appeal but am not even sure how to do it. I've been told it will take four to five months for them to even look at it again. Social Security required me to see a psychiatrist who diagnosed me with situational anxiety. I was a very stable person before. Now, every time I have to go to IMEs, or if I get letters or anything about my injury, I have this anxiety. I don't take anything for anxiety. I already take narcotics for pain and muscle relaxants and don't want to add any more medication.

A difficult part about being injured is that you might gain weight. I went from working 12- and 14-hour shifts to doing nothing. I didn't gain weight but see that it would be easy. I didn't eat and actually lost weight. I sort of lost my appetite for food altogether and only eat because I need to, which I think is part of the anxiety and depression. Being injured has affected me so much emotionally. It nags me all the time. I still have the stigma of someone trying to rip off the system. That is as hard as to deal with as the physical pain.

I saw two nurses I hadn't seen for 18 months. They thought that I was fine, that I had decided to not come back to work. Nurses seem to be shuffled off somewhere when they're hurt. I never knew what happened to some of them, but always thought they were happier, wherever they went, but it wasn't true. Some of them had to change complete lifestyles. One nurse who was injured is still working. One had back surgery and received an $8,000 settlement, $3,000 of which went to the attorney. This was not her first injury at work. She filed for Disability and Social Security but was turned down. She's been working on it for two years and feels pretty frustrated. One nurse who had fusion surgery is now looking at fusion of the next higher level.

My hospital has made no accommodation to put me back to work. I don't know what I can do anymore. I've looked into teaching and assessments in skilled nursing facilities. Workers' compensation vocational rehabilitation didn't go anywhere. They won't send me back to college to become a teacher because they said I can work as a nurse. I still have trouble driving for 30 minutes. It just continues. When a friend called yesterday, I told him I would be on psychotropic medication if I had not found some people who validated my experience. My physician says part of this is mind and body and you need to get yourself together. I went six times to a counselor. I cried for the first hour that I talked to the counselor about all the stuff I was going through. She validated me.

In reality, the hospital knew all along that we were lifting dangerous amounts of weight – lifting, bending, and twisting. Obstetrics patients weigh in excess of 200 lbs., lifting them off the commode, and so forth. There is so much heavy lifting that it would be insane to return to hospital nursing. The hospital is only trying to keep nurses with strong backs. All the doctors agree that I'm permanently disabled but some of them say I could still work at a clinic where I might be required to transfer people from wheelchair to table, lift patients' legs up, etc. In reality, I know that I can't. It's not that I don't want to. I'd love to. People say there's so much nurses can do, but who is hiring injured nurses? When I have interviews, I don't take my cane. When they ask why I left and I say, "I'm leaving the hospital for less strenuous work," they don't hire me.

People look at me and say, "You can do something." I've been warned about this all along. They say if you can't work as an RN, anybody can still have a job doing something. Social Security said I could do "unskilled, critical-thinking labor," for minimum wages. Every once-in-a-while I feel so beat up that I'd do anything they say. But, I'm going to try to stay strong and keep myself up, in order to keep myself going. I was running out of steam, felt that I was a crumb, and could find a job if I tried hard enough. The mental and physical pain weighs heavy on you, like having a dead leg or something. I keep thinking I'll wake up tomorrow and everything will be fine.

My husband goes to these appointments with me. When I look at what he thinks, he has a better understanding of my situation than I do. My husband keeps asking, "What is it worth?" Because I tell them I'll get a job, I'll do what they say, he says, "You know everything you do causes pain. Why do you want to do this?" I say, "I'll do everything they say," trying to make sure everything is fine. In the end, we need to take care of ourselves. When do we get to that point, that we realize that we're disabled? If I didn't have such a supportive husband, I don't know what I'd do. I know some injured nurses are forced back to work because they don't have any support. The only reason I can be home on my little pension is that my husband has a stable job and we've paid everything off except our house payment. The caveat is that my husband could have retired already, but was not able to because of my injury.

If I move around, it pinches a nerve and it hurts. Lifting is the worst activity. What do you do when your 18-month-old grandson comes to you with his arms up? You sit down and take him into your lap. He looks at you kind of funny because everyone else just picks him up. I used a long tire brush for washing out the toilet and a long car brush for washing tubs and showers. I've learned to adapt with some of these things. When I learned about the things you can get at the auto parts store, it was like "Eureka!"

Then, the funny part – my friend said everything has to have humor – none of the doctors ever asked how I manage to have sex. I think they believe that 50-year-old women don't have sex. In my 20s, I would have questioned it, but now that we're in our 50s… So, it's really funny. I've asked a couple of doctors, "Just tell me, exactly, do you have any ideas?" They said, "No, but maybe at the pain clinic." The pain clinic gave me a sheet of something copied from the physical therapist. None of them will talk about it. None of the doctors asked how I shave my legs, either. My husband does my toenails. He'll paint them for me, too, because he knows it makes me feel good. I don't know what I would have done without him.

More of the nurses won't lift patients by themselves anymore, because of what has happened to me. They're looking more at the lifting, but I think some of them don't understand the real risk. I keep telling them, "It might be you next. You need to talk about getting more lift devices." One nurse gets tired of asking for help so she goes to the head of the bed and pulls the patient up by herself. I used to do that, too, but it's crazy. I want nurses to know that an injury could happen to anyone. I was always strong. One physical therapist said that being twisted and bent with heavy weight stops blood flow to the spine, so a process of deterioration begins and progresses.

With the last nursing shortage, the American Medical Association said they were going to replace nurses with unlicensed assistive personnel (UAP). So, if nurses are out with back injuries, what is going to happen? UAPs may be hired to replace injured nurses, and when the UAPs are injured, it won't cost as much to replace them with more UAPs as to replace RNs. You wonder, then, whether people believe this is happening or if they make excuses in order to save money on nursing staff.

If the court determines that I am permanently disabled, I will be eligible for a pension from the hospital. So, the hospital attorneys will be fighting against it. They don't want a precedent set for cumulative trauma with my case or the hospital will have to pay, not just for me, but, probably, for other people as well.

I've heard that if the hospital offers you a job, and you don't take it, you lose all of your rights. If they give you a job, and it makes your condition worse, that would be in the record, and it would make a better case for disability. If you go for unemployment, and unemployment won't pay you because you're injured, that makes a better case. But, if I were to receive unemployment, I'd lose my union pension. It's a catch 22. It looks like the hospital might offer me a job next or the attorneys will string it out so long that I'll just get tired and go away. It could be a long, drawn-out battle.

I had a future when I worked as a nurse. I don't have that future now. I've been to a counselor and a psychiatrist. I was a strong pillar of the community. I volunteered in many areas and did all kinds of things. Now, I have herniated spinal discs, degenerative disc disease, chronic pain, narcotic use, situational anxiety, depression, and a ruined nursing career. Who would have guessed it two years ago?

I want other nurses to know what could happen to them and that there is no advocate. I loved being an L&D nurse. It's hard to be away from nursing but it's also hard because my job was the cause of my injury. I would advise anyone considering nursing to beware. Nursing is a heavy lifting job which can cause permanent injuries.

We live in a small community. I see women in town who I've been with when they've had their babies. I was the nurse for some of them for two or three babies. Now, they're pregnant again and they say, "I'll be there in three months. Will you be there with me?" It tears my heart out because I won't be there.

Injured Nurse Story #8: Fine When I Entered the Room

by Want it Over

I've been transferring patients since I was a 19-year-old nursing assistant in a nursing home. I received mostly hands-on training. We had an old Hoyer lift but the hydraulics only worked part of the time. It was the kind you had to pump with your foot. We used it on super-heavy people, but, it didn't always work. We had one really large lady that was actually too heavy. We never got her out of bed because she was just too large. That was years ago when I worked in the nursing home.

For many years now, I've worked on a general medical floor in a hospital where we have a lot of stroke patients, cancer patients, and a lot of really ill people. I work as a secretary but I am cross-trained as a nurse tech. Since I was crossed-trained, they called me when they needed extra help lifting. I've always helped lift patients. It just depended on how many patients I'd lift each day.

The hospital had new lift equipment, just nothing that would lift the patient I was trying to help lift when I was injured. She weighed over 400 pounds and was maybe 5 feet 2 inches tall. She was quite compact, very large through the middle. The way she was proportioned, they had to order a special bed, but only got the right bed after several were injured on her.

There were five of us using a draw sheet to pull her up. Two of us had the draw sheet on the left side. I had fat from her belly hanging over up to my arms. She was just short and so large that much of her body hung over the bed. When we went to lift her, I felt my neck pop and twisted my back. I immediately felt it. It's been a fight ever since. I've never had any trouble lifting patients before. In the case of my injury, it just depended on how the patient's weight was distributed.

The young CNA that was lifting on the side with me immediately complained about her back. She ruptured two discs in her lower back and had to have fusion surgery. A total of six were hurt on that one patient. One of them didn't fill out an incident report. I almost didn't myself. Boy, am I glad I did. So many got hurt that the nurse manager and the employee health nurse did a study to see what we were doing wrong. They decided it was the way the patient was proportioned. The injuries were unavoidable since we had no equipment to lift her.

I feel like I'm being punished because they didn't have the right equipment to lift her. Workers' compensation is making me go to two different doctors. I saw some before my surgery. Now they're making me see two different ones. It's very frustrating after working there all those years.

The first doctor asked, "How many weeks have you been on the job?" When I answered 28 years, he said he had been doing workers' compensation claims for that many years. They made me feel like they thought you just go to work for a few weeks in order to get workers' compensation.

The injury broke an old fusion in my neck. C4 and C5 in my neck had been fused after a car accident over 15 years ago. This time it was C4 to C7. I continued going to work while my neck just kept getting worse. The radiologist read the MRI as a "tilted disc" and, so, it took 2 months to see the doctor.

I'd go to work and the charge nurse said, "You can't work in this shape," and they would send me home. I told the internist I couldn't work. It got to the point that I couldn't stand it anymore. I could only work 4 hours a day and ran out of leave time. It took so long to get scheduled with the doctor because of workers' compensation delays. The orthopaedic doctor did an x-ray with my head tilted and you could see the break all the way across. Workers' compensation denied that my

1-56670-631-9/03/$0.00+$1.50

injury was work-related saying they were only responsible for a "neck strain." That took 2 or 3 months. Then, they had to deny my claim for surgery and get approval from the hospital. There was a form to sign saying if I won my workers' compensation claim, they would be reimbursed.

I had surgery about 3 months ago. When I had been home less than a week following surgery, my leg started hurting and it got so bad I couldn't walk. I went to the ER at 3:00 a.m. My right hip graft site had broken; about a quarter-size piece broke off. When they went to x-ray my hip, the radiology tech pulled me with the pull sheet trying to get me onto the gurney. He just jerked me instead of asking if I could help. Jerking me tore my rotator cuff in my right shoulder. The x-ray guy hurt me really bad. I felt really sick and just wanted to go home.

A couple of days later, I complained to the hospital that he hurt me really bad. They said they did an incident report and didn't call me back. The physician's assistant called and checked for me. The risk control officer wanted to do an employee incident report. I said I didn't come in as an employee. I came in as a patient and he hurt me. I even had to wait for approval for an MRI of my injured shoulder.

I had my first release to go back to work about a month ago with restrictions, but, they wouldn't let me come back because of my shoulder. They couldn't use me anywhere that I wouldn't have to lift my arm and lift charts that sometimes weigh over 10 pounds.

They've given me just a few hours when I could have worked for 2 months. I usually work 9 days a pay period. My savings went. My husband became ill and passed away about a year ago. Now, I've had to use my credit line and credit cards to pay the bills. I'm a survivor. I'll do okay but it's very frustrating. I have no more leave time. My insurance gave me a waiver through the union so I didn't lose my insurance. You have to have 80 hours work a month to keep it. I received $184 a week for a few weeks, which helped a lot. A lot of extra things have come up and I just want it all over with. I'm getting really frustrated with the hospital. I feel they ignored that my shoulder was injured while I was getting the x-ray of my hip, like they would have paid more attention if I had not been an employee, and that they could have put me to work somewhere.

My right arm from the shoulder injury bothers me more than my neck. I've never had trouble with my shoulder before and, now, the doctor says I'll eventually have to have surgery on it. I went to PT and really worked it but still have to cope with pain from the 60% tear in it.

I feel angry, like I'm being punished. I would have had no financial problems at all if they had let me go back to work. The holidays are coming up, they didn't let me work, and I had no income. I requested release with no restrictions and the doctor okayed it. About a month later, they gave me just a few hours but won't let me start for another month with partial days. It will be 3 months before getting back my full schedule.

My coworkers have been wonderful. My unit manager called me. But, after the incident with my rotator cuff, they should have been on it better. Risk control should have contacted me. I feel like I'm getting put off. I think the hospital should be responsible for my lost wages from my shoulder, my lost leave time, and doctor's bills from the shoulder injury. They said for me to put it through my insurance until they look into it. They haven't responded on that.

For my neck, the hospital bill was over $18,000. The ortho doctor bill was over $16,000. The ER, anesthesiology bill… It just goes on and on. My shoulder didn't hurt until he jammed me into the gurney. I told him my leg really hurt but he was really rough. When I reported it, they said, "I'm just glad it didn't happen to someone in the community." I am someone in the community. I didn't go in as an employee. I went in as an orthopaedic patient with post-op complications.

It's been a nightmare. I don't want it to go on for 2 or 3 years. I just want it to be over. I'd really like to take a vacation. Vacation is when you go someplace. Being off work in pain wasn't a vacation.

I've learned what it's like to take pills with pudding because with neck surgery, you're all swollen because they go in through the front and clamp structures in your neck to the side. I learned to walk with a walker. I had to use the walker even after the rotator cuff was torn because I could do no weight-bearing on my right leg.

When the fragment from the graft site was broken, I had no weight-bearing for about two weeks. The first week, I couldn't put my foot down on the floor. And, when I went to the ER for the x-ray of my hip, that was when my rotator cuff was torn. So, it really hurt to use the walker with my shoulder. I had to have help getting my leg out of bed. It really hurt. They don't do anything about it when a fragment breaks off; it just absorbs back into the tissue, taking three or four months to completely heal. The doctor said it's rare to break off like that, but, it does happen sometimes because the bone is weakened when they saw it off.

The CNA that was injured lifting with me and had surgery is now working temporary light duty as a hospital greeter, like a Wal-Mart greeter. No hands on, just kind of a PR thing, helping direct people where they need to go. They told her they would only hold her job a couple more months but she might not be able to go back then.

When patients are so very large, it's hard on everyone. The family wanted to take her home but the doctor said there was no way. It took six people to move her. She went from the hospital to the nursing home. Now, they have the right equipment but I haven't been trained on it.

They want me to get another appointment, to see what any permanent restrictions might be. If they remove the nurse tech aspect of my job and make me just a secretary, I'll lose a dollar an hour.

We had one patient lift, which we used at times, though it was hard to work with, and was really inconvenient to go get, plus, it wasn't always there. When you're really busy, you don't always have time to run and get the lift. And, if another unit was using it, it wasn't available. They have more lifting equipment now than they used to; I saw an in-service on the ceiling lift in one room. Makes sense that they should get one for every room for really heavy patients or stroke patients that are limited on one side.

My neck claim was originally approved but, then, they denied it. I'm getting all the flack from workers' compensation because I had the prior fusion. My attorney is still working on it. My case manager, the nurse advocate who works between you and workers' compensation, called and said, "You need surgery immediately, but, I just called to let you know that we're not paying for it." I yelled at her, "How would you feel if you were fine when you went into the patient's room and weren't when you left the room?" She sent me a letter saying they'd send me to a stress management class. I don't feel they are there to help me.

The ortho and neuro specialists for workers' compensation claims, if that's all they do, how could they be that good of doctors? The ortho doctor moved my neck, made me squat, and do other movements. The neuro doctor sat at the table and wrote what the ortho doctor said. They kept talking over my head. I asked them to talk so that I could understand what they were saying. One said, "Well, this isn't about you."

They agreed with my orthopaedic doctor's letter. I think the dictation may have been unfinished because he wrote that the need for surgery was due to my previous injury. When I told him they denied my claim because of what he wrote, he rewrote the letter clarifying that my injury was from lifting patients. The judge will talk with the doctor and attorney and will decide whether my claim will be accepted.

Everybody says, "Don't worry" but it's not their money and their life. That workers' compensation is absolutely the worst for denying people. I will fight it because they are wrong. I was supposed to have the hearing, but the judge postponed it for 60 days.

We get treated like dirt after getting hurt doing your job like you're told to do. Then, they want to push it under the rug. They sure don't want to talk about it. I just want it all to be over.

Injured Nurse Story #9: They Let Me Go

by Down but Not Out

My story began in January 2000 when I was hurt on the job as a night shift nurse in our small town hospital. I sustained a herniated disc injury of C5 through C7 (neck area) while having to turn several patients on one night in particular. There were only two nurses for eight patients and most of them needed lifting, turning, and assistance with their personal care.

There was one mechanical lift for the whole hospital, but we never knew where it was. We were too busy to leave and go look for it, so we just lifted and moved the patients ourselves. Most patients who have bone surgery also have bone stabilizing machinery attached to them necessitating manipulation of the machinery as well.

When I got off shift, I had a nap and when I woke up later that day, I had excruciating pain in my neck and down my right arm with accompanying muscle weakness. Over a course of time, I had physical therapy, medications, and spinal injections, all to no avail.

Nine months after my injury, on September 6, 2000, I had cervical neck surgery. Bone was taken from my right hip to place in my neck where the discs were removed and, then a metal plate was inserted to add stability to the region.

I did well after the initial surgery with resolving symptoms until a week later when the donor site in my hip cracked. I was brought to the emergency room in severe pain and anxiety. I was in the hospital for approximately three days and was then transferred to an extended care facility for further pain control and physical therapy. This was no easy feat because of all the pain I was in. I was on a tremendous amount of pain medication including Soma, Neurontin, OxyContin, Valium, Morphine, and Lortab.

Luckily, I don't remember most of what happened but that made it very difficult on my 17-year-old daughter. She didn't realize that I had no memory of our conversations. As a matter of fact, I didn't know myself until a few days before I went home that anything was amiss with my memory. I didn't realize that I wasn't remembering things until someone would ask me a question, like about the surgery and, then I would vaguely recall having surgery. My poor recall of events was from requiring so much medication.

On October 25, 2000, after being in the nursing home for approximately 36 days, the surgeon felt that I was on too much medication and abruptly stopped all the aforementioned medicines. As if my body hadn't gone through enough already, I now had a week of withdrawal pain to deal with. I had tremendous pain and spent a week pacing up and down the halls, night and day, trying to get away from the pain. I was nauseated and hardly ate throughout withdrawal. I was in so much pain and was also on an emotional roller-coaster going between inappropriate crying and laughing spells. I was then diagnosed with fibromyalgia, the result of all the trauma to my body, and now respond painfully to any touch, no matter how light.

When I went back to light duty about 6 months later, the workers' compensation insurance carrier wanted to know my status. Reluctantly, my internist signed a "medically stationary" release, which meant that I probably wouldn't get any better and that I could not go back to hospital nursing again but I could start retraining for another position.

1-56670-631-9/03/$0.00+$1.50

Through workers' compensation vocational rehabilitation, I was retrained to work in the hospital business office under the assumption that I would have employment when the training was completed but that also fell through. So, now after 20 years of faithful service to the hospital, I am out the door! I have had to make some lifestyle changes and also help pay for my daughter's college expenses.

To this day, I continue to have right shoulder, right arm, and right hip pain which has become worse since April 2002 when I started full-time training in the business office. I am still under doctor's care and don't know where or how it will all end. It has been a long frustrating time for me, my daughter, and my internist.

August 16, 2002, was my last day of business office training at the hospital. It was a sad moment in my life when I walked out the door. I thought for sure someone would stop me as my hand hit the door handle shouting, "No, wait! You're too valuable to let go!" but nothing happened as I continued to walk away.

Thank you for the opportunity to tell my story. I feel that if there had been adequate staffing and appropriate equipment to lift and turn the patients that night I might not have been hurt and my world turned upside down. I can't possibly express here all the emotions I've been through in the last 2 years but I think my story will give you a good idea.

Injured Nurse Story #10: In Pain and Out of Work

by Down Right Mad RN

I became a nurse in my mid-thirties when my children were teenagers. I wanted to be with them while they were small. When they were older, I wanted to be able to support myself as a single mom.

After 20 years as an RN, I was working full-time on a general medical/surgical floor in a hospital. I usually had four to five patients and lifted my patients on a daily basis, plus assisted other nurses with lifting their patients. We didn't have equipment, or equipment available, to assist with lifting, so we just asked for assistance from the other nurses.

When I injured my back the first time, I was lifting a patient with a partner and felt a pain in my back. We were lifting a surgical patient, an average-sized man of about 170 pounds, up to the head of the bed. He was unable to help himself so we were pulling him up in bed. All we had to use was the draw sheet. We had no friction-reducing sheets and didn't even know about such things.

We put the bed down flat. Then, holding the edges of the rolled-up draw sheet, we lifted and pulled on the count of three, just as we were trained to do. At the instant, I knew I was injured, but couldn't identify anything unusual about the lift. We did the "one, two, three, lift" and seemed to lift pretty much in unison, but I still got hurt. I had not really twisted or turned out of the ordinary with this lift.

I filled out the required forms and was sent to the ER and had x-rays. I was off duty a few days or a week. I continued working on and off for intervals as my back would allow. I felt like my back wouldn't tolerate working any longer, like I couldn't do it anymore.

I started drawing workers' compensation because of the back injury. After the length of time I was eligible to draw, they created a modified-duty job for me as a hospital "lobby assistant." At that time, they gave me a new name tag which didn't have my title of RN on it. That was very degrading and upsetting to me, that they even took my title away. They didn't want me to work in the capacity of a nurse because my lifting was still restricted.

They reduced my wages by fifteen dollars an hour which was supposed to be reimbursed by workers' compensation. And, while drawing workers' compensation, they somehow reduced me to three days a week. In that position, I was only permitted to work four hours a shift or less, if my back didn't feel like I could work any longer.

I felt really out of place as a lobby assistant because every time someone came in, I wanted to help them as a nurse and I couldn't. I worked that position for a few months until they gave it to another injured employee. Basically, it was just greeting people when they came in, seeing that they signed in, directing them where they needed to go, a basic nonnursing job.

After that job ended, I continued to draw workers' compensation after filling out an aggravation form because of continued and worsening back pain, until I drew the maximum amount. They paid me a settlement. It was supposed to have been around $5,000, but they took out workers' compensation wages from that amount. So, I was paid a check of $3,500 and didn't expect to receive any more until I filled out the aggravation form at a later date for a re-injury. I was re-injured just a number of months after my initial injury.

Then, when I drew that money again as long as I could draw, they did another Determination Order and Closure. At that time, I was awarded around $8000. Then, again, workers' compensation

1-56670-631-9/03/$0.00+$1.50

wages were taken out of it. I don't remember the exact amount that time. There was a total of about $13,000, if you count the wages and the settlement, but I was left with the back pain and being unable to return to work.

When I was re-injured, five of us were lifting a female patient who probably weighed 300 pounds from the chair back to bed. Three of the five of us were injured on that one lift. The patient had somehow been placed in the chair but was just too weak to bear weight or assist at all with getting back to bed. We never used the Hoyer, so it never occurred to us to go looking for it. We lifted patients all the time; so, the nurses just gathered and lifted her.

The bed was in its lowest position, but we still had to lift her up onto the bed. It was really awkward. One nurse lifted her lower legs and feet. With two nurses on each side, one on each side lifted her torso and one lifted an upper leg and buttock. It was scary when we lifted her and shoved the chair out of the way because then there was nothing beneath her. We barely got her up onto the bed.

Right away, three of us said, "Oh, my back!" Since I had had a previous injury, I filled out the forms but the others didn't report their injuries and said, "I'll be okay; I don't think it'll last." Luckily, none of the others were injured seriously that time, but, within a matter of months, one of them had a ruptured disc from lifting a patient. She had surgery, which was really successful, and she was back on the job in 6 months.

My diagnosis was herniated nucleus pulposus of L4/L5 with right radiculopathy. The doctor said I had lifting restrictions of 50 pounds, which was way more than I could handle.

After having a BAK implant fusion at L4/L5, where a BAK cage — a hollow porous, titanium cylinder — is filled with autograft bone and placed in the intervertebral space, the pain continued. The hospital kept my position open until it was determined that I could not return to my former position and, at that time, they terminated me. I was then left with no income, no job, and continuing back pain, in spite of having the surgery.

I continued to see my doctor. Eventually, I told him I wanted to have a discogram done. When I asked a previous doctor, when I was first injured, about having a discogram, he told me that the MRI is the best test we have. I told him statistics prove that 25% of the time MRIs don't show correct results. When I asked the second time for a discogram, from a different doctor, I was referred to an out-of-town doctor to perform the discogram.

The discogram showed that I had three other discs involved, and one that would require surgery. I had complained of pain on both sides, with pain on the right being more severe. After seeing the workers' compensation independent medical examiners (IMEs) four different times, those complaints are documented, even in their notes.

Now, at the present time, I'm waiting for an intradiscal electrothermal therapy (IDET) surgical procedure to be approved by the hospital's insurer. I believe that all those discs were involved at the time of my injury although they weren't all recognized at that time because the MRI failed to show injury to them. So, I'm still suffering, still waiting to have the IDET approved, and wondering if they will. They leave you in limbo for months before they finally acknowledge anything.

In the interim, I've depleted my savings and have been overwhelmed by bill collectors calling. All I can say is, "I'm sorry. I have no income. I have no job. I have no way of paying," and explain the situation with my back. I've been sued a couple of times because of failure to pay bills and lost. I knew that I owed the money. I didn't deny that I owed the money. I just didn't have any way to pay. I went from $25 an hour to zero dollars an hour.

The whole situation has been so degrading and humiliating. It was something I never imagined I'd have to go through. I thought the hospital, being a healthcare provider, would at least be empathetic about my injury and my situation. But, in spite of applying for three different jobs, in different areas of nursing, I was told, "Oh, that job shouldn't be posted. It should have been taken down two weeks ago" or, else, just no response. They said one job had been filled but I found out that it wasn't filled. No one ever called me for an interview.

So, basically, I couldn't return to my previous position and there wasn't going to be any other job available to me at that hospital. The occupational health nurse never let me know of any positions available. Several would have been appropriate with less stress, less lifting. If you've had an injury, it's just a finality. They will not hire you back.

During this whole period I've been in, like, a major depression. I don't even like to leave the house unless absolutely necessary or to contact anyone. It's just really affected me adversely. I used to be pretty out-going. It's just changed my whole life. It's affected every area of my life. It's affected my ability to do housework, my grocery shopping. It's affected my relationship with my significant other as far as being real intimate sexually because of pain. It affects everything.

I take an antidepressant; have gone through half a dozen because of the side effects. One gave me auditory and visual hallucinations. Another gave me horrible nightmares. They cause dry mouth, lots of bad side effects. Initially, my ortho doctor ordered them for me. All the other doctors continued to try to get one that works for me without unbearable side effects. Methadone caused stasis dermatitis because it's a circulatory depressant. I'm kind of moody and don't see the bright side as often as I'd like to. It all looks kind of bleak. My family is supportive but is a two-hour drive away. My faith in God is my major support.

You just don't realize how much everything you do affects your back until you have an injury. Like washing windows — you'd never think it would hurt your back, but washing windows or mopping the floor are some of the hardest things to do. I used to enjoy all kinds of work around the house and yard but I just can't do it anymore.

Years ago in nursing school, I had minimal basic training on transferring patients, perhaps a couple of hours. Things like that weren't really addressed much at seminars. Maybe transferring patients was mentioned one time at a seminar, not lifting, but transferring patients.

There was one Hoyer lift available on the floor. It went unused because it was cumbersome and awkward and required more time. Time was of real value in order to get your work load done. So, it was easier to just get someone to help lift. You always rush, rush, rush and try to save as much time as you can so you can get done with the things you have to do. And, you still go home feeling like you haven't done enough, haven't spent enough quality time with the patients.

Some of the nurses weren't very supportive. They just don't understand — not until they have a back injury. You just can't understand it until you have it. Some act like they think you're faking. I had never tried to shirk work. I always enjoyed my job. Some suggested that I should get out of nursing and get into the administrative aspect of hospital work.

One time I contacted my union rep. I needed to talk to her about my injury and things that were happening with the hospital because of my injury. She seemed really harried that day, so I asked her to get back to me when we could talk because I really needed to talk with her. She never called me back. It made me feel like you were just not worthwhile.

They are putting nurses out of commission and don't want to replace them when they can't work any longer. It's all I can do to keep up with everything required — jumping through their hoops to remain eligible to keep drawing workers' compensation, going through board reviews, and everything that they require you to do, seeing all the doctors they want you to see, and getting your own doctor to keep you in pain medication.

I had to be referred to a pain specialist so that I can feel half-way normal. Even taking pain medication, I have to do work around the house at times I'm covered and rest while it's waning until it's time to take more 12 hours later. I don't like taking medications, have never liked taking them. It's hard to get used to having to take medicine in order to function.

At night you wake up hurting. You have to turn over and reposition to try to get some relief. Sometimes you have to reposition five minutes later because of the pain. A back injury disrupts your sleep pattern terribly.

I had to go on the state health plan in order to be covered by insurance, to pay for my medication and procedures that weren't covered by the hospital's insurance. I had a cage exploration where they inject dye all around the cage and do a CT scan to see if the bony growth has made a solid fusion. It appears to have been a good fusion but I still have so much pain on the right. The pain may be caused by scarring or by the other discs that need to be fixed now. If that's the case, the discs are what have caused so much pain all along. That's why I need to have the IDET done, to determine whether the injured discs are the source of my pain.

I don't know if I'll work again until I know the results of any surgeries in the future. If I ever do get pain free, and get my back into shape, I'll go into some aspect of nursing but not back at the hospital. I had a Preferred Workers card and made them aware of that when I applied for the jobs.

I felt terrible when I got that termination notice. A period of time had gone by. I'd had no verbal contact. I just got this letter saying I was terminated as of that date. You might think your supervisor would contact you, to see how you're doing, or when you're coming back and it doesn't happen. The hospital occupational health nurse doesn't help you. She takes down information as you fill out papers and forms and does nothing to help you.

You have everyone on every side saying you need to re-educate. I'm thinking all the time, I can get my back fixed, I can get back into nursing. They had this vocational rehabilitation assessment of what your other abilities might be. I had an associate degree in nursing and wanted to earn a bachelors' in nursing under vocational rehabilitation. I was thinking that if I could get a BSN, or higher degree, I could get into the administrative side of nursing, if they would approve it for vocational rehabilitation but they denied it. They said they wouldn't pay for it.

So, they got me enrolled into classes, small business, accounting classes to be re-trained away from nursing. They tell you way early in the game that you're not getting back into nursing. So you throw away your whole career.

I never wanted to be an accountant. You had to take 18 hours of classes to qualify for vocational rehabilitation. I was acing the classes, even with the stress and pressure, but all of the pain made it hard to concentrate. The pain was so distracting. It became so stressful that I felt like I would buckle under. I finally dropped out because I didn't feel like I could do my best in the classes. It was too hard to even get up and get around to be there in the classes.

The emotional pain is as bad as the physical pain. It's not really healthcare; it's a business. They're supposed to be caring, loving, understanding, supportive.

My claim was denied. The hospital appealed everything. I went in front of the workers' compensation board with a judge there. The judge determined that I was injured and qualified to draw workers' compensation. He was even in disagreement with the IME saying that the IME was biased in his opinion. The IMEs are working for the hospital and aren't going to do anything in your favor. They see you for 5 or 10 minutes and make a determination. They are paid to not approve you.

I did three rounds of physical therapy and two of chiropractic with pool therapy included with PT. I did PT for work-hardening in order to try to get back to work. None of this was successful because you are limited in what you can do in PT because of the pain. I still had the same pain after surgery as I had prior to surgery. Something that needed to be fixed had not been fixed.

I had several MRIs when I experienced increased pain. The first doctor said that MRI is the best test we have. The other doctor said MRI doesn't show everything. I think they want you to wait so long to have surgery. Eventually, the doctor said to have the surgery and said there was 50 to 75% chance of improvement. I was that 25% that didn't improve with surgery. But, when I had surgery, they didn't know about the other injured discs which were discovered by the discogram when that was finally done. Another injured nurse who got information off the internet about discogram told me I needed a discogram. I had my back injected twice with steroids with no help.

Pain meds make you nauseated. You don't want to take that much Tylenol. I'm now on OxyContin. When I had the cage exploration out of town, the doctor gave me a prescription for pain medicine after the procedure. When I tried to fill the prescription, they said the state health plan wouldn't cover it out of town. Back home, in town, they wouldn't cover it because the prescription was not written by my primary doctor. I told them I was referred by my primary doctor but they didn't relay that to the pharmacy.

My doctors were out of town until Monday — my ortho, internist, and pain doctors, all out until Monday. I called the state health plan and told them I couldn't get my prescription filled. They said, "Just have the pharmacy call the state plan." I drove back to the pharmacy where they told me I couldn't fill the prescription because the referral was made over three months ago. In order to get the prescription, I would have to pay for it myself. So, following the painful procedure, I had my regular pain medication but have gone without medication for breakthrough pain.

I've had to deal with more pain and have gone uncovered because of their rules. All of these things add up and they are so frustrating. I've run back and forth so much that my gas tank is on low. It is destitution time now. My significant other is having his own physical problems. He's self-employed and is presently unable to work.

I applied for Senior and Disabled assistance. She told me I might as well not apply because, "Backs are never covered." I applied anyway because what do I have to lose? They finally called and said I am eligible. It's only $300 a month and I'm still waiting to receive the first check. I filed for Social Security Disability, too. They said that since I qualified for Senior and Disabled, they sent info to Social Security, boosting my chances for drawing $700 a month, which is a drop in the bucket compared to what I was accustomed to earning as a nurse.

My worker's compensation case has been closed, unless I have surgery. My attorney has told me I can't sue and get money. Workers' compensation does their thing and makes their determination. You have no power over it. The board makes the decision and you can't even be present. It's all open and shut. You have no power to make a difference.

I would warn anyone thinking of going into nursing about the risks and to take care of their backs. I still think nursing is a great profession, a worthy profession, but would warn them. You spend a lot of hard work and years of training and suddenly it's snatched away from you and you have no where to turn.

8 Prevention of Back Injury to Healthcare Workers Using Lift Teams: 18 Hospital Data

William Charney

CONTENTS

INTRODUCTION

The incidence, human and financial costs of back injury in the healthcare industry all point to a need for a major attack on the problem.[1] Jensen, in 1987, in a comparative study of 24 occupational groups using annual incidence data from four U.S. states, demonstrated that nursing aides, orderlies, and attendants ranked first for back injuries due to moving patients.[2] In the state of Oregon, nursing aides rank first for back injury and in Florida they are third.[3] In a survey conducted by the American Nurses Association, 39% of respondents reported that the fear of a severe back injury may make them leave the profession.[4] SEIU/1199, a healthcare union in New Jersey, surveyed its membership and found that 40% of CNAs leave the profession due to back injury.[5] Fuortes et al. found that nurses' aides were more than 3 times more likely to suffer a low back injury compared to registered nurses.[6] Owen found that patient transfers were perceived to be the most stressful tasks leading to injury.[7] Fuortes also reported after reviewing workers' compensation records in a large university hospital for a 2-year period that nurses aides had the highest rate of injury of any occupational group in the hospital.[8] Overexertion (including lifting patients) is the number one cause of injury for health services with almost 74,000 cases (about 45% of all health services cases) in the year 2000. That is a much higher rate than private industry, which had a 27% rate.[9]

Manual lifting of patients has been shown to be the trigger cause for back injury to the population charged with moving or repositioning patients in healthcare.[10] Marras concluded that there is significant risk when manually transferring the patient with either one or two patient handlers and that up to 46% of two-person lifts resulted in exceeding tolerance limits.[10] Marras also concluded that the 20% of the manual lifts exceeding the NIOSH limits of 6400 N exposed healthcare workers to risk of vertebral endplate microfractures even during the two-person transfers. Different authors (Knibbe, 1996; Smedely, 1995; and Marras, 1999) have found that repositioning of patients is a high risk for back injury.[10–12]

Intervention strategies ranging from training on safe lifting techniques to using good body mechanics have shown little or no influence on injury rates.[13] The lift-team method was first

1-56670-631-9/03/$0.00+$1.50

reported on in 1991(Charney)[14] at San Francisco General hospital where a one-year study showed injury rates were reduced by over 90%. It was reported on again in 1997,[15] in a ten-hospital study with aggregate data (23.5 combined years of lift team experience) of all ten hospitals reporting a 69% reduction of injury rates, 62% reduction in incidence rates, 90% reduction in lost days, and a 72% reduction in workers' compensation dollars. The lift team method is a design approach to remove nurses from the exposure to back injury by using a team of professional lifters, mandated to use mechanical lifting devices, who respond to the total body transfers in their facility. The method incorporates the risk-management axiom of putting risk where it can be controlled by limiting the exposed population, and by controlling the variables that create back injury: primarily the lack of mechanization.

DESCRIPTION/METHOD

Many hospitals in the United States are running lift teams. The exact number is not known at this time. Included in this article are 12 hospitals in the United States, which have implemented lift teams that have allowed the author to present their data on injury rates and cost-benefit figures pre- and postimplementation up to and including their latest year retrievable. This data is the latest data available on this intervention method. All the reporting hospitals are acute care tertiary hospitals. Some of the reporting hospitals are running teams 7 days a week, 24 hours a day; others are running teams 7 days a week, 12 hours a day. All the teams are responding on beeper system or portable phone systems. All hospitals are of different sizes with daily patient census ranging from 150 to 500 beds. For the purposes of identification and anonymity all hospitals are labeled with a letter. Each hospital's data was retrieved using workers' compensation data. The data and rates if provided are not formatted in the same way for each hospital. Instead the data is presented as it was provided by the individual institution as seen in all the tables. All data is presented so that the results can be self-explanatory.

DISCUSSION

All reporting hospitals have realized significant reductions in their back injury rates for severity and frequency of injury, where the associated cause was transferring a patient. All reporting hospitals have shown large reductions in workers' compensation and medical costs associated with musculoskeletal (MSK) injury due to transferring of patients. Different systems compute their cost-benefits differently, but the bottom line for all reporting hospitals is that even with the costs of running multiple teams a profit calculation has been determined. The injury that was captured in the data is primarily back injury. Knee and shoulder injury, unless specifically delineated to be patient transfer related, is not captured.

Hospitals B and H are still experiencing repositioning injuries by healthcare workers as their lift teams were not mandated to do these manipulations. Hospital B hired six additional lift team members in order to get the repositioning need covered.

Commentary with patient care staff at Hospital C reveals some interesting perspectives:

1. Patient care staff (PCS) feel that the lift teams are much better trained and equipped to do routine and high-risk maneuvers. As a resource the lift team enhances the quality of patient care by allowing PCS to direct more attention and/or time to clinical patient care activity. Many interviewees stated that the lift team "frees PCS personnel to perform more clinical care."
2. PCS were asked to rate the lift team performance in the various patient transfers (on a scale rating 1 as lowest to 10 as highest). All responses were either a 9 or 10 for overall rating.

3. Response times were calculated as between 5 to 10 minutes. All response times on the day shift were excellent but some respondents said the night shift needed improvement.
4. Utilization reviews were defined as the number of times the lift team was called. Predominately it was described as frequently. During the period of 1996 to 1998 the combined total of logged calls for the two hospitals was 141,400 calls.
5. Caring for heavier and obese patients increases the risk of injury. Most PCS recognized the increase in obesity and acuity and therefore an increased reliance on the lift team.

CONCLUSION

Lift teams, as an intervention method, have reduced the number of lost time injuries, the claims in medical dollars and compensation dollars, and the number of lost days in all the hospitals surveyed and who have contributed data to this study. Many lift teams are considered "untouchable" and the anecdotal satisfaction data is extremely supportive of the lift team method. All cost benefit data also supports using lift teams to prevent unnecessary injury dollars from flowing out of the healthcare system. Some hospitals are realizing zero transfer injuries on the shifts that lift teams cover.

Some hospitals are still seeing injuries during lift team shifts. Some of these injuries are repositioning injuries that the lift team does not cover. Some hospitals are administrating their teams better than others. The hospitals that have consistent administrative oversight of the teams are realizing a lower injury rate data picture than the hospitals that do not have good administration of the teams.

In this age of acute nursing shortages, prevention of injuries to healthcare personnel must be a priority. This method now has a good history of success and these successes have been published in the science and the data has stood up to a combined 30-year data history. For hospitals with more than 200 beds, this might be the method of choice.

TABLE 8.1
Hospital A

500-bed hospital, 2 year data
Reduced injury claims: 69%
Reduced lost days: 95%
Reduced restricted days: 88%
Reduced average cost per injury due to patient handling from $9894 to $1099
Medical cost savings direct and indirect cost by $690,504

TABLE 8.2
Hospital B

400 beds, running team 7 days/week day shift and night shift
Zero transfer injuries on shifts covered
Turning and repositioning injuries occur
Hired six additional FTEs at $170,000 to reduce reposition injuries by 30% for a net savings of $54,000/year
3-year annual average of workers' compensation cost and medical $242,000 prelift teams
Reduced to 14,000 after 1 year implementation
Lost days prior to lift team averaged 788
Reduced to zero

TABLE 8.3
Hospital C

Exhibit I

Patient Transfer Injury Claims
Comparison of Equivalent Three-Year Periods
1993-95 vs. 1996-98

					Combined	
	Number Claims	**Incurred Cost**	**Number Claims**	**Incurred Cost**	**Number Claims**	**Incurred Cost**
1993-95 Valued 9/30/96	113	$914,073	134	$1,334,744	247	$2,248,817
1996-98 Valued 9/30/96	65	$655,269	57	$896,604	122	$1,551,873
Reduction		$258,804	77	$438,140	125	$696,944
% Reduction	42.5%	28.3%	57.5%	32.8%	50.6%	31.0%

Exhibit II

Patient Transfer Injury Claims vs. All Other Claims
Comparison of Equivalent Three-Year Periods
1993-95 vs. 1996-98

	1993-95		Reduction
Patient Transfer Injury Claims	$2,248,817	$1,551,873	$696,944
All Other Claims			$305,993
Total =			$1,002,937

Exhibit III

Loss Development Illustration
Patient Transfer Injury Claims vs. All Other Claims
Comparison of Equivalent Three-Year Periods
1993-95 vs. 1996-98

	1993-95 Valued 9/30/99		Reduction
Patient Transfer Injury Claims	$2,950,692	$1,921,219	$1,029,473
All Other Claims		$5,910,576	$1,169,216
Total =			$2,198,689

*The 1996-98 incurred costs are developed using development factors applicable to the 1993-95 period.

Exhibit I

- Reflects both the 3 years prior to and the 3 years immediately following the implementation of the 24-hour lift teams, both the number of claims (frequency) and the total incurred loss costs (severity). Each of these outcomes have been reduced. The resulting reduction in costs is $696,944 and in number of claims is 125 claims between these two 3-year periods. At first review of this data does not reflect savings over costs of $400,000 annually. Therefore, we further analyzed the data to ensure every facet was taken into consideration.

Exhibit II

- Illustrates an incurred cost comparison of patient transfer injury costs versus all other claims and total incurred for equivalent three periods combined. These periods reflect pre- versus post 24-hour lift team implementation. The following factors further contributed to the understated reduction of costs:
- During the 1993 to 1995 period, only day shift lift team coverage was provided. Thus, the "total" reduction of $696,944 is understated due to the fact that the day shift lift team reduced claims costs by 50% during this period. Using a conservative reduction of only 40% (not the 50% achieved above) patient transfer costs would have been $366,000 higher in the absence of the lift team. *By including this cost factor, patient transfer claims would increase to $2,614,817, resulting in savings of $1,062,944.*

TABLE 8.3 (CONTINUED)
Hospital C

- Since inception, the lift team provided services to selected patient care units only. While these units represent only 2% of total lift team calls, units outside of the selected units account for 37% of patient transfer claim costs during the 1996 to 1998 period. These units did not derive the injury reduction benefit through use of the lift team. If we subtract the incurred costs generated by these units of approximately $5799,000, the incurred costs for the 1996 to 1998 period are reduced to $972,944 resulting in a revised reduction of $1,275,873, or a reduction of $425,000 annually.
- By combining the above two factors, a more equitable comparison of the 3-year periods is achieved, resulting in patient transfer costs in 1993 to 1995 of $2,614,817 and in 1996 to 1998 of $972,944, *with a resulting net revised reduction of $1,641,873, or $547,218 annually.*

Exhibit III

- This document reflects the "volatility" of workers' compensation claims for the 1993 to 1995 period from $7 million in 1996 (shown in Exhibit II) to $10 million in 1999 (shown in Exhibit III). By assuming the same loss development factors applicable to the 1993 to 1995 period, we have projected the "ultimate" cost of claims, including patient transfer costs for the 1996 to 1998 period. Comparatively, the value of "All Other Claims" increased by 49% while patient transfer injury costs increased by 31%. This comparison suggests a lower level of severity associated with patient transfer injury claims. This can be attributed to the lift team's role in performing a large ratio of the "high risk" lifts, which minimizes the PCS's injury severity exposure and potention.

TABLE 8.4
Hospital D

Lift team start: February 1995

150 bed hospital

Lift teams now working 8 a.m. to 7 p.m., 7 days a week

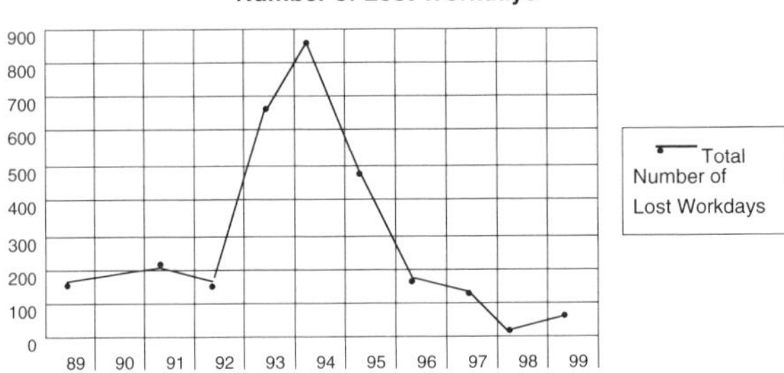

1989-1999

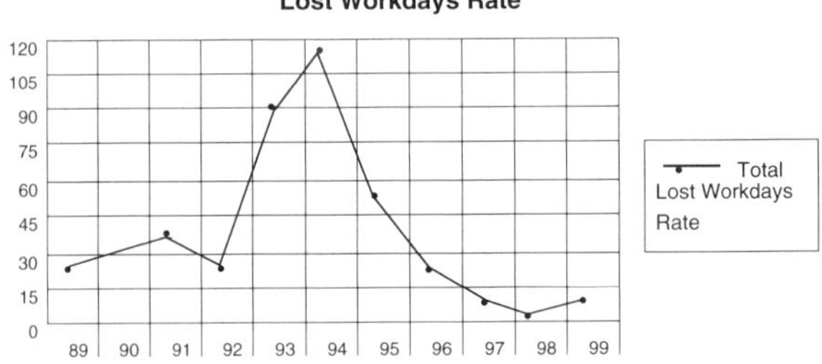

Lost Workday Savings

1989–1994 average lost workday rate per 100 = 60.6 shifts

1995–1999 average lost workday rate per 100 = 20.0 shifts

"Saved" workdays per 100 FTEs = 40.6 per year

$$\frac{40.6}{100\,\text{FTE}} \times 12{,}000 \text{ FTE} = 487.2 \text{ "saved" workdays in CY-1999}$$

Lost Workday Cost Savings

487.2 "saved" workdays in 1999 including peak years.

255.6 "saved" workdays in 1999 excluding peak years.

Assume $20 per hour gross wages and 8-hour shifts.

487.2	"saved" days/year
×$160	per shift
=$77,952	saved wages
+2196	medical costs/year
=$80,148	direct costs
×4	indirect costs
=$320,592	estimated saved costs

TABLE 8.5
Hospital E

Unit	Year	Number of injuries/rates per 100 FTEs	Using team
Telemetry	1999	7.4	Yes
		2000	1.9
		2001	10.1
		2002	0.0
Neurology	1999	11.1	No
		2000	17.1
		2001	17.5
		2002	7.0
Med/ICU	1999	7.2	Yes
		2000	4.9
		2001	15.7
		2002	0.0
Renal	2001	6.2	Yes
		2002	0.0
Telemetry	2000	2.3	Yes
		2001	2.1
		2002	0.0
Inpatient	1999	9.8	Yes
		2000	7.3
		2001	9.2
		2002	0.0
Oncology	2002	6.3	Yes[a]
ICU	1999	6.9	No
		2000	1.4
		2001	5.4
		2002	2.7
ICU	2000	6.7	Yes
		2001	4.4
		2002	0.0
Med	2000	3.6	Yes
		2001	13.7
		2002	0.0
		Ortho	
Bariatric	1999	1.7	Yes[b]
		2000	6.7
		2001	3.2
		2002	3.2

Note: Running two teams: Teams implemented October of 2001: Pre-data up to 2001: Post-data 2002. Data entered by unit. Units not using the teams are having injuries. Units utilizing the teams are down to 0 injuries.

[a] One injury accounted for this data point.

[b] Injuries occurred when lift team off shift.

(*continued*)

TABLE 8.5 (CONTINUED)
Hospital E

Lift Team Data: Jan. to May 2002, Unit Frequencies

2002	Jan.	Feb.	Mar.	Apr.	May	Total/Unit	%/Unit
C 6200	5	1	3	3	1	13	1.4
C 6100	11	7	38	3	6	65	7.0
C 5200	11		28			39	4.2
C 5100	50	7	138	11	27	233	26.3
C 4200	2	5	7	6	1	21	2.3
C 3100	1		1			2	0.2
W 4700	14	1	55	8	12	90	9.8
W 4200	3		3		1	7	0.8
W 3700	5	11	15		1	32	3.5
W 2700	2					2	0.2
WES G					3	3	0.3
S 6200	3		2			5	0.5
S 5400	17	7	27	10	12	73	7.9
S 4400		3	1			4	0.4
S 3500	2	1	5	1		9	1.0
S 3400	10	2	24	3	3	42	4.6
S 2500			1			1	0.1
S 2400	54	20	93	19	34	220	23.9
S 1400	35	1	24			61	6.6
	226	66	465	64	101	922	
Days/Slips	7	2	14	2	5		
Avg. Lifts/Day	32	33	33	32	20		

TABLE 8.6
Hospital F

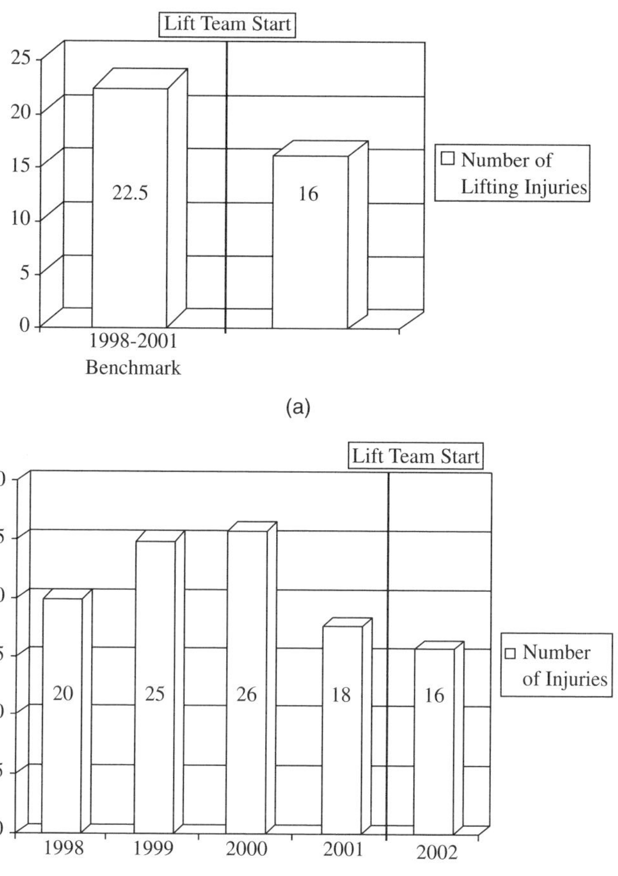

TABLE 8.7
HOSPITAL G[a]

1500 beds
Over 60,000 lifts performed without injury to staff or lift team members on the shifts covered
Restricted days were reduced 361%

[a] Hospital has discontinued use of lift teams in 2002 despite excellent record of results.

TABLE 8.8
HOSPITAL H

350 beds
Running two teams covering 7 a.m. to 2:30 p.m., 20 hours a week, 7 days a week.
Pre-lift team years of 1999 and 2000 showed 34 injuries and 36 injuries, respectively, with an average of 822 lost days.
Post-lift team data for transfer injuries showed 40 injuries with 394 lost days[a]
2002 data from January to June show 22 injuries with 114 lost days.

[a] 115 additional lost days due to repositioning injuries for a total of 509.

TABLE 8.9
HOSPITAL I

300 beds
Cross-trained transport team to a lift team.
Two-year prior data showed 151 and 171 lost and restricted days.
First year post-lift team lost and restricted days were reduced to 2.

TABLE 8.10
HOSPITAL J[a]

350 beds

Running lift team on day shift only 5 days a week

Reduced lost time injury on that shift from 16 to 1 the first year of implementation and workers' compensation and medical costs on average from $144,000 to less than $10,000 per year for the day shift.

Experienced 9 years of implementation reducing lost time injuries on the day shift to between 0 and 1 during the 9-year period.

[a] Hospital discontinued use of lift teams in tenth year despite excellent success rates.

Patient Handling Injuries
Nursing Units
2002

80
60
40
20
0
68
74
64
20
1999
2000
2001
2002

(a)

Patient Handling Injuries
Lateral Transfers
2002

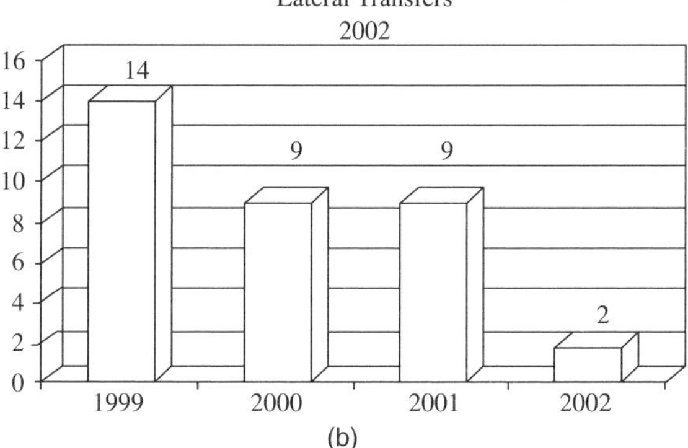

(b)

Patient Handling Injuries
RN
2002

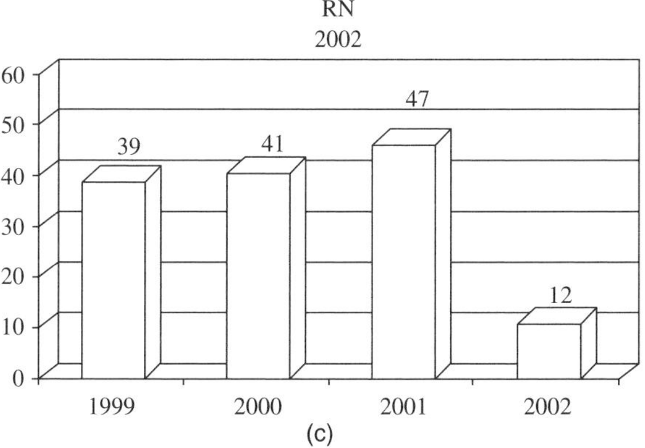

(c)

TABLE 8.11
Hospital K — Workplace Safety Metric Rates 2002 — Second Quarter Update Adult Acute Care Nursing — All Nursing Claims (Adjusted[a])

North Medical Centers	Lift Team Implemented	2001 Rate	2002 Rate (To Date)	Change
A	4Q 01	10.95	10.35	-5%
B	1Q 00	15.95	13.35	-16%
C	3Q 01	17.12	16.42	-4%
D		7.43	4.18	-44%
E		11.53	13.27	15%
F		16.98	15.36	-10%
G		12.96	25.62	98%
H		7.51	6.61	-12%
I		14.69	23.50	60%
J		17.25	10.18	-41%
K	4Q 01	11.10	9.35	-16%
L		11.93	17.04	43%
M		12.85	11.80	-8%
N		9.97	4.57	-54%
O	4Q 01	9.52	8.98	-6%
	.	9.30	9.78	5%
North TOTAL		12.31	12.08	-2%
South Medical Centers	**Lift Team Implemented**	**2001 Rate**	**2002 Rate (To Date)**	**Change**
A		11.08	15.24	38%
B		14.28	10.27	-28%
C		9.51	14.83	56%
D		8.83	14.29	62%
E		3.62	3.69	2%
F		13.27	14.96	13%
G		6.42	14.99	134%
H		7.39	22.97	211%
I		6.38	1.75	-73%
J	4Q 01	7.28	6.31	-13%
K	4Q 01	9.79	9.49	-3%
South TOTAL		9.65	12.77	32%
GRAND TOTAL		11.20	12.33	10%
Facilities with lift teams		12.13	11.23	-7%
Facilities without lift teams		10.87	12.75	17%

Notes: Injury rate based on workers' compensation indemnity and medical only claims opened per 100 productive FTEs (Number of claims opened)(200,000 hours)/(total productive hours). Productive hours do not include PTO, holidays, etc. 2001 baseline rate - all injuries: 22.38. Adult acute care nursing - all nursing claims.

[a] "Adjusted" means only patient handling lift related injuries were included (e.g., lifting, transferring, repositioning).

TABLE 8.12
Hospital L

Patient lift history summary

Lift team implemented: August 4, 1999

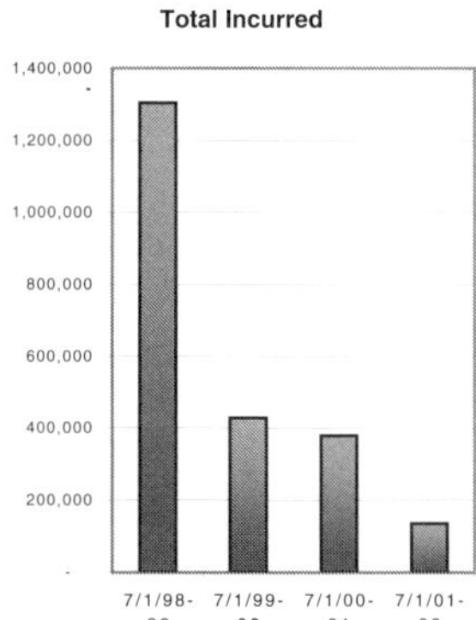

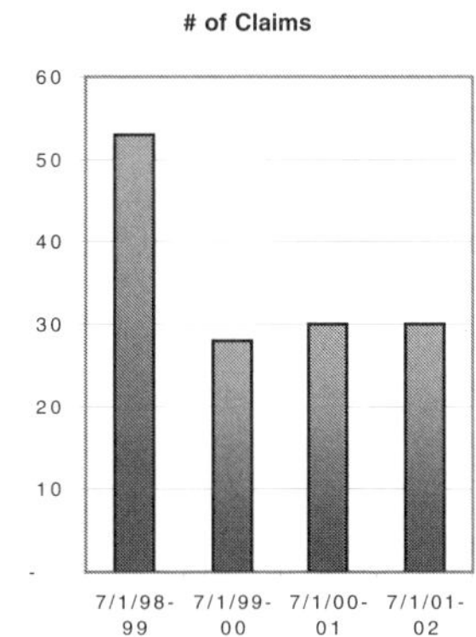

Policy Year	Total Incurred	# of Claims	Valuation Date	Carrier
7/1/98–99	1,303,862	53	5/2/2002	Self insured
7/1/99–00	426,804	28	5/9/2002	Zurich
7/1/00–01	378,566	30	5/9/2002	Zurich
7/1/01–02	133,839	30	5/2/2002	Alpha fund

Incurred $$

$4,484

$156,144

$188,294

$63,835

During Lift Team Shift

M-F, 18:00 to 8:00

Weekends, 9:00 to 17:00

Weekends, 17:00 to 9:00

# of Claims

4

5

16

6

REFERENCES

1. Mcattaminey, L., Ergonomic workplace assessment in a healthcare context, *Ergonomics*, 35(9) 963–978, 1992.
2. Jensen, R., Disabling back injuries among nursing personnel, *Res. Nursing Health*, 10, 29–38, 1987.
3. Oregon Department of Labor, Florida Department of Labor Data, 2001.
4. American Nurses Association: Health and Safety Survey, 2001.
5. SEIU/1199. Health and Safety Survey, New Jersey: 2001.
6. Fuortes, L., Epidemiology of back injury in a university hospital, *J. Occ. Med.*, 36, 1022–1026, 1994.
7. Owen, B.D. and Garg, A., Four methods for the identification of most back stressing tasks, *Int. J. Ind. Ergonomics*, 9, 213–220, 1992.
8. Fuortes, L., Epidemiology of back injury in nurses, *J. Occ. Med.*, 38(9) 1994.
9. Hodiewicz, T., Healthcare hazards, *Ind. Hyg. Saf. News*, August, 2002
10. Owen, B.D. and Garg, A., Four methods for indentification of most back stressing tasks in nursing homes, *Int. J. Ind. Ergonomics*, 9, 213–220, 1992.
11. Marras, W., A comprehensive analysis of low-back, disorder risk and spinal loading during the transferring and repositioning of patients, *Ergonomics*, 42(7), 1999.
12. Knibbe, J., Prevalence of back pain and chararcteristics of the physical workload of community nurses, *Ergonomics*, 39, 186–198, 1996.
13. Smedley, J., Manual handling activities and risk of low back pain in nurses, *Occup. Environ. Med.*, 51, 160–163, 1995.
14. Charney, W., The lifting team; a design method to reduce lost time injury in nursing, *AAOHN J*, 39(5), 1991.
15. Charney, W., The lift team; 10 hospital study, *AAOHN J*, 45(6), 300–304, 1997.

Injured Nurse Story #11: After Years of Service

by Judy Sims

Here is my story. I have been a registered nurse for 18 years, thirteen years have been in critical care. Every now and then I had a sore back but nothing that an anti-inflammatory and a heating pad could not fix. In June 2001, I began having pain in my lower back and left groin area. I related it to my weight and age and went to my physician. We talked and decided to start medication and an increase in exercise. I walked and rode a stationary bike. I lost 20 pounds but the pain persisted.

I was working at St. Vincent's in Portland at the time as a critical care float. Long hours, heavy pulling, turning, pushing, etc., were a given in my job. I returned to my physician and we started the noninflammatory, narcotic analgesics, and physical therapy. I let the charge nurse know about my back but nothing was said. I kept on working. Of course the staffing was always short and they would ask me to work more. Like the good girl I was, I did because the patients and hospital needed me. Who would care for the patients?

The physical therapist was the one who suspected it was a disc problem. By September, the leg pain was excruciating. Around my birthday on the 22nd, I had two near-syncopal episodes. Getting up from a sitting position was extremely painful. My physician finally sent me for an MRI on the last day of the month which showed (of course) degenerative disease and a bilateral disc herniation of L4-L5, worse on the left.

After attending a pacemaker class in the morning of the day after my MRI, I reported to the cardiac recovery unit at 3 p.m. for my shift. At 3:15, my doctor called to inform me of the MRI findings. She told me to cease work, start steroids, and left strict restrictions faxed to me.

My colleagues on the unit were angry I had to leave. I finished my charting, told the charge nurse, and went to employee health, which was closed. My manager was gone so I left a voicemail. The next day, I came back to employee health and gave them my paperwork. I explained my physician verbally told me this "was not related to my biking but repetitive nursing tasks such as pulling, pushing, turning, etc." The employee health personnel stated that, since it did not involve one specific patient, it was not a workers' compensation case. End of story in her mind. I again called my manger but she was in Seattle; so, I left a voicemail message.

Ceasing the 16-hour days; pushing beds; CPR for many minutes; patient pulling, pushing, and turning; carrying heavy supplies; hours of walking and running; quick decision making; short staffing; and lack of support from colleagues and management didn't cease the pain though. Neither did medications.

One morning in October, I nearly passed out again from the pain, scared my daughter tremendously, and almost called 911. Now, I truly know what someone experiencing a heart attack goes through. Cold clammy skin, tachycardia, fear, pain, and anxiety are the very real feelings. I saw my physician that day and was referred and sent to a neurosurgeon.

The neurosurgeon could not see me for over a month. I explained I was an intensive care nurse at the hospital and the wonderful woman at the end of the telephone in that office understood. I then became an emergency consult and saw the surgeon on November 22. This was the same man I made rounds with on his patients who I cared for. He did pediatric and adult patients. I had

1-56670-631-9/03/$0.00+$1.50

worked with him in two different medical systems. I kidded him at one time, telling him he operated on patients from "womb to tomb," never dreaming it would be me at one time.

At the same appointment, after reviewing the films, he questioned me on symptoms related to tethered cord problems as he saw some fatty and enlarged filum. After examining me, he also detected a systolic murmur. I asked him to stop giving me such good news — an operation, a possible spinal cord defect, and a heart murmur in less than 3 minutes!

My life was falling apart. My left leg was on fire and painful. The lower back pain was uncomfortable. I got dizzy most of the time I stood up. I was separated and had four young children to care for. I could not work to support us and now I needed surgery. I have no family in town.

I called employee assistance and explained my situation. The gal suggested an attorney and returning to employee health and demanding workers' compensation. I did and was invited in to "start the paperwork." The gentleman called from the compensation company. After explaining my inability to walk at times, having to crawl on all fours to the bathroom, not being able to bend and tie my shoes or wipe my bottom after using the bathroom, he said it looked like a real claim and progressed to tell me the name of a vocational person who would be calling me after surgery to help transfer to another job. Boy, was I relieved.

Surgery was set for November 16, 2001. The workers' compensation company wanted an independent medical exam to verify. I saw Dr. Coulter. Our conversation stared out talking about Los Angeles where we both had come from to Oregon. Since he was a neurosurgeon from that area, and I knew several neurosurgeons from that area, since I worked in the ICU where they practiced, I asked him if he knew a certain group I worked with, mentioning them by name. Of course, he did recognize the names I mentioned. We also discussed my surgeon's comments on my films. He asked if he could take them home to analyze and I said yes. After examining me, he agreed with the need for surgery as he had just had himself.

Still no call from nursing management. The next thing I know is a call, the afternoon before surgery, to tell me my workers' compensation claim was denied because it wasn't work-related; and, since I had mentioned other neurosurgeons, I must have seen them as a patient, before, in previous jobs. WHAT?!

So, while trying to recover from surgery, I'm looking for a lawyer. The holidays are here, very little sick time income and I'm looking for a lawyer. Years of service for a hospital, hours of overtime, over and above extra work for an organization, and they said it wasn't work-related, contradicting my primary care physician and surgeon.

Well, it got worse. I developed severe pain, burning, etc., post-op down my left leg to my foot and worse back pain than I started with. My leave of absence was extended because I couldn't even pick up a 10-pound turkey or my 5-year-old son, let alone a 200 to 300 pound comatose patient.

The medications were making me sleepy and I couldn't concentrate. There was no way I could calculate my checkbook let alone a vasodilator drug dose. The attorney I retained missed the appeal date on my compensation case so it was a done deal. I ended up on public assistance and food stamps because I had no income. Employee assistance helped us one month with my mortgage, helped with Christmas as a "needy" family. Still no call from nursing management.

Physical therapy, pool therapy, acupuncture, magnets, narcotics, counseling, epidural injections, prayer…I did it all but to no avail. I was passed on from one physician to another. The pain management docs, physiatrist, and neurologist just concluded it was a severe neuropathy. Drugs and more surgery were in the cards and, of course, no work. I tore a lateral meniscus in my left knee apparently while trying to compensate for the back. With pain each time I took a step, I was reminded of my work as a nurse and how I came to this.

I was terminated from my job in June 2002 because my physician wrote I could not return as an ICU nurse due to my back. I asked for a different position in the organization to accommodate; they declined. I was amazed they could not find a position for a nurse with 18 years experience, 13 in critical care, and various charge positions. I have a Bachelor of Science degree in Nursing, also. The interesting thing was, they had head-hunted me to come work for them when I was recovering in the PACU after some surgery on my leg, I guess, as long as I was okay to work 8 to 16 hours in various settings on day, evening, and night shifts, I was okay; but, damaged goods don't sell.

I was hurt and angry. This was not fair. Another nurse hurt in the line of duty.

One more nurse making the nursing shortage shorter. I called the nurses' union, Bureau of Labor and Industries (BOLI), and Equal Employment Opportunity Commission (EEOC) because I felt discriminated against because of my disability. By now, it was not only the neuropathy, but depression and post traumatic stress syndrome. Depression and chronic pain go hand and hand. I could not work to support myself or my family. I was in debt. My children and I were traumatized by a system that did not care.

Working I was fine, was called all the time, but hurt I was not. My self-esteem was in the toilet. Not only physical pain but emotional.

The only one who seemed to care and help was my vocational rehabilitation counselor, Susan Guentner, from the state of Oregon office. She tested, listened, and analyzed. She recognized my strong desire to return to work and my willingness to search for solutions, transferable skills, and my shortcomings of wanting to fix and make it work right now. I cried a lot and learned a lot.

This blessed angel found a job as nurse case manger for an insurance company in downtown Portland. I was hired immediately and have been working there for a month. I still have chronic pain. I wear my transcutaneous electrical nerve stimulator (TENS) unit to work, down pills three times a day, and exercise. I had to give up the pool as there is no time left in my day. My spirits have improved as I work with wonderful caring people. They have accommodated me with some office equipment but my nursing skills have been maximized.

My legal suit against the workers' compensation lawyer still lingers. I've learned nothing is stat in the world of government, law, or life, often as it is in hospital critical care. I have networked with other nurses injured on the job for support and to unite to prevent this from happening to someone else. I would not wish this on my worst enemy. The devastation is overwhelming.

I think of all the years I helped ill and injured people. The hours, physical effort, love, and kindness I showed to patients, colleagues and families. It was not there for me. To this day, not one nursing manager has called to see how I was doing. God bless whoever reads this.

Injured Nurse Story #12: The Tub Bath

by CNA on Blue Ice

I was a 24-year-old CNA working evening shift in a facility for developmentally disabled adults. There were about 50 residents in the home with about a third of them needing to be lifted. We had touched on lifting and transferring in the CNA course and didn't get any training in the facility once I was hired. There were no mechanical lifts at all, no Hoyers, nothing to help with lifting. We just got help from each other.

I hurt my back that evening bathing a resident. She was wheelchair-bound and non-weight-bearing. We had to position the wheelchair up beside a deep tub, like an old-fashioned claw foot tub. We had to put the wheelchair right up to the tub and lift her, with a coworker one on each side, each of us having an arm and a leg and then lift her up and slide her down into the tub. Then, we squatted down to bathe her. She couldn't walk but she could sit up. We had to kind of hold her up for balance, for safety, though. I believe, when she was in the wheelchair, we had her harnessed in with a safety belt to help her sit up.

We washed her and rinsed her off and then let the water out and dried her off while she was still in the tub. Then, we put a towel in the wheelchair, where it was beside the tub, because when we lifted her up, her bottom was still wet. We couldn't turn her side to side in the tub. I know this stuff; I live it. We could wash and dry the entire front of her, every inch, but not her buttocks because the tub was slippery and we couldn't turn her side to side. You get people in a tub with a bar of soap and they just slip around like a soapy baby slipping around.

I can see this person in my mind as I talk about her. I can even remember what her voice sounds like and her eyes, the way she looked at you, the way she communicated. I have hundreds of them in my head, from the places I've worked. And, they never go away. You always wonder what became of them.

So, then, with my coworker, we had to bend over and both of us get an arm and a leg and, on the count of three, heave-ho, to lift her straight up and put her into the wheelchair that had the towel in it.

When we lifted her up, I twisted, I turned to get her up into the wheelchair, but my feet didn't go along with my body because the weight in my arms was too heavy to reposition my feet without losing my balance. I had had to bend down and turn. I was in an unsafe position to try to lift and turn my feet along with my body without losing balance. You don't have time to think about it but, typically, when your body moves in a direction, your feet do, too. You keep your body aligned. We do that without thinking. But, when the weight that you're transferring is too heavy, you just strain and do whatever it takes to move the patient. So, that's how we got her into the wheelchair. Then, we draped her in the big robes and towels.

The point is the twisting and unsafe environment with the low tub. I would never think of doing it now, but when you're young and lack experience, and they tell you, this is how to do it, then you do it. And, when that job provides rent and food because your husband's a college student, you do it.

I knew I was injured. Within a few minutes my lower back was burning and painful. They sent me home. Well, I basically told them I was going home. They acted concerned; one of them was real nice. I took a couple of days off, or had a couple days off. After that, my back was still as painful and I went to the doctor.

1-56670-631-9/03/$0.00+$1.50

They diagnosed it as lower lumbar strain and gave me lots of pain medication. They didn't do an x-ray or anything — just kind of poked and prodded to see where the pain was coming from. The doctor just said lower lumbar strain, like a cord that gets stretched out and never tightens up and goes back in, like a wire that gets stretched.

I went home with pain medicine and was just kind of dopey and druggie. They had me put ice on my back, heat to cold, and lie around. That lasted about a week and then, somehow, I filled out the forms at work. When I went into the facility, they said, "You're going to be off work for workers' compensation," and were real nice about having me fill out the forms.

After a week or so of doing almost nothing, they started me in with the work-hardening program. Actually, before that, they had me sit in a Jacuzzi and they also gave massage to my lower back. That was actually the treat, if you had to look for one in the whole thing. And, so, I had the Jacuzzi and massage and then had to hit the gym and do different exercises.

Doesn't that seem asinine? I've hurt my back and can't come back to work and now they're going to make me exercise? I remember being so doped up. It was pretty hot and the air conditioner was on. I had a little box of bonbons and watched the soaps. I watched them for a while and would then just doze off because the medication was so strong. Then, I'd just kind of wake up and lie there and watch TV, dozing back and forth. And, oh, yes, I gained 12 pounds.

So, after the work-hardening program, after I'd ended my exercise, I was released to go back to work. I did become stronger but I was still in back pain. That's kind of a no-brainer. If you do work, you will get stronger, but it doesn't eliminate the pain in my back.

I tried to go back to work. And, I did for a week or so but it was too painful. I just couldn't do it. I ended up quitting because there was no other work there for me. I heard about an opening in a mental hospital where there was no lifting so I applied and went to work that night, eleven at night to seven in the morning. Most of the residents slept until morning time. The residents in this facility were mentally ill rather than developmentally delayed. They didn't have physical limitations. They all ambulated, so didn't need to be lifted.

Looking back, I still put myself in jeopardy because mentally ill patients are unpredictable. So, my safety could have been jeopardized, with my back already injured. Anybody else working there also put themselves in jeopardy but I feel like I wasn't thinking what kind of jeopardy I was in because they could have picked up a chair and whacked my back.

I've slept on Blue Ice every night after that for over 15 years. Quite frankly, when you share a bed with somebody, you'd better keep your Blue Ice on your side of the bed. It totally affected my sex life because some nights my back was really painful and, you know, if you hurt, it doesn't make even thinking about sex very pleasant.

I mean, you try lying on your left side, you try lying on your right side. The injured person shouldn't be on the bottom because when somebody is on top of you, they have to partially lift their weight off of you and they have to be in really good shape to do that and also perform. Having the weight of somebody on you is not a good thing. It really is part of having a back injury and it was difficult because pain always affects your mentation and your outlook.

It is depressing, also, because pain attacks your mind. I've worked in healthcare long enough with enough pain control seminars. It controls your day and your activities and, then, to think that exercise is good for depression. But, what comes to your mind is lying down and putting ice on your back and soothing yourself. They say exercises, body stretches, are good for strengthening and back pain and, also, for depression, but what makes it difficult to act on is that when you're in that much pain, pain becomes fatiguing to you. Pain tires you so you want to lie down and curl up. It's one of those damned if you do and damned if you don't things. Either way is a loss.

Everyday things, packing groceries, along with your attitudes, become very poor. And lifting children, "Mommy, pick me up. Mommy, carry me. Mommy, up, up, up." And, Mommy just can't. It went for over 15 years this way, before I'd just had enough. Blue Ice every night and, on long trips, packing extra Blue Ice to make sure you have it or else you hit that ice bucket.

Most of my effort was functioning at work, getting through the days, and trying to get over it every night. I never had a diagnosis other than lower lumbar strain though the pain just went on and on and on.

I moved and changed doctors and went to work again as a CNA but I was real careful where I went to work. Because there isn't too much lifting in home health and, if the clients are on the heavy side, you get family members to help you. And, if it still isn't manageable, you leave them where they're at. Then, you just let your supervisor know that you're not going to injure yourself.

When I applied for this job, they didn't ask me, and I didn't tell them, that I'd had a back injury and that I had chronic back pain because most everybody that works in healthcare has some kind of pain — neck or shoulder or back or knees from the bending and lifting. Think about what we do and what's expected of us to do.

That pathetic body mechanics thing they do at the annual training, as if that would protect us from the lack of adequate lifting devices. Not a very good match is it? Body mechanics isn't a very good replacement for them not having proper lifting devices for us. It shouldn't even be considered to take the place of their responsibility to keep us safe physically. The body mechanics booth that you go through — what is that? Are we morons? I used to play Twister when I was a child. It was easy then, but not when you're lifting such heavy ill patients.

They may be 380 pounds and you need to sit them on the side of the bed to cleanse or get them dressed. They lift up their arms like little wings and go, "You'll have to help me up; I can't do it alone." Then, when you try to help them sit up, they don't give even 10 per cent effort of their own. In your attempt to lift them up, you start to lift and find that you have their whole upper body weight in your arms. So, you have to lower them back down.

Isn't it amazing? You can already feel the strain and have to tell them, "I'm not going to be able to lift you by myself." And, then, they look at you like, "You're the healthcare people; you should do the work. You're being paid and I'm sick." They look at you like they think you're judging them or making fun of them because they're too heavy when it's just common sense. They're simply too heavy to lift.

I mean fatigue. Go into Safeway and try to pick up a 300-pound sack of potatoes. Even the grocers have this figured out. That's why they ask you if you need help out. They don't want you injuring yourself on their property. It has nothing to do with being courteous. It comes down from management.

In healthcare, someone must be making a very good income who doesn't have the relationship with the patients and with what's expected of us. They don't have the first-hand experience, that relationship. Quite frankly, I don't think that they even have the desire for that relationship, no feeling for caring or for people. It's just a business. I do this job because I came with the desire to work in the medical field. It's part of my makeup; it's who I am. And, it's who they are. They have a closer relationship with the numbers and the business. They aren't healthcare providers.

During orientation, management should have to go into the field and see and hear and feel what we go through. So, maybe, they could feel beyond their spreadsheets.

I've worked at a number of places now as a CNA and injuries to coworkers happen so often. Many are off work and then come back and still work in pain. I was on workers' compensation but it's not the same as your salary. It's just another loss.

A young woman who took the CNA course had a patient get hurt at the nursing home after 2 or 3 weeks on her first job. There were so many people to take care of. Every night she was just exhausted. All she could do was try to keep them dry but wasn't able to give all the good care they needed. She had a patient on the toilet and called for help, rang the call bell, for the coworker to come help get the patient up, but they never came. She stayed with her patient. Then, when she stepped out to flag down help, the patient slid off the toilet onto the floor.

Of course, the new CNA then felt horrible. She was so upset. She went into the break room and was crying. The manager said, "You need to get it together. Go take a break." She was still tearful. The manager said, "Get yourself together." There was no apology for lack of staff or safe lift equipment. The CNA was still so upset that she just got into her car and left. She just quit. She called them when she got home to tell them she had to go home; the supervisor hung up on her. A very caring, kind-hearted young woman, that was her experience as a new CNA.

I, honest-to-God, wouldn't even recommend anyone I care for going into nursing, like this one I've been talking to. I know that sounds bad but I have feelings for her. I'd recommend her doing occupational therapy in a non-lifting environment or social work. I've told that to several people. I'd never choose nursing unless you work with the babies or on the teaching side. I'd never recommend it to anybody.

If I wanted to invest in the coursework I could be a good nurse, but I know too much. I'd never do it to myself. I'd never earn a nursing degree. I've watched the stress they're under, counting out the medications at the end of shift. I've watched them, with the people on the mental health unit. I'd never take on that responsibility. It changes you. You come home and your family doesn't even know who you are. That's my gut feeling. You're maxed out. Look across the country at what really has gone on in the nursing profession. There's a lot of turmoil, always turmoil over something.

My back will always be this way, with pain management and lifestyle changes. You don't realize it, but, with groceries, you get help at the store, but, once you get home, you still have the groceries to deal with. It takes a lot of energy to just do my everyday domestic things at home. Like, you look at the groceries you have to bring in and say, the eggs are in this one, the canned food is in this one. So, you take one canned food bag and one bread bag only, hoping to balance out your load. I think, "I should flip the mattress on the bed, but, I can't, so, I won't." And, I really do think that's the reason why I let my child do her own thing. I don't wash her feet for her. I put some bubbles in and let her bathe. It's just that general feeling of what's been taken away.

Another time, I pulled or strained myself lifting patients and missed work. I was only off a few days. Once, I hurt myself lifting at work and had to go to the physical therapy department, to go through the body mechanics pamphlet with a physical therapist. It's irritating and demeaning to have to sit down and go through that pamphlet. She goes over it almost like you're illiterate. She reads it: "And, you know, when you lift, bend your knees. Use your legs, not your back." Isn't that great? That whole body mechanics thing just burns me.

Don't get an SUV — they're bouncy. All those trips, having hot flaming feelings shoot up my back, like somebody took a lighter and was teasing me, getting the flame up close to my skin. In the car, I'd say, "Don't hit that bump!" I finally had enough of constant pain and not sleeping at night. I went to the doctor and insisted that I needed relief from the pain. Now, I'm on Celebrex and a muscle relaxer and it's a lot better. I'm not pain free just because I'm on the medications. I'm in pain just sitting, but it's more manageable with medication.

I told the CNA, "You are young and have a young child. Protect yourself. Be very, very careful and lift no one. To keep from being hurt, do not lift anyone alone. Do not lift anyone who is heavy, even with help." It's not worth it, I'd say.

9 Equipment for Safe Patient Handling and Movement

Audrey Nelson and Guy Fragala

CONTENTS

INTRODUCTION

There is no safe way to manually lift and transfer physically dependent patients. Simply applying principles of body mechanics and using good lifting techniques are insufficient for safely performing high-risk patient handling tasks. One key to safe patient-care work environments is technology. The purpose of this chapter is to:

1. Identify high-risk patient handling tasks
2. Delineate five categories of patient handling equipment
3. Describe the best use of equipment by category
4. Discuss the limitations of the equipment by category

1-56670-631-9/03/$0.00+$1.50

5. Consider barriers to use of technology
6. Identify gaps in technology for high-risk patient handling tasks

HIGH-RISK PATIENT HANDLING TASKS

Few would argue that one of the highest-risk patient handling task is a patient transfer. Patient transfers can start with the patient in a sitting position (vertical transfer) or when the patient is supine (lateral transfer). However, high-risk tasks are not restricted to vertical and lateral transfers. Other high-risk patient handling tasks include repositioning a patient in bed, repositioning a patient in a chair, and transporting a patient in a bed or stretcher. Further, risk for injury extends beyond tasks that involve patient movement. Patient handling tasks can be designated as high-risk if the tasks are performed in a forwardly bent position with the torso twisted, such as feeding, bathing, or dressing a patient. It is the combination of frequency and duration of these high-risk tasks that predispose a caregiver to musculoskeletal injuries. Since many patient-handling injuries are cumulative in nature, efforts to reduce musculoskeletal stress can prevent and reduce the severity of injuries.

Owen and Garg (1990) identified 16 stressful patient handling tasks in nursing homes. The most stressful tasks identified in rank order included: (1) transferring patient from toilet to chair, (2) transferring patient from chair to toilet, (3) transferring patient from chair to bed, (4) transferring patient from bed to chair, (5) transferring patient from bathtub to chair, (6) transferring patient from chair lift to chair, (7) weighing a patient, (8) lifting a patient up in bed, (9) repositioning a patient in bed side to side, (10) repositioning a patient in a chair, (11) changing an absorbent pad, (12) making a bed with a patient in it, (13) undressing a patient, (14) tying supports, (15) feeding a bed ridden patient, and (16) making a bed while the patient is not in it.

Furthermore, Nelson and colleagues (2001) identified the following nursing tasks as high-risk: (1) bathing patient in bed, (2) making an occupied bed, (3) dressing a patient in bed, (4) transferring a patient from bed to stretcher, (5) transferring from bed to wheelchair, (6) transferring from bed to dependency chair, (7) repositioning a patient in a chair, (8) repositioning a patient in bed, and (9) applying anti-embolism stockings (TED hose). Many of these high-risk tasks will be briefly described and appropriate technology solutions will be discussed.

EQUIPMENT TO FACILITATE LATERAL PATIENT TRANSFERS

A lateral transfer is defined as movement of a patient in a supine position from one surface to another. Lateral patient transfers include transfers to and from bed, stretcher, prone cart, or bathing trolley. Lateral transfers necessitate use of the weaker muscles of the arms and shoulders as primary lifting muscles, rather than the stronger muscles of the legs. Technology solutions for lateral transfers rely on strategies to reduce friction as the patient is transferred from one surface to the other. Lateral transfer aids can be grouped into three categories: (1) friction reducing lateral sliding aids, (2) air assisted lateral sliding aids, and (3) mechanical lateral sliding aids. Each will be briefly described and evaluated.

Friction Reducing Lateral Sliding Aids

Description. Friction reducing lateral sliding aids can assist with lateral patient transfers. These fabric devices are positioned beneath the patient and provide a slick surface for the patient during transfers (see Figure 9.1). The key is in the friction-reducing properties of the fabric. These simple, low cost devices are foldable and easy to store. While not all friction reducing lateral sliding aids have long handles, this feature significantly improves the safety and ease in performing the task by reducing horizontal reach.

Best use of equipment. This device is best used for patients that can offer caregivers limited or no assistance during a lateral transfer. To perform the task safely, the two transfer

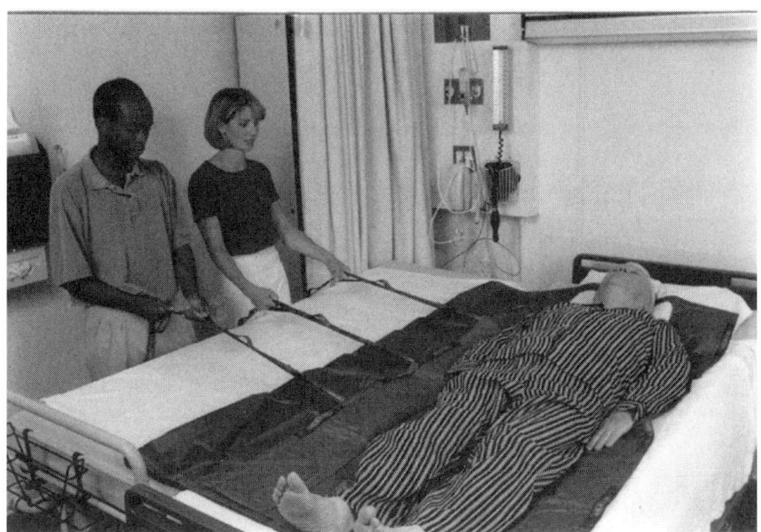

FIGURE 9.1 Friction reducing lateral sliding aid with long handles. (Courtesy of Phil Strong, ERGO-IKE Associates.)

surfaces should be at the same height, preferably at elbow level of the caregiver. When the task needs to be performed frequently, it is advantageous to store the device at the bedside for ready access.

Limitations. Some effort is required to place the device under the patient. Even with the device, two caregivers are needed to perform this task safely for patients up to 200 pounds; for heavier patients, three or more caregivers are needed. For patients with stage III or IV pressure ulcers, care must be taken to avoid shearing force when inserting the device as well as when moving across surfaces.

Air-Assisted Lateral Sliding Aids

Description. Air-assisted lateral sliding aids can also reduce risk associated with lateral patient transfers. A flexible deflated mattress is placed under a patient; next, a portable air supply inflates the mattress. Air flows through perforations in the mattress and the patient is moved on a cushioned film of air allowing staff members to perform lateral transfers with less effort (see Figure 9.2).

Best use of equipment. This device is best used for patients who can offer caregivers limited or no assistance during a lateral transfer. To perform the task safely, the two surfaces should be at the same height, preferably at elbow level of the caregiver. This technology is particular suitable when performing lateral transfers involving patients with specific comorbidities, such as pressure sores or severe musculoskeletal pain. Another advantage is that there is no weight limit, so the device is also useful for obese patients.

Limitations. While air-assisted lateral sliding aids are effective for lateral transfers, using this device takes more time than manually performing the task or using a friction reducing lateral aid. Much of the extra time is associated with bringing the device to the bedside and setting up the equipment. Furthermore, the mattress has to be deflated and the devices carried away from the bedside and stored after the task is completed. Further, this device requires more storage space than friction reducing lateral sliding aids. Due to its size, it cannot be kept at the bedside for easy retrieval.

FIGURE 9.2 Air-assisted lateral sliding aid. (From HoverTech International. With permission.)

Mechanical Lateral Transfer Aids

Description. There are several types of mechanical lateral transfer aids. Stretchers are available that are height adjustable and have a mechanical means of transferring a patient on and off the stretcher from a bed (see Figure 9.3). Some mechanical lateral transfer aids are motorized while others use a hand crank mechanical device. Not all mechanical lateral transfer aids are attached to the stretcher; some can be moved to the bedside to perform lateral transfers (see Figure 9.4). Each of these devices eliminates the need to manually slide the patient, minimizing risk to the caregiver.

Best use of equipment. Key advantages of mechanical lateral transfer aids are ease in moving in tight spaces and performing the lateral transfer with minimal caregiver effort. These devices are particularly useful for patients who are physically dependent on staff for performing these tasks.

Limitations. Frequently these devices are more costly than other lateral transfer aids and require significant storage space. This device is best used when job requirements involve a high frequency of lateral transfers.

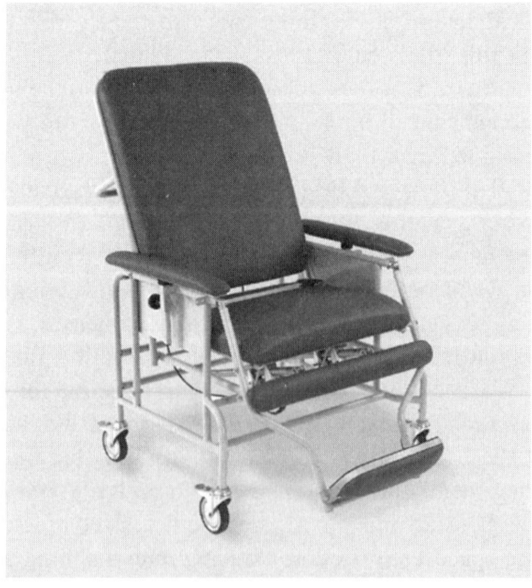

FIGURE 9.3 Mechanical lateral transfer aid: chair/stretcher combination. (From StretchAir. With permission.)

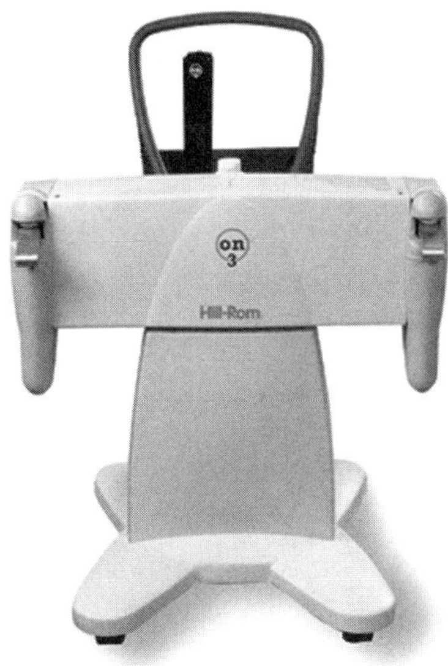

FIGURE 9.4 Mechanical lateral transfer aid: motorized and brought to bedside. (From Hill-Rom Services, Inc. With permission.)

EQUIPMENT TO FACILITATE VERTICAL PATIENT TRANSFERS

A vertical transfer is defined as movement of a patient from one surface to another, where the movement begins with the patient in a seated position. Examples of vertical patient transfers include transfer to and from bed, wheelchair, toilet, or shower chair. Key technologies to support safe vertical transfers include sliding boards, gait/transfer belts, stand-assist lifts, floor-based mechanical lifts, and ceiling-mounted lifts.

Sliding Boards

Definition. Sliding boards are usually made of a smooth rigid material with a low coefficient to friction. The lower coefficient of friction allows for an easier sliding process. These boards act as a supporting bridge when seated slide transfers are performed. Some, but substantially reduced, manual lifting is still required to move the patient, however, sliding boards do offer considerable improvement at a minimal cost (see Figure 9.5).

Best use of equipment. This low cost device maximizes the patient's functional level, promoting patient independence. Sliding boards are most appropriate for patients with limited to no weight bearing capability that have strong upper body strength, e.g., a person with paraplegia. The patient may need limited caregiver assistance or may be able to perform the task independently.

Limitations. Sliding boards should not be used for patients who cannot offer any physical assistance, with obese patients, or patients who have cognitive deficits and/or difficulty following instructions.

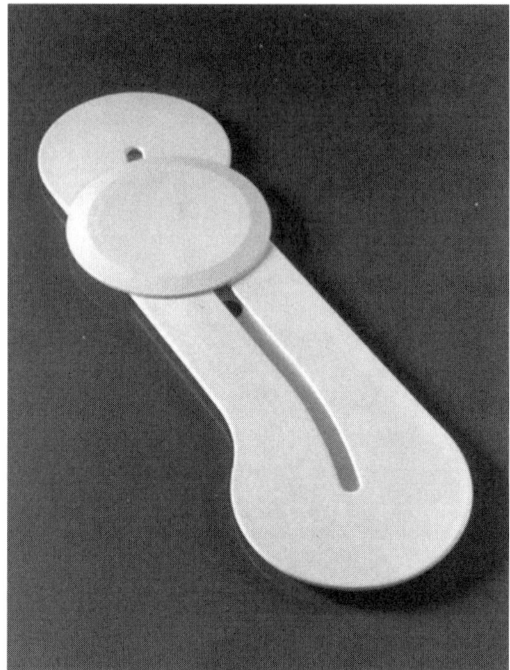

FIGURE 9.5 Sliding board. (From Beasy Trans. With permission.)

Gait/Transfer Belts

Definition. Gait/transfer belts can wrap around the waist of a patient providing handles for a worker to grasp when assisting or transferring a partially dependent patient, as shown (see Figure 9.6). Small hand-held slings that go around the patient can also facilitate a transfer by providing handles. An object with handles improves the grasp opportunity for the worker and thereby reduces the risk.

Best use of equipment. Gait/transfer belts are low cost and are useful for a patient with weight bearing capability that needs only minimal assistance. If this is the transfer device of choice for a patient, the belt should be stored at the bedside for easy access.

Limitations. These belts should not be used with patients who are at times combative, unpredictable, or have cognitive deficits. Further, the belts are not useful for bariatric patients.

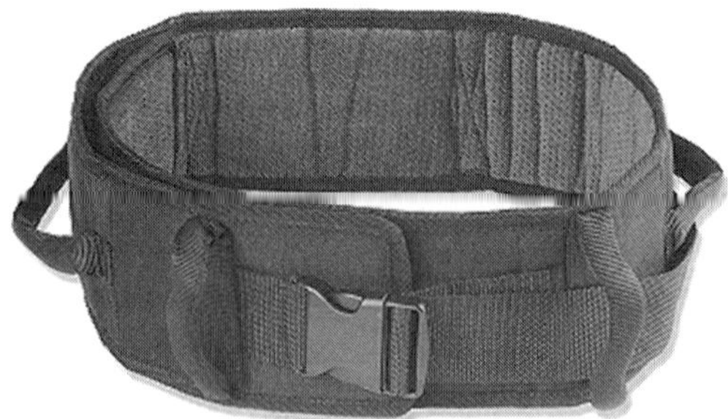

FIGURE 9.6 Gait belt with handles. (From Posey. With permission.)

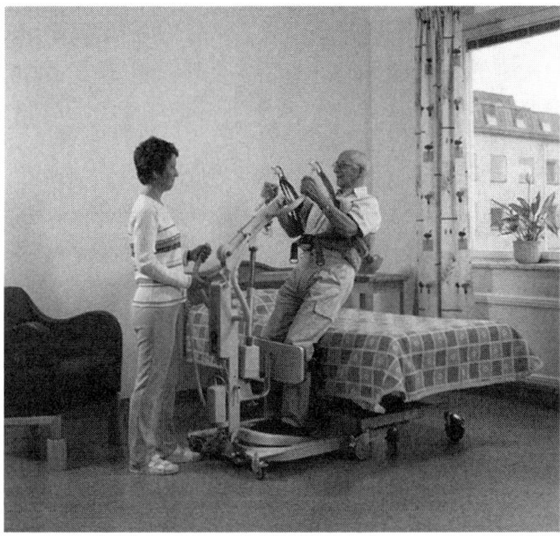

FIGURE 9.7 Stand assist lift. (From Arjo, Inc. With permission.)

Stand Assist Lifts

Definition. Stand assist lifts provide an alternative to full body sling lifts. There are some variations in the sling design, but the basic concept is of simple design, as illustrated in Figure 9.7, and very easy to place around the patient.

Best use of equipment. This type of lift is very useful when patients are partially dependent but have some weight-bearing capabilities. Stand assist lifts are excellent for moving patients in and out of chairs and for toileting tasks, since they are easily maneuvered in restricted areas, such as small bathrooms.

Limitations. Stand assist lifts should not be used with patients who are at times combative, unpredictable, or have cognitive deficits. Further, the equipment is not suited for patients with limited weight-bearing capability.

Floor-Based Mechanical Lifts

Definition. Probably the most common lifting aid device in use is a floor-based mechanical lift (see Figure 9.8). A number of models and configurations are available. These lifts are available with many features and there is a wide variation in the types of slings available. The newer sling designs are much easier to install beneath the patient.

Best use of equipment. Sling lifts are usually used for highly dependent patients. They can be used to move patients out of bed, into and out of chairs, for toileting tasks, bathing tasks, and for any type of lift transfer. These lifts can be used safely for patients who are at times combative, unpredictable, or have cognitive deficits.

Limitations. Floor-based mechanical lifts take longer to use than a manual lift, but the task is much safer for both the patient and caregiver. Most of the added time in this task is for finding the lift and bringing it to the bedside. This time can be reduced if there are adequate numbers of lifts on a unit and the lifts are conveniently stored. Many of the newer lifts are battery powered, necessitating that the battery packs be recharged.

Unusual situations in which mechanical lift devices cannot be used should be identified in advance and specific lifting procedures should be designated for those situations. For example, when a patient weighs in excess of the safe lifting capacity for the standard mechanical lifts at a facility, appropriate mechanical lifts able to accommodate higher weight limits should be obtained. Standard mechanical lifts will normally accommodate

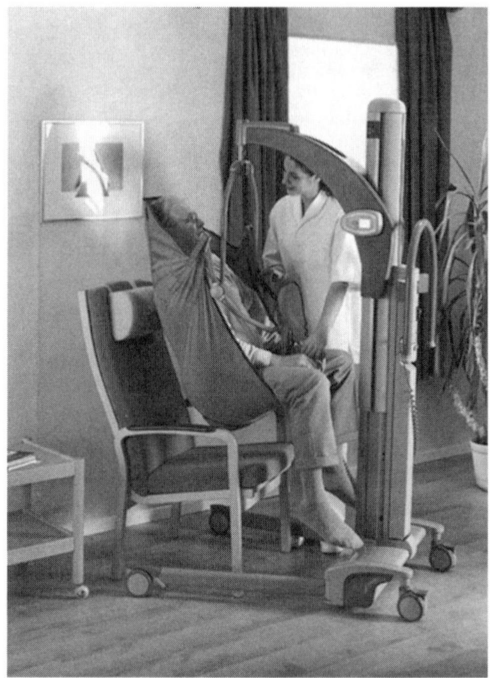

FIGURE 9.8 Mechanical full sling lift. (From Arjo, Inc. With permission.)

350 to 400 pound patients. More substantial lifts will accommodate 600 pound patients, and bariatric lifts are available that can lift patients up to 1000 pounds.

Ceiling-Mounted Mechanical Lifts

Definition. With a ceiling-mounted device, there is no need to maneuver over floors and around furniture. The lift is mounted overhead on tracks installed in the ceiling (see Figure 9.9). These lifts are available with many features and there is a wide variation in the types of slings available. The newer sling designs are much easier to install beneath the patient or resident.

Best use of equipment. Sling lifts are usually used for highly dependent patients. They can be used to move patients out of beds, into and out of chairs, for toileting tasks, bathing tasks, and for any type of lift transfer. These lifts can be used safely for patients who are at times combative, unpredictable, or have cognitive deficits. Unlike the floor-based lifts, ceiling mounted lifts can accomplish lifts safely in tight spaces with fewer caregivers.

Limitations. These units are quite easy to use, however, transfers are limited to where overhead tracks have been installed.

EQUIPMENT TO FACILITATE REPOSITIONING IN BED

Repositioning a patient in a bed is also a high-risk task that occurs frequently in physically dependent patients. Bed repositioning includes side-to-side repositioning as well as pulling a patient up to the head of the bed. Technology to support bed repositioning is still in the early stages of development. Current bed technologies incorporate many ergonomic improvements. For example, one bed feature, shearless pivot, reduces the need to constantly reposition a patient in the bed by minimizing the amount of slippage down to the foot of the bed experienced by the patient when raising the head

FIGURE 9.9 Ceiling-mounted lift. (From Hill-Rom Services, Inc. With permission.)

of the bed. Further innovations with bed mattress surfaces can aid rotation and move a patient as needed by utilizing air bladders incorporated into the mattress surface.

Shearless Pivot

Definition. TotalCare™ Bed System offers a unique feature named Shearless Pivot®. This unique mechanism combines the articulation of the frame, the surface, and the patient in a fashion that minimizes patient migration towards the foot end of the bed when raising or lowering the head section of the bed (see Figure 9.10).

Best use of equipment. TotalCare Shearless Pivot is used with all patients, but can be most effective with patients who have limited mobility and need assistance for repositioning. This unique feature enhances caregiver safety and productivity, as well as patient comfort.

Limitations: For maximum impact of Shearless Pivot, the patient should be aligned with the hip placement indicators, which are marked on the side rails.

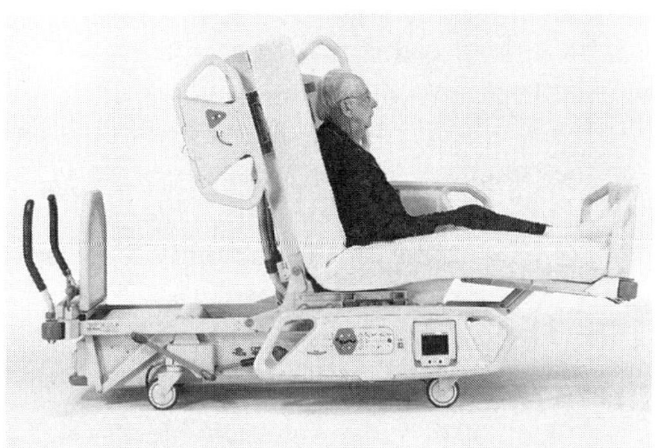

FIGURE 9.10 Shearless Pivot. (From Hill-Rom Services, Inc. With permission.)

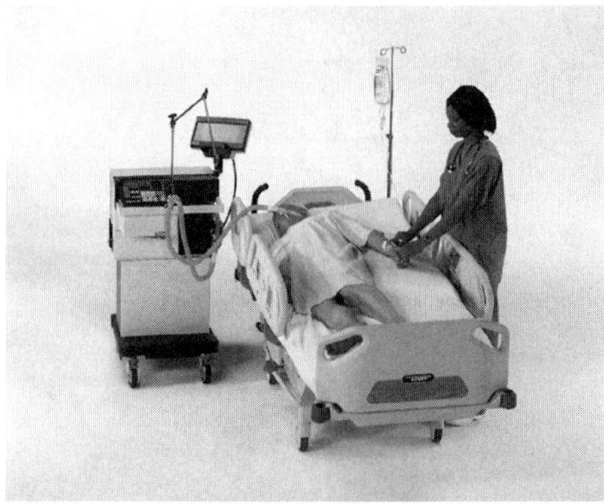

FIGURE 9.11 Turn Assist. (From Hill-Rom Services, Inc. With permission.)

Mattresses to Support Bed Repositioning

Turn Assist

Definition. TotalCare SpO_2RT™ uses the inflation and deflation of air bladders in the mattress to assist with patient turning (see Figure 9.11). With the simple press of a button, one nurse is able to turn the patient to the right or left hand side.

Best use of equipment. TotalCare SpO_2RT Turn Assist provides enhanced patient positioning for easier back care, linen changes, and routine nursing procedures, thereby reducing the number of caregivers needed and the risk of caregiver injury.

Limitations. TotalCare SpO_2RT Turn Assist should not be used with patients that have unstable spinal injuries or long bone fractures.

FlexAFoot™

Definition. TotalCare Bed System offers a unique bed frame design that allows the caregiver to customize the overall length of the bed in both the bed and chair position (see Figure 9.12). The retractable footboard provides support to the patient's feet, reducing the need for additional foot support devices. The overall surface can be retracted up to 12 inches, which also aids the caregiver during transport (i.e., small elevators, smaller turning radius).

Best use of equipment. FlexAFoot can be used for patients that required foot support, as well as those patients who need leverage to adjust themselves in the bed. By positioning the footboard to the individual needs, the nurse may decrease the number of times they will need to reposition the patient.

Limitations. Do not use ankle restraints when activating this feature. Extremely short patients may not be supported with the foot section fully retracted. Move the patient down in bed until the patient's feet contact the footboard.

EQUIPMENT TO FACILITATE PATIENT TRANSPORT

Transporting a patient in a bed or stretcher requires significant effort, particularly over uneven terrain, carpeting, or long distances. This task typically requires two or more staff to perform safely.

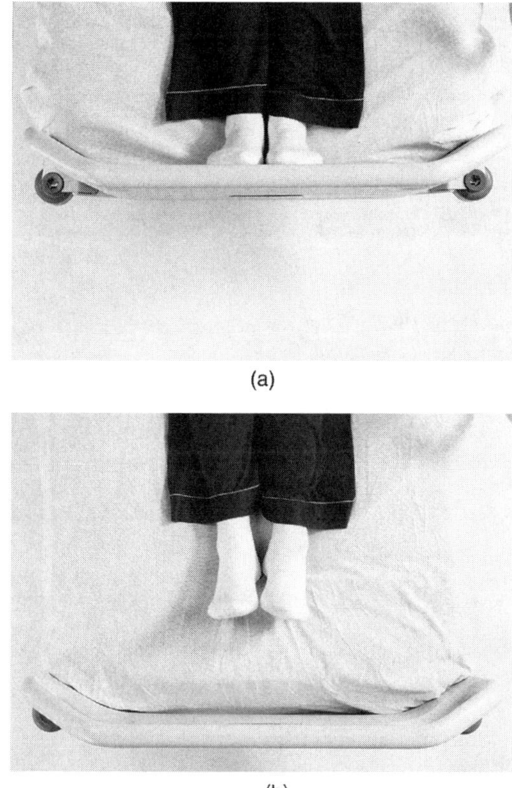

(a)

(b)

FIGURE 9.12 FlexAFoot™. (a) Patient needing repositioning. (b) After repositioning. (From Hill-Rom Services, Inc. With permission.)

Powered Transport Devices

Definition. Powered transport devices, like StatMover (see Figure 9.13) have universal clamps that attach to a bed or stretcher. A motor propels the occupied bed/stretcher as long as a caregiver pushes a switch. The device fits on standard elevators.

Best use of equipment. This low-cost device can be used for patient transport throughout a hospital or nursing home, requiring only one caregiver to perform the task.

Limitations. The device needs to be stored and may be less effective in transport outdoors.

Power Driven Beds

Definition. Power assisted transport system is integrated into the bed to allow for one caregiver to effortlessly transport a patient (see Figure 9.14). The nurse simply unplugs the bed, releases the brake, grips the handles and propels the bed. The variable speed drive wheel is controlled through the push handles.

Best use of equipment. Power assisted bed transport can be used for transporting dependant patients throughout the hospital. The benefits of potentially reduced risk of injury to patients and staff, as well as restoration of caregivers' time may be affected since the TotalCare IntelliDrive™ control helps minimize the need to transfer patients to the stretcher.

Limitations. The cost of this device may be prohibitive.

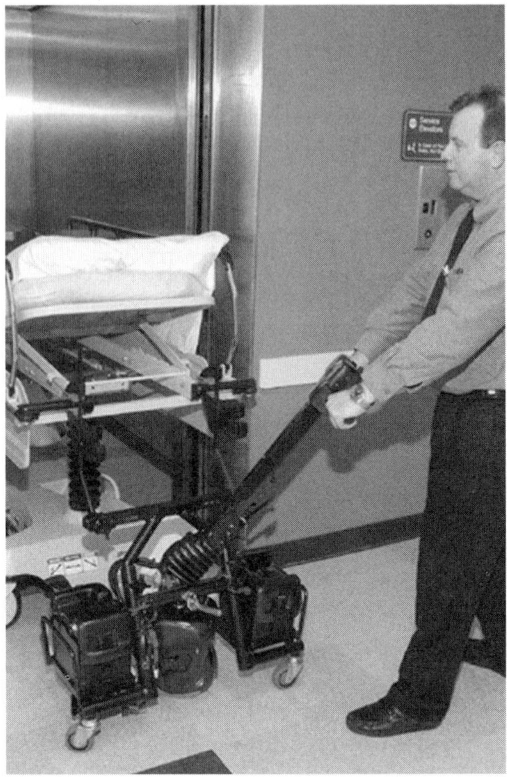

FIGURE 9.13 StatMover transport device. (From StatMover. With permission.)

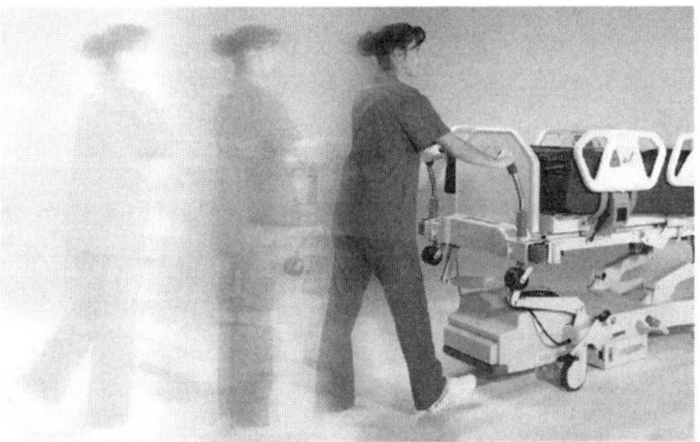

FIGURE 9.14 Power driven bed. (From Hill-Rom Services, Inc. With permission.)

EQUIPMENT TO ELIMINATE AND MINIMIZE THE NEED TO LIFT AND TRANSFER PATIENTS

Moving a patient out of bed to a chair position or standing position is strenuous on the nursing staff as well as the patient. This task often requires two or more nurses to complete, but because of limited time or nurses, many transfers are attempted with an insufficient amount of staff. This can result in injury to the staff and the patient.

Current bed technologies allow one nurse to adjust the bed into positions such as full chair and chair egress while the bed supports the weight of the patient. This allows the nursing staff to move the patient efficiently, safely, and comfortably.

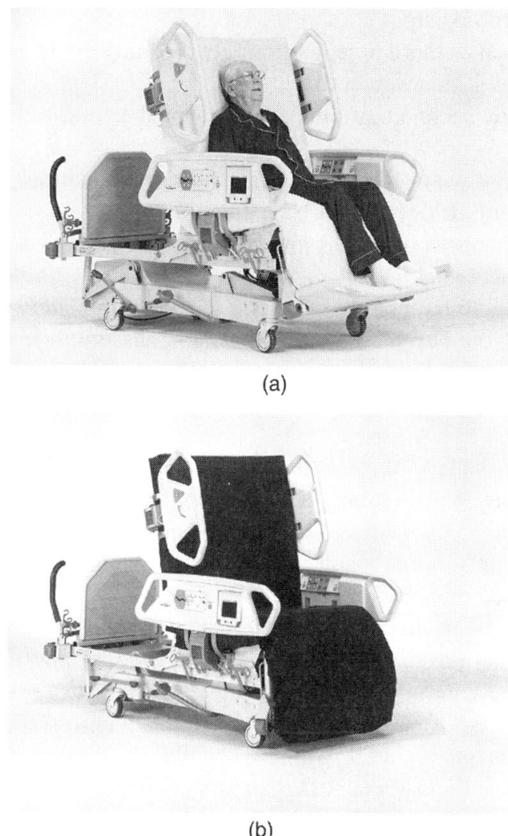

FIGURE 9.15 (a) Patient in bed converting to chair without transfer need. (b) Bed that converts to chair. (From Hill-Rom Services, Inc. With permission.)

FullChair® Mechanism for Up-in-Chair Positioning

Definition. The chair feature allows the caregiver to place the patient in a fully seated position without having to remove the patient from the TotalCare Bed System (Figure 9.15). The chair feature also provides a means to support the patients feet for comfort and security. TotalCare FullChair position is achieved by pressing the chair button, thus raising the head section of the bed to 65°, knee section to 10°, and the foot section lowering to 70°.

Best use of equipment. As patient acuity permits, one caregiver can easily and frequently position immobile patients in the FullChair position per physician and nursing "up-in-chair" orders. The FullChair position allows a nurse to place a patient into an upright chair position, which has been noted to improve pulmonary mechanics and gas exchange.

Limitations. Chair position should not be used with patients who have unstable spinal injuries or long bone fractures.

Barriers to Use of Technology

While use of technology has been shown to be far safer for nurses and patients than manual patient handling (Harvey, 1987; Owen et al., 1995), several barriers have been identified as to why technology is not used (Prezant et al., 1987; Venning et al., 1987; Bell, 1987). Key reasons why equipment is not used for patient handling include:

- The equipment is too costly.
- Use of technology takes more time, particularly if equipment is purchased in insufficient quantities or is stored inconveniently.
- Patients dislike new technology and prefer manual lifting rather than use of lifting devices.
- Equipment is not feasible in confined spaces, such as in bathrooms where two caregivers, a patient and a lifting device cannot be accommodated.
- Equipment is not intuitive for care providers and/or is hard to operate.
- Equipment is not accessible.
- Equipment is available but not charged (e.g., no back-up battery pack available).
- Equipment is available but there are insufficient slings for wide spread use on a unit.
- Equipment is poorly maintained.
- The equipment purchased does not meet patient handling needs on the unit.

A mistake commonly made is the purchase of manual equipment rather than slightly more expensive powered versions. When making decisions about whether or not to use a lifting device, nurses balance the amount of effort required with the amount of extra time it will take. Slight improvements to minimize effort can result in an increased number of staff members that use the equipment, making powered devices more cost effective.

Other common mistakes are purchase of insufficient quantities of devices, locating the lifts inconveniently, or failing to adequately maintain equipment. The way that nurses organize their work assignments must be carefully considered. Patient lifting tasks are not evenly distributed throughout a 24-hour period. Often, there are peak periods where staff must compete for lifting devices. If the expectation is that staff will use equipment to reduce risk, there should be a commitment to purchase sufficient quantities so this is feasible. Furthermore, few healthcare facilities have adequate and conveniently located storage space. Developing a plan for placement of equipment is critical to success. In addition, a plan for routine service/maintenance is needed. This includes not only the motor and frame, but cleaning of the equipment, laundering of the slings, and a plan for sling and battery replacement.

Furthermore, the hospital or home environment adds to the complexity of patient handling and movement tasks. Access to patients can be very difficult because of clutter around a bedside or lack of space, such as a bathroom. It can be very difficult for nursing staff to position themselves properly when trying to assist a dependent patient with toileting activities. Patient rooms are often crowded and awkward postures are often required when trying to gain access to a patient in a bed. The environment in which nurses care for patients can be very unpredictable and is constantly changing.

Selecting the right equipment. Operation of some lifting devices can be as stressful as manual lifting. Equipment needs to be evaluated for ergonomics as well as user acceptance. In a study conducted to redesign at-risk nursing tasks, Nelson et al. (2001) found that lifting devices were not intuitive and staff had difficulty using some equipment as it was designed. Furthermore, sling attachment mechanisms varied and some were significantly more stressful than others to use. Lifting devices that require manual pumping to raise the lift can be stressful to shoulders and may be more stressful than a two person manual transfer. Specialty hospital mattresses, designed to reduce patient risk for pressure ulcers, have been shown to increase caregiver exertion by 17%, by allowing the patient to sink low into the mattress and reducing access to the patient (Nelson et al., 2001).

Cost/benefit. The long-term benefits of proper equipment far outweigh costs related to nursing work-related injuries. In nine case studies evaluating the impact of lifting equipment in healthcare facilities, the incidence of injuries decreased from 60 to 95%, workers' compensation costs decreased by 95%, insurance premiums dropped 50%, medical and indemnity costs decreased by 92%, lost work days decreased by 84 to 100%, and absenteeism due to lifting and handling was reduced by 98% (Bruening, 1996; Fragala, 1993; Fragala, 1995; Fragala and Santamaria, 1997; Logan, 1996; Villeneuve, 1998).

As these studies show, the purchase of lifting devices benefits the facility, patient, and nursing staff. A higher quality of work life for healthcare workers results from occupational injury risk reduction, which translates into improved quality of care for the patient due to higher staff productivity and reduced turnover. Costly mistakes have been made in selecting equipment that is inappropriate for the patient population or that staff do not use.

Job-related injuries that occurred during the performance of patient care activities cost the VHA over $23 million in the year 2000. Approximately 31% of injuries to nurses consisted of upper extremity injuries; 25.5%, back injuries; and 19.1%, lower extremity injuries . The vast majority of these injuries were related to patient transfer and repositioning tasks. Back injuries, although not the most frequent injury, do result in the most lost workdays. The importance of developing reliable approaches to injury prevention is obvious.

FUTURE DIRECTIONS

Technology holds much promise for creating safe work environments in healthcare. Unfortunately, we do not yet have technology solutions for every difficult patient handling task. Opportunities for technology improvement exist for the following high-risk tasks:

- Repositioning a patient in bed
- Repositioning a patient in a chair
- Applying TED hose
- Efficient bathing
- Feeding dependent patients
- Dressing/undressing a patient
- Changing an absorbent pad
- Making an occupied bed

Nurses can serve as partners with industry in identifying needs for new technology or improving existing designs. These collaborative efforts to improve technology for safe patient handling and movement hold promise for the creation of a safer work environment for nursing staff.

REFERENCES

Bell, F., Ergonomic aspects of equipment...patient lifting devices, *Int. J. Nursing Studies*, 24(4), 331–337, 1987.

Bruening, J., Keeping healthcare workers healthy, *Ergonomics News*, Mar/Apr, 20–21, 1996.

Fragala, G. and Santamaria, D., Heavy duties? *Health Facilities Management*, May, 22–27, 1997.

Fragala, G., Injuries cut with lift use in ergonomics demonstration project, *Provider*, Oct, 39–40, 1993.

Fragala, G., Ergonomics: the essential element for effective back injury prevention for healthcare workers, *Am. Soc. Saf. Eng., Mar*, 23–25, 1995.

Harvey, J., Back to the drawing board: training in correct lifting techniques may even increase the amount of back injury, *Nursing Times*, 83(7), 47–48, 1987.

Logan, P., Moving and handling, *Community Nurse*, April, 22–24, 1996.

Nelson, A., Lloyd, J., Gross, C., and Menzel, N., Redesigning patient handling tasks to prevent nursing back injuries, Research report #95–1502 to the Veterans Health Administration, 2001.

Owen, B. and Garg, A., Assistive devices for use with patient handling tasks, in Das, B. Ed., *Advances in Industrial Ergonomics and Safety*, Taylor & Frances, Philadelphia, PA, 1990.

Owen, B., Keene, K., Olson, S., and Garg, A., An ergonomic approach to reducing back stress while carrying out patient handling tasks with a hospitalized patient, in *Occupational Health for Health Care Workers*, ECOMED, Landsberg, Germany, 1995.

Prezant, B., Demers, P., and Strand, K., Back problems, training experience and use of lifting aids among nurses, *Trends in Ergonomics/Human Factors IV,* 839, 1987.

Venning, P.J., Walter, S.D., and Stitt, L.W., Personal and job-related factors as determinants of incidence of back injuries among nursing personnel, *J. Occup. Med.,* 29(10), 820–825, 1987.

Villeneuve, J., The ceiling lift: an efficient way to prevent injuries to nursing staff, *J. Healthcare Saf., Compliance Inf. Control,* 19–23, 1998.

Injured Nurse Story #13: Is That What a Nurse Is?

by Worth More

I am a Canadian nurse who has been working in the active nursing field for 28 years. It was always understood that lifting and moving of patients was part of the job and injury was a possibility we accepted.

In the 1970s, when I started nursing, we had orderlies on the wards to do the heavy lifts. But by the 1980s, the orderlies disappeared and all the lifting and moving of patients fell on the nurses' shoulders. If a nurse is injured, she is expected to suffer quietly and try to carry on. On the whole, I find nurses are discouraged from filing workers' compensation claims, and if they do, are made to feel guilty and rushed back to work doing modified duties. What are modified duties? You are put back on an active ward in situations where you're at great risk of being injured again.

Over the years, I have been scratched, punched, kicked, and numerous times have cushioned a patient's fall with my body. Remarkably, I have lost very few days work in all these years due to injuries. My luck ran out in June 2002 when I began to experience cervical pain and pain and numbness in my right arm. Our ward had been especially heavy for several months and our work loads had increased. Many of our patients were total care. This meant more lifting.

I was trying to use the mechanical lift on as many of these patients as possible but found I got resistance from at least 60% of the staff. They seemed to think using the lift showed weakness and it was the "lazy" way out. I also found that physical therapy discouraged using the mechanical lift. Their reasoning is that manual lifting encourages patients to increase their mobility. Even some doctors don't like the use of mechanical lifts for similar reasons. So, it seems those of us who choose to use the mechanical lifts face opposition on many fronts.

I first went to my doctor on June 28, 2002, and was diagnosed with cervical strain and sent for massage therapy. Because of the pain and weakness in my right arm, I tried to lift using my left side as the dominant side. I was assisting another nurse in lifting a 250-pound woman up in bed with a lifter sheet. I heard a tearing sound in my upper left arm.

When I saw the massage therapist, she advised me not to do so much heavy lifting and to take a break from my job. She compared my condition to someone suffering from a severe case of whiplash. She could also feel a tear in my bicep in the left arm. I happened to be going off work to have abdominal surgery and would get my long needed rest from lifting. So, I was sure that my arm would heal with rest.

Unfortunately, resting my left arm did not help. The pain in my left arm and shoulder increased while mobility decreased. I have been to physical therapy, which is a slow process. Despite recovering from my abdominal surgery, I have not been able to return to work because of the shoulder injury. My health nurse has been supportive thus far, but, she is anxious that I get back to work as soon as possible. My doctor has put in a claim for workers' compensation on my behalf. However, they are still to approve my claim. If they do not approve the claim, then I will have to pay for all the related therapy for this injury.

At the present time, my range of motion is limited. I have difficulty doing many of my household chores. My sleep is very disrupted by the pain because it is difficult to find a comfortable sleeping

1-56670-631-9/03/$0.00+$1.50

position. Simple tasks, like doing up my bra, are impossible. I am lifting two pounds with my left arm in physical therapy and was rather shocked that it seemed to take effort.

I am beginning to think that my career in nursing may be over and all just because I can't lift heavy weights. It makes me wonder: is that what a nurse is? Someone who has the strength to lift?

I have been trying to advocate for a no-lift policy at work with my coworkers but have met opposition. I have been told if I can't lift, then it is time for me to leave because I can't do my job anymore. This makes me angry because I feel I have more worth. I can only hope that this attitude changes and things will improve for the next generation of nurses.

10 Bariatrics: Considering Mobility, Patient Safety, and Caregiver Injury

Susan Gallagher

CONTENTS

INTRODUCTION

Bariatrics is a term derived from the Greek word *baros,* and refers to the practice of healthcare relating to the treatment of obesity and associated conditions. The specialty of bariatrics is increasingly important as the number of obese and overweight Americans is increasing. The implication for caregivers is that in healthcare facilities, activities such as: turning, lifting, and repositioning very heavy patients can predispose caregivers to physical injury. In addition, failure to provide adequate patient activity and mobility leads to issues of patient safety.

A number of studies reveal the increasing incidence, cost, and number of back injury claims associated with patient care. More than half of strains and sprains can be attributed to manual lifting tasks while assisting dependent patients with their mobility needs. Injuries that result from manual lifting and transferring of patients are among the most frequent causes of nurse-related injuries. These and other patient care tasks are becoming increasingly more difficult for patient care providers as the size and weight of patients continually increase.

Overweight and obesity are common health conditions, and their prevalence is increasing globally. Recent estimates suggest that two in three adults in the United States are overweight,

1-56670-631-9/03/$0.00+$1.50

defined by a body mass index (BMI) of higher than 25. Of all Americans between the ages of 26 and 75, 10 to 25% are obese. This is an increase of more than 25% over the past three decades. These dramatic increases have occurred across racial and ethnic groups, and include both sexes.

This chapter serves to describe the common immobility-related complications associated with obesity in the healthcare setting, to identify the specific risks of injury related to mobilizing the very dependent patient who is overweight, to present ideas for reducing or preventing caregiver injury and promoting patient safety, and to outline some of the challenges clinicians face when introducing change. A variety of strategies are available to reduce or prevent caregiver injury, and to promote patient safety, of these, transfer teams, appropriate equipment and criteria-based protocols are discussed herein. The philosophy of continuous quality improvement is presented as a model for change. Techniques for mobility are described.

UNDERSTANDING OBESITY

The simplest explanation of weight gain is that it occurs when caloric intake is greater than the energy expended to maintain bodily functions and perform physical activities. However, because of the mounting concern about the increasing prevalence of obesity, researchers are trying to more fully understand the metabolic, psychologic, and genetic factors that lead to excessive weight. The traditional view that obese people gain weight because they either eat more or exercise less than normal-weight people is only part of the explanation. For example, there are striking differences in energy requirements among individuals. In terms of weight theories, there is only one point of agreement, and that is that obesity is a complex disorder with multiple etiologies and numerous comorbidities. Excess body weight is associated with an increased incidence of cardiovascular disease, type 2 diabetes, hypertension, stroke, dyslipidemia, osteoarthiritis, and some types of cancers.

To fully understand the meaning of current statistics related to overweight and obesity it is important to know how overweight and obesity are defined and measured. Overweight refers to an excess of body weight compared to set standards. The excess weight may come from muscle, bone, fat, and/or water. Obesity refers specifically to an abnormal proportion of body fat. One can be overweight without being obese, as in the example of the body builder or other athlete who has greater than average muscle mass. However, many people who are overweight are also obese.

A number of methods are used to determine if the patient is overweight or obese. Some of these methods are based on mathematical formulas, which relate height to weight. Other methods rely on measurements of actual body fat. There are a number of ways to measure the presence of body fat. Historically, the standard method is to weigh a person underwater. This procedure, unfortunately, is limited to certain laboratories with specialized equipment.

Other simpler methods for measuring body fat include skin fold thickness measurement and bioelectrical impedance analysis (BIA). Skin fold thickness is measured over the subcutaneous fat at specific sites on the body. To measure BIA, a harmless amount of electrical current is sent through the body. The body's ability to conduct an electrical current reflects the total amount of water in the body. Generally, a higher percent of body water indicates a larger amount of muscle and lean tissue. Mathematical formulas are used to calculate the percent of body water into an indirect estimate of body fat and lean body mass. Despite the presence of these methods, neither is widely used currently to track short-term changes in body fat brought on by diet and activity. In addition, BIA may not be accurate in severely obese individuals.

BMI, is the most common and widely accepted method of measuring overweight and obesity. BMI, which describes relative weight for height, is significantly correlated with total body fat content. The BMI can be used to assess obesity. Caution must be used when interpreting BMI in a patient with edema, ascites, in pregnant women or persons who are highly muscular, as an elevated BMI will not accurately reflect excess adiposity in these instances. Normal BMI measurements fall in the range of 18.5 to 24.9 kg/m^2. A patient with a BMI of $\geq$ 25 kg m^2 is considered overweight. Obesity is defined as a BMI of $\geq$30 kg/m^2. Within the obese classification are grades I, II (BMI > 35 kg/m^2) and III (BMI > 40 kg/m^2) (see Table 10.1).

TABLE 10.1
Calculating BMI

Height in inches (cm)	Obesity = BMI 30 kg/m² in pounds (kg)	BMI = 40 kg/m²
58 (147.32)	143 (64.86)	191(86.8)
59 (149.86)	148 (67.13)	198 (90)
60 (152.40)	153 (69.40)	204 (92.7)
61 (154.94)	158 (71.67)	211 (95.9)
62 (157.48)	164 (74.39)	218 (99)
63(160.02)	169 (76.66)	225 (102)
64 (162.56)	174 (78.93)	232 (105.4)
65 (165.10)	180 (81.65)	240 (109)
66 (167.64)	186 (84.37)	247 (112.7)
67 (170.18)	191 (86.64)	255 (115.9)
68 (172.72)	197 (89.36)	262 (119)
69 (175.26)	203 (92.08)	270 (122.7)
70 (177.80)	207 (93.90)	278 (126.36)
71 (180.34)	215 (97.52)	286 (130)
72 (182.88)	221 (100.25)	294 (133.6)
73 (185.42)	227 (102.97)	302 (136.27)
74 (187.96)	233 (105.69)	311 (141.36)
75 (190.50)	240 (108.86)	319 (145)
76 (193.04)	246 (111.59)	328 (149)

Notes: BMI is calculated as:

Metric Conversion Formula = weight (kg)/height squared $(m)^2$

Example: A person who weighs 78.93 kg and is 177 cm tall has a BMI of 25 or: 78.93 kg/$(1.77\ m)^2 = 25$

Nonmetric Conversion Formula = [weight (pounds)/height $(inches)^2] \times 704.5$

Example: A person who weighs 164 pounds and is 68 inches tall has a BMI of 25 or: [164 lbs./(68 $inches)^2] \times 704.5 = 25$

Conversion table: selected BMI units categorized by inches (cm) and lbs (kg)

Source: From *Clinical Guidelines on the Identification, Evaluation, and Treatment of Overweight and Obesity in Adults — the Evidence Report*, The National Heart, Lung, and Blood Institute (NHLBI), National Institutes of Health, 1998.

When hospitalized, clinicians should recognize that most very overweight patients are at risk for certain hazards of immobility. When physically dependent, the obese patient is more inclined to develop complications resulting from a long hospitalization. Common immobility-related complications could include skin breakdown, pulmonary concerns, pain management challenges, and depression. The fear of falling has implications in caring for larger patients in that patients and caregivers may be reluctant to ensure adequate mobility; therefore, the patient becomes at risk for certain immobility-related complications. Skin breakdown such as pressure ulcers, candidiasis, and incontinence dermatitis are related to immobility, and aggravated by obesity. Each one of these complications is exacerbated by poor wound healing. Immobility also contributes to pulmonary complications such as pneumonia, and exacerbates pre-existing conditions such as overweight hypoventilation syndrome or sleep apnea. Immobility can lead to a prolonged hospitalization, feelings of powerlessness and subsequently depression. Mobilizing the patient early and safely can reduce some of these immobility-related complications of hospitalization.

UNDERSTANDING CHALLENGES OF IMMOBILITY

In addition to the safety hazards of obesity and immobility, there can be hazards to caregivers when mobilizing the patient. Even the most compassionate caregiver's intervention can be colored by the realistic fear of physical injury. Consider the numerous court cases involving injured caregivers, and the caregivers inability to provide care, which entails lifting or transferring. Hospitals are not legally mandated to ensure assistance to the caregiver that is providing these activities. The Americans With Disabilities Act requires an employer to offer reasonable accommodation to any employee who becomes disabled. However, any accommodation that places the burden of performing one employee's job responsibilities on other employees substantially impinges on the rights of other employees and is inherently considered unreasonable (see Table 10.2). In addition, this failure to accommodate the needs of an injured caregiver further perpetuates the potential for injuring other healthcare workers. In other words, injured caregivers are removed from their roles, others are added as replacements. The process for care does not change, only the people — the noninjured are substituted for the injured. This problem not only affects the injured but the noninjured, or those who are likely to be injured in the future; this is a mounting healthcare concern.

Safety specialists, insurance carriers, administrators, and caregivers are increasingly aware that healthcare is fast becoming one of the most dangerous jobs in the United States. Injuries are reaching epidemic levels. For example, 89% of back injury claims filed by hospitals are related to patient handling, and the direct costs associated with these injuries exceed $15,000 per claim. It is estimated that $50 billion per year are spent on treatment of back injuries. Workers' compensation back injuries cost 255% more than nonwork-related back injuries and hospitalization is twice as likely for these individuals. Caring for the obese patient places the caregiver at particular risk for injury.

TABLE 10.2
Back, Neck, Shoulder Injuries: Nurse's Disability Discrimination Case Dismissed

A staff nurse worked in a university hospital's hyperbaric and vascular surgery unit where most of the patients were bedridden, weak, wheelchair-bound amputees or severely disabled individuals. Many are morbidly obese.

Nursing tasks on the unit are very demanding physically. Patients often had to be physically lifted from stretchers or wheelchairs, on and off examining tables and on and of bathroom commodes. The hospital expects nurses on this unit to be able to work alone.

The nurse in question injured her back, neck and shoulder in a car accident. She returned to work and injured her back, neck and shoulder again lifting a heavy patient. When she was released to return to work this time she had medical restrictions on lifting that prevented her from performing her job. The nurse wanted to remain in her current job, but wanted assistance in moving patients.

According to the Supreme Court of Iowa, the University Hospital System fulfilled its legal obligation to offer the nurse reasonable accommodation. Human resources assigned an accommodation specialist to help her find another position for which she was qualified and physically able to perform with her limitations. She also got priority status for any vacancy that was open or would become open for which she was qualified.

The nurse was told to check the university hospital's job line frequently. The accommodation specialist would act as her advocate in securing a suitable position, and many were available.

In her lawsuit the nurse claimed disability discrimination because the hospital refused to let her return to her old job with other nurses being instructed they had to help her with any lifting tasks in excess of the lifting restrictions her physician had imposed.

The court ruled it is inherently unreasonable to expect other employees to perform the essential physical tasks of a disabled employee's job.

An employee has no right to accommodation beyond what is reasonable. The court ruled this nurse had no basis to sue for disability discrimination.

TABLE 10.3
The Story of Ann

Ann is a 43-year-old registered nurse. She has worked in intensive care for 15 of her 20 years as a nurse. Fifteen months ago while manually lifting a critically ill patient she was overwhelmed with pain and numbness. She had sustained a significant back injury, yet attempted for work 13 more months until she could no longer focus on her work because of the massive doses of medication that were needed to control her pain and other emerging symptoms.

The purpose of this interview is to illustrate the noneconomic implications in caregiver injury, to more fully examine the meaning of quality of life when caregiver injury occurs. During the interview, Ann explains that she is really angry about her situation. She carefully explains that this is not an economic issue for her, it is something much bigger.

First of all there are role changes that must be dealt with, she is no longer employable. So much of our self-esteem and self-image is reflected in our work and when asked "what she does, " her response is "I'm unemployed." Ann explains that her years of education, training and experience are now unimportant or meaningless in a work-oriented culture. "I'm not asked to use my brain any more, and its killing me." She explains that, "No part of my life is unaffected!"

In addition to the issues of self-esteem, there are actual physical issues that she never imagined she would need to deal with. First of all, consider the family's shock when it was discovered that she could no longer rock climb, hike, or mountain bike with them. The vehicle she had driven for 3 years needed to be replaced because she could no longer get into it comfortably. "I can't shop, I can't dress myself comfortably … and imagine the idea of sex when the back pain is constant"! Ann explains that life revolves around her back and it is trying for her friends and family. "The only member of my family who is happy with me anymore is my dog because the only activity I can do comfortably is walk briskly. It's really a pathetic existence — I won't wish this on anyone." "Something must be done about this."

It is important to recognize that the economic costs are only a portion of the real cost of occupational injuries. Injured caregivers are faced with lost time from work, emotional and physical distress, job and career changes, and role changes at home (see Table 10.3). There are also hidden costs for the organization such as lost revenue due to a loss in productivity, decreased employee retention, costly orientation of new staff, diminished staff morale, and added administrative time for investigation and paperwork.

The degree of difficulty in transferring patients will depend on the functional status of the patient. In other words, not every obese patient will pose the risk of injury. However, when the patient is obese and uncooperative, sedated or in pain risks do exist. Some tasks are more difficult than others. For example, Fragala (1999) and others explain that one of the most frequent and difficult tasks to accomplish is the bed to chair transfer. The challenge of this task varies depending on how physically dependent the patient is. With a totally dependent patient, in order to have access to the patient, the caregiver must reach across the bed. This is especially true in the case of caring for the morbidly obese patient. The bed serves as an obstacle. The caregiver will probably not be able to bend his or her knees because of leaning up against the bed. If the patient needs to be physically lifted, the weight of the load involved in the lift is unacceptable for a typical caregiver. In addition, the transfer into a chair requires moving the patient to a different height level; some carrying is usually involved. Therefore, the risk factors of this particular task include reaching, lifting a load using suboptimal lifting postures, and carrying a load a distance. These types of transfers have traditionally been identified as high-risk activities. Organizations best serve their employees and their patients, especially the obese patient, when patient safety and caregiver injury is addressed; however, most agree that this is not forthcoming.

IDENTIFYING RISK

Perhaps insufficient attention to issues of obesity, immobility, patient safety, and caregiver injury is more a product of inadequate identification of those patients who pose risks. The obese patient may or may not be identified by an admitting diagnosis and/or list of clinical problems. The

TABLE 10.4
Preprinted Orders for Care of the Bariatric Patient

A private room is assigned to the patient
Skin care consultation
Monitor skin for breakdown every 12 hours
Nutrition consultation
Pharmacy consultation
Physical therapy consultation
Occupational therapy consultation
Respiratory therapy consultation
Pulmonary medicine consultation
Notify Radiology one hour before all procedures
Case management/social service consult for specific needs and discharge planning equipment
Based on the BMI, bariatric bed 39 inches if BMI = or < 50 (*Bari-Rehab Platform Bed)*
48 inches if BMI > 50 (*Bari-Rehab Platform Bed*)
Optional rotational therapy (*Big Turn*)
Alternating/low air loss mattress with airflow chux (order only if spine stable)
Weigh patient (circle one) qd, qod, every week (weight scales incorporated into beds)
Over the bed lift with *commode slit sling* (*Bari-Lift Transfer System)*

OR

Bed trapeze
Bariatric wheelchair with regular back or high rise back that reclines
Bariatric walker
Bariatric commode
Bariatric gowns

caregivers and patients are best served when the bariatric patient is properly identified even before admission to the hospital. Or at least upon entrance into the healthcare setting whether it is in the emergency department, or the inpatient setting. Preplanning, based on risk assessment is best initiated for the patient with a BMI that is 50 or greater, or for the patient is 100 pounds overweight, with a potential for special skin care needs, respiratory issues, mobility difficulties, or additional needs (see Table 10.4).

FINDING SOLUTIONS

CONSIDERING MOBILITY

Physical and occupational therapists (PT/OT) are often responsible for making recommendations for durable medical equipment, in-patient mobility, and subsequently, discharge needs. PT/OT therapists are instrumental in evaluating the patient for strength, functional mobility, and safety issues.

The following equipment should be considered in mobilizing the obese patient (see Figure 10.1 to Figure 10.5):

- Specialty beds both 39-inch or 48-inch beds
- Wide front wheeled walkers
- Wide wheelchairs (28 inches to 34 inches)
- Wide room chairs
- Wide beds that lower closer to the floor
- Patient lifts
- Transport stretchers

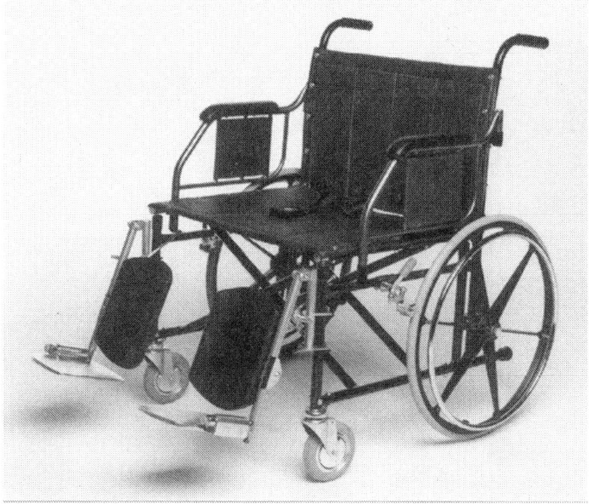

FIGURE 10.1 Bariatric wheelchair.

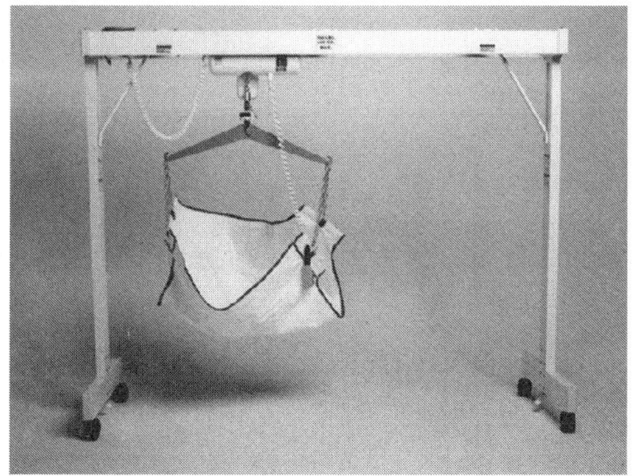

FIGURE 10.2 Bariatric lift that fits over bed.

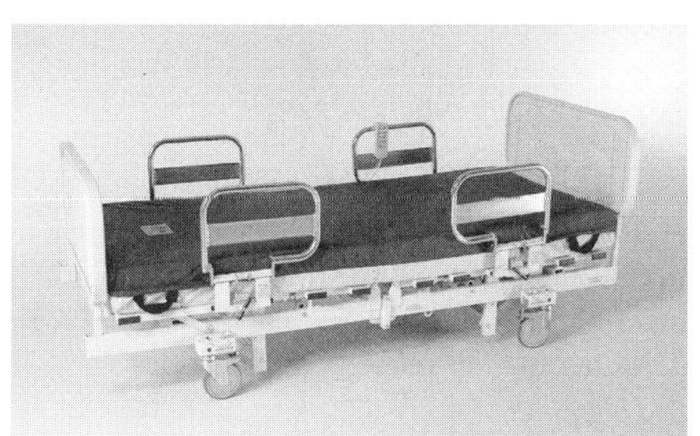

FIGURE 10.3 Bariatric bed.

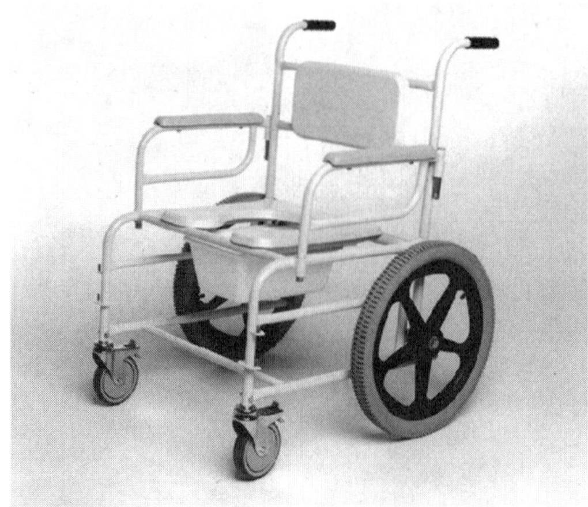

FIGURE 10.4 Bariatric commode.

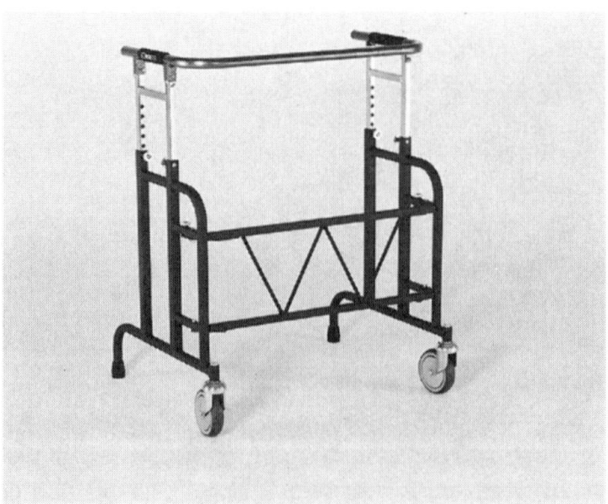

FIGURE 10.5 Bariatric walker.

- Gowns large enough to cover the patient when out of bed
- Wide bedside commodes
- Scales to weigh the patient (are on the specialty beds)
- Bed trapeźe (appropriate for weight of patient)
- Oversized stretcher in ED (appropriate for up to 650 lbs)
- Bariatric beds in operating room (appropriate for up to 1000 lbs)

Although fall prevention is imperative, in order to fully appreciate the issues related to caring for larger patients it is important to recognize other factors that influence patient safety. These factors are more than falls, a number of immobility-related complications that influence safe management of the patient. In addition, managing caregiver injury is essential in subsequently appreciating the prevention of immobility-related complications.

Considering Patient Safety

The concept of patient safety entails fall prevention but it also encompasses preventing nosocomial complications such as skin injury or respiratory complications. In addition, pain management can lead to falls, and emotional concerns can further lead to immobility-related complications.

Fall Prevention

Falls pose a serious risk for hospitalized patients, and are typically thought to be a problem of the elderly population. However, falls are increasingly common among heavier patients, regardless of age. And when a large, heavy patient does fall the consequences can be profound for the patient and caregiver. Further, the fear of falling can also affect the transition from hospital to home (see Table 10.5).

Falls impact patients, caregivers, and organizations. Patient falls are extremely costly in terms of injury and disability with a subsequent impairment of functioning and quality of life for the patient. Caregivers, too, can suffer injury and possible disability with subsequent loss of wages, function, and even earning power. Additionally, caregivers are likely to suffer emotional pain. Patient falls can be costly to healthcare organizations, as they can be a source of litigation, loss of experienced employee hours, workers' compensation claims, and increased length of stay with inadequate reimbursement. The bariatric patient poses greater challenges in delivering safe quality care.

Even patients who are able to maintain the highest level of functioning and independence at home may be compromised upon entering the hospital environment. For example, the home

TABLE 10.5
The Story of Anthony

The transition from hospital to home is often challenging. Certainly, patients look forward to moving back to their own homes, but following a prolonged hospitalization, there can be a number of anxieties. One is the fear of falling, which is more intense at home because there may not be sufficient resources, either people or equipment, to assist the person up off the floor. This was Anthony's fear.

Anthony had been in the hospital for 62 days, following a fall at home that resulted in a fracture. His hospitalization was rocky at times, which was exacerbated by his weight. At 480 pounds, caregivers had a difficult time mobilizing Anthony after surgery, especially when he was having pain or lightheadedness. Tomorrow, Anthony will be discharged from the acute rehabilitation facility to his own home.

Arlene Orhon Jech describes precautions for the elderly at home. Many of her ideas apply to Anthony, as well. Jech explains that because dehydration is a risk factor due to postural hypotension, Anthony ought to drink enough fluids. In addition, sitting on the edge of the bed before getting up may prevent dizziness when standing. Some medications can cause postural hypotension. Other medications can cause sedation, cognitive impairment, blurred vision or impaired balance, and postural hypotension. A pharmacist can help determine if either the prescription or over-the-counter medications that Anthony takes place him at risk for falling at home. Occasionally, patients will take multiple medications from multiple providers, which was the case with Anthony. In reviewing his medication, the rehabilitation nurse discovered that he was planning to take both digoxin and lanoxin at home. The nurse asked Anthony to make a list of all the medication that he planned to take once he returned home. He listed prescription medications, over-the-counter drugs, vitamins, and any herbal medicines so that he and the nurse could review the actions and interactions of each.

Along with medication-related falls, hazards in his home might exist. If a home safety evaluation reveals such hazards as loose electric cords, high wax floors and throw rugs, the situation is easily remedied. Electric and telephone cords can be stapled or taped (with 2 inch electric or duct tape) to the floor so they won't be tripped over. Nonskid wax should be used on floors, and throw rugs can be eliminated. Adequate light is important. Nonskid appliqués in the shower and a shower chair that accommodates Anthony's weight may be helpful early in the transition period.

Although home care planning is essential for all patients, it becomes more critical in planning for patients such as Anthony.

environment may be modified to allow for the patient's size and unique needs. There may be wide pathways through rooms with handrails or heavy furniture that the patient may use for balance. The patient's bed, chairs, even bedside commode may be oversized to allow for adequate support and balance. The patient and/or caregiver may have devised certain workable routines for daily activities such as bathing, toileting, and exercise. However, in the hospital, the patient is presented with equipment that is often too small or lightweight, preventing safe activity and mobility. Most furnishings in the hospital environment are on wheels, this is not safe or adequate to support a large person attempting to transfer in and out of bed or chair, or to ambulate around the room or restroom.

In addition, some restrooms are too small to accommodate the patient comfortably. Many wall-mounted commodes will not support the weight of the obese patient. Bedside commodes, wheelchairs and walkers are too narrow for the patient. This inappropriate equipment may give the impression of being too small to accommodate the patient's weight. The patient may be in a weakened condition with decreased balance or function due to illness, medication, pain, immobility, or dietary changes associated with a hospital admission. Turning, transferring, or ambulating the obese patient places the caregiver and patient at risk for injury. Often the patient is not only fearful of falling and incurring injury to self, but also of falling on or otherwise causing injury to a caregiver.

Appropriate equipment, along with PT/OT intervention and education are instrumental in preventing falls. The actual size of the patient room and access to timely nursing care are essential in preventing the fear of falling and actual incidence of falls.

Skin Injury

Pressure ulcers typically occur over a bony prominence and develop because of immobility and the inability to adequately reposition the patient. Pressure ulcer staging is dependent on depth of damage to the underlying tissue. In addition, obese patients can be at risk for atypical or unusual pressure ulcers, which can occur due to pressure within skin folds, as a result of tubes or catheters, or from an ill fitting chair or wheelchair.

Pressure within skin folds can be sufficient to cause skin breakdown. Tubes and catheters burrow into skin folds, which can further erode the skin surface. Pressure from side rails and armrests not designed to accommodate a larger person can cause pressure ulcers on the patient's hips. One way this atypical skin breakdown can be minimized is by using properly sized equipment. In addition, the patient needs to be repositioned at least every 2 hours, as do tubes and catheters. Tubes should be placed so that the patient does not rest on them. Tube/catheter holders may be helpful in this step. In the event that the patient has a large abdominal panniculus, it too, must be repositioned in order to prevent pressure injury beneath the panniculus. Patients who are alert are able to physically lift the pannus off of the suprapubic area. The dependent, weak or unconscious patient could be placed in the sidelying position and the nurse can lift the pannus away from the underlying skin surface allowing air to flow to the regions, while relieving pressure. Use of rotation therapy is often regarded as the standard of care for certain pulmonary situations, however, it can serve to ensure sufficient repositioning for a very large patient, who otherwise may pose a realistic challenge to frequent turning. Despite the value of rotation therapy in prevention and treatment of skin injury among the obese patient, it is necessary to take precautions to prevent friction and shear. Correct pressure settings, fitting the patient to the appropriate-sized surface, and assessment for skin changes can provide these precautions.

Incised and sutured wounds are expected to create a water-tight seal within 24 hours; however, wound healing can be delayed in some obese patients because of intereference with the normal wound healing process. Wounds are prone to dehiscence. In addition, blood supply to fatty tissue may be insufficient to provide an adequate amount of oxygen and nutrients. This can interfere with wound healing. Wound healing may also be delayed if the patient has a diet that lacks essential vitamins and nutrients. Wound healing can be delayed if the wound is within a skin fold, where excess moisture and bacteria can accumulate. Further, the excess body fat also increases the tension at the wound edges. To reduce the occurrence of abdominal wound separation, some clinicians use a surgical binder to support the area. The binder will need to be large enough to comfortably fit the patient, such as the Dale Abdominal Binder, which can accommodate waist sizes of up to 94 inches. Binders are especially important when the patient ambulates. Some patients find the binders so comfortable that they ask to leave the binder on at all times. Patients report that using a properly fitted binder when ambulating provides an added sense of security and comfort. Careful assessment under the binder will reveal any signs of early pressure-related breakdown where the edges of the binder meet the skin.

Candida albicans is a yeastlike fungus that thrives in a dark, moist environment, such as within skin folds. It is a normal inhabitant of the mouth, gastrointestinal tract, and vagina. Most fungal infections among humans are from *Candida albicans* or *Candida tropicalis*. *Candida albicans* is one of the most, if not the most common perpetrator of human disease. Factors that contribute to this condition include: immunocompromised states, diabetes mellitus, infection, chronic steroid use, hyperhidrosis, and obesity. This is characterized by scaling erythema, and in some cases small pustules, pus-filled lesions, may appear. Patients often complain of itching or burning. Without intervention, this condition can lead to fissuring and maceration. Further, in the face of associated pruritis, patients may scratch the skin surface, further compromising skin integrity. This can lead to a secondary bacterial invasion.

Candidiasis is manageable in a number of ways. If the patient complains of a moist skin surface, initially, an antifungal powder can be applied to clean, completely dry skin. For a dry, flaking surface, an antifungal cream can be helpful. To help soothe and cleanse affected skin, a soak or compress of Burrow's solution (aluminum acetate) can be applied for 15 to 20 minutes twice a day. Others suggest use of a 1% acetic acid (10 ml of vinegar to 1 quart of water) solution as a soak or compress. If the condition does not improve within 24 hours, consider reassessing the condition because many skin conditions mimic one another.

Moisture is a risk factor in skin breakdown, therefore, incontinence can complicate skin care. Many continent people develop short-term incontinence when physically dependent in the intensive care area. This may occur because of medication, a delay in locating enough caregivers to place the patient on the bedpan, or simply because the patient can't reach a commode in time to prevent an incontinent episode. Also, patients are frequently reluctant to ask for assistance with hygiene. Patients need to be reminded that our goal is to serve their needs. Maintaining clean, dry skin is our objective and if the patient needs assistant in this effort, caregivers are available to help in this respect.

After each incontinent episode, clean the entire affected area with an incontinence cleanser, and then rinse and dry the area. Patients report that drying the buttocks, perineal area, and between folds within institutionally approved blow dryer on the cool setting is more comfortable than towel drying. This technique may be less traumatic to the outer most layer of skin.

If, despite preventive efforts, skin breakdown occurs an aggressive plan of care is indicated. A moisture barrier ointment can serve as a protective barrier to chemicals in urine or stool (BAZA, Sween Company; Calmoseptine, Huntington Beach, California). Few moisture barrier ointments adhere to weeping or moist areas of superficial breakdown. A light coat of protective powder applied to the moist areas may increase adherence of the moisture barrier ointment, thus more completely protecting the skin surface from the irritating chemicals found in stool and urine.

Although it is important to understand the principles of local treatment, the most cost effective strategy is prevention. Candidiasis and other moisture-related skin breakdown are exacerbated by pressure related injury. Timely, appropriate mobility, which allows for regular commode use, can reduce incontinence-related skin injury. Early activity and mobility are imperative in preventing this sort of skin injury. Prevention of skin injury is especially important in the bariatric patient who may have a poor capacity for healing once the injury occurs.

Any nonintact skin (wounds, pressure ulcer, or procedural skin invasion such as a tracheostomy) is influenced by the negative effects of poor perfusion, tension to the wound edges, intra-abdominal pressure, inadequate oxygenation, and protein malnutrition. Thus healing may be seriously delayed.

Immobile, dependent patients should be assessed for risk for skin breakdown at least every 12 hours and more frequently as indicated by their condition. Specifically, moisture, pressure, and friction must be addressed in the obese patient. Criteria-based protocols are available to facilities interested in incorporating skin assessment and early mobilization into their policies.

Preventing Respiratory Complications

Morbidly obese patients tend to have pulmonary problems, two in particular: obesity hypoventilation syndrome (OHS) and obesity sleep apnea (OSA). OHS is an acute respiratory condition wherein the weight of fatty tissue on the rib cage and chest prevents the chest wall from expanding fully. Because patients are unable to breathe in and out fully, ventilatory insufficiency can occur.

Sleep apnea usually occurs when the patient is asleep in the supine position. The weight of the excess fatty tissue in the neck causes the throat to narrow, severely restricting or even cutting off breathing for seconds, or even minutes at a time. At home many patients manage the problems of nighttime sleep apnea with the use of a continuous positive airway pressure (CPAP) machine. However, in the hospital, some patients use BiPAP for a short time especially after extubation. Breathing can be made easier by keeping the patient in the semi-fowlers position, which takes some of the pressure off the diaphragm for reasons described earlier. Mobilizing the patient as early as possible will also help.

If long-term ventilator support becomes necessary, performing a tracheostomy can be especially challenging if the trachea is buried deep within fatty tissue. A large wound may be needed in order to locate the trachea. This larger wound can lead to complications such as bleeding, infection, or damage of the surrounding tissue. Postoperative tracheostomy care, therefore, includes steps to protect the persistomal skin, manage the tracheostomy, and contain wound drainage. To compound this dilemma, standard-sized trach tubes may be inadequate for use with patients with larger necks. A normal size endotracheal tube will usually accommodate the bariatric patient; however, a special tracheostomy tube may be required. The special tracheostomy tube for the obese patient may not have an inner cannula so safely maintaining a patent airway is important to address. In addition, narrow cloth trach ties can burrow deep within the folds of neck further damaging the skin. The thicker or wider ties have been used by clinicians to prevent this sort of damage.

Immobile, obese patients should be observed, particularly at night, for decreased oxygenation on pulse oximetry, snoring with sleep, and apneic periods. Remember pulse oximetry does not tell you if the patient has CO_2 retention. Patients with hypoventilation may have significant CO_2 retention with pulse oximetry readings of 90%. Patients should also be observed for confusion, especially upon awakening, and hypersomnolence during the day. Hypersomnolence typically manifests as the inability to stay awake, falling asleep during a conversation or other inappropriate times.

Therapy often includes positioning in a reclined or semi-fowlers position when possible to move abdominal contents away from diaphragm. CPAP or bilevel ventilation (pressure varies between inspiration and expiration) is usually applied via a nasal mask. The mask should not be applied too tightly. The CPAP and bilevel machines actually work better with a small leak; be sure to observe for skin breakdown. Slow deep breaths followed by inspiratory holds and yawning, as well as activity and mobility are essential in controlling respiratory complications.

Early mobilization of the critically ill patient can be especially difficult; however, equipment exists to promote movement. Lift and transfer systems are available to move the patient from bed to chair/wheelchair or bed to commode, slide boards and hover-type products can move the patient from bed to stretcher/cardiac chair. Full body lateral rotation therapy is available to provide kinetic therapy thus protecting the skin while improving pulmonary status with little to no risk to caregivers. Education on the methods of preventing injury and use of the equipment is an essential part safety in the critical care setting.

Pain Management

Pain can interfere with patient mobility for physiologic and psychologic reasons. In most cultures, pain serves as a warning that something is wrong. Patients experiencing pain often respond with reluctance to move or participate in activity and mobility. The patient may be reluctant to reposition because of sustained discomfort. Often this reluctance is misinterpreted as noncompliant or health defeating behavior.

One of the greatest challenges in healthcare is to ensure the physical, emotional, and spiritual comfort of our patients. Management of pain is an important factor in patient comfort, and all patients are entitled to the best pain relief that can safely be achieved. Yet the problem of pain is pervasive, and the myths and misconceptions surrounding the pain experience and the assessment of pain often preclude adequate comfort and quality care. This is especially true among bariatric patients where little evidence-based practice is available to make decisions about patient care and pain management. Misconceptions not only affect clinical decisions, but patients may also hold these misunderstandings, further interfering with pain control. Pain must be managed adequately as research now shows that unrelieved pain can inhibit the immune system, increase oxygen demand, respiratory dysfunction, decreased gastrointestinal motility, and confusion.

In addition to the challenges that all patients face, bariatric patients have additional concerns. More questions than answers arise when dealing with pain in patients whose bodies are a greater percentage of adipose tissue. For example, is the medication of choice water or fat soluble, and what are the clinical consequences of either? Will a 1-inch needle deliver a medication into the muscle or into the fatty tissue and should intramuscular injection even be attempted? What is the effect of opioids on sensorium or already compromised pulmonary function? What role has long-term chronic pain had on the surgical patient's interpretation of acute pain, and how is this assessed preoperatively? Finally, is postoperative nausea among morbidly obese linked to the type of surgery performed or a side effect of the medication? These largely unanswered questions further serve to complicate pain management among the obese patient. In many cases, pain can lead to behaviors that appear to be health-defeating activities.

Pain is a completely subjective experience. A widely used definition is that pain is whatever the experiencing person says it is and exists whenever he or she says it does. The self-report of pain by a patient should be considered sufficient evidence to establish pain as a nursing diagnosis. Overcoming the myths and misinformation that abound regarding pain assessment and treatment is a challenge. The acceptance of pain as a subjective experience is difficult for many healthcare professions. Often these old paradigms continue to pose barriers to adequate pain control. Inadequate documentation tools fail to communicate pain and pain ratings. A pharmacist and pain specialist are essential in determining the most appropriate drug based on absorption mechanisms and rates, which can be complicated by the patient's adiposity. The unique needs of the obese patient are especially important in making a drug selection. This clinical perspective also reinforces the need to clinically interpret the consequences of the chosen medication.

In many settings, use of the intramuscular route is widespread, despite the fact that research suggests that the intravenous or intraspinal route for analgesics is usually safer and more effective. In the bariatric patient, it may be impossible to deliver an intramuscular (IM) injection because of the presence of a thick layer of fatty tissue. The IM route may not provide predictable levels of drugs, and therefore is not recommended.

Respiratory depression is a potentially life-threatening side effect among all patients but can be especially serious among morbidly obese patients. Respiratory depression is thought to be preventable by the clinician's careful monitoring of sedation levels and easily treatable if it occurs. It is critical to assess for sedation levels and respiratory status when starting opioids on a patient who has moderate to severe pain and has not been receiving opioids regularly. When an opioid causes the patient to be so sedated that he or she has difficulty staying awake, the dose should be decreased to prevent respiratory depression. The likelihood of respiratory depression decreases the longer the patient has been on opioids because tolerance to respiratory depression develops and information about the patient's response to opioids is known. The antidote for respiratory depression is naloxone (Narcan) administered intravenously. Narcan is a pure opioid antagonist that can reverse both analgesic effects and respiratory depression. Sufficient amounts of narcan should be given to decrease sedation and increase respirations to an acceptable level without completely reversing analgesia. Giving too much naloxone can also precipitate hypertension and ventricular dysrhythmias. Therefore, dilute 0.4 mg of naloxone in 9 to 10 ml of saline and administer 0.5 ml over 2 minutes.

Each patient's situation should be reviewed separately to determine the acceptable level of sedation and to assess the potential for harm from decreased sensorium. Some patients choose to endure more discomfort if it means less sedation. If the sedation level is still unacceptable after a few days of adaptation, another opioid can be substituted until a satisfactory one is found.

Many obese patients suffer with chronic pain. Research suggests that the body adapts to this state of continuous pain and after a period of time, and vital signs normalize. This adaptation distorts the patient's perception to pain. It does not necessarily mean pain has been controlled adequately. The patient may minimize expressions of pain for a number of reasons. A patient may wish to be seen as a good patient or may place a personal or cultural value on a stoic response to pain. The patient may become too exhausted to respond vigorously to pain. Sometimes patients use distraction techniques to move the focus away from pain, especially when intense and unrelieved.

One way of screening for pain among bariatric patients is to include questions on the clinical admission form, such as "Do you have pain now or have you experienced any pain recently?" The 0 to 10 (0 = no pain, 10 = worst pain) pain rating scale usually is used. If chronic pain is identified on admission, a more comprehensive pain assessment should include one that includes location, quality, onset, frequency, and intensity. These items are self-explanatory, and it is easy to ask the patient about them. Location can be assessed by asking the patient to point to the site of pain on his or her body or on a figure drawing. To assess quality, you may need to give the patient some examples, using terms such as burning, aching, dull, knifelike, or shooting.

With this assessment, a comfort/function goal is established by asking the patient what pain rating would make it easy for him or her to perform specific activities required for recovery or quality of life. For example, the patient in this case situation may need to ambulate three times daily to prevent skin and pulmonary complications. The patient would be asked, "What pain rating will make it easy for you to walk with the physical therapist three times daily?" He may respond that 3 on a scale of 0 to 10 would be sufficient. The comfort/function goal would be documented as "3/10 to ambulate with PT." When establishing comfort/function goals, keep in mind that research has suggested that pain ratings above 4 significantly interfere with activities and mood.

In summary, the value of an interdisciplinary team should not be overlooked. Mobility is profoundly affected by uncontrolled pain. The pain specialist, pharmacist, physician, physical/occupational therapist, WOC nurse, bariatric CNS, and other interested professionals are important members of the team.

Psychosocial Issues of the Bariatric Patient

Morbidly obese people often experience prejudice, discrimination, humiliation, and embarrassment simply because of their weight or body configuration. Many people believe that obesity is due to a lack of discipline or poor self-control. Although research suggests that a distorted sense of self-image and certain eating disorders are common among morbidly obese people, the incidence of psychological disorders, such as depression, anxiety or substance abuse is no greater than among those of a healthy weight. However, situational depression and obesity are common.

Bariatric patients who are hospitalized report a lack of privacy and a loss of control. Unfamiliar surroundings interfere with mobility and activity. Some patients report that they overhear caregivers discussing them and their care, and that some caregivers hold a negative tone toward them. It is inappropriate to refer to someone by his or her weight or size. It is a violation of patient privacy to speak about a patient's size/weight with other staff members not involved in care. Staff members need to be aware of their own feelings and opinions about obesity and their possible effect on patient care. Because of experiences in past hospitalizations, some obese patients are reluctant to accept care. In addition, assessment can be difficult because of inappropriate equipment; scales or even a blood pressure cuff may be insufficient in size for an accurate assessment. Still other patients may have many issues related to a prolonged hospitalization.

Many patients who experience sustained pain become depressed and anxious. Depression and anxiety can be exacerbated if it is difficult to establish a physical cause for the pain. These patients may begin to question their judgment, fear they will be perceived as troublesome, and worry that pain relief will be withheld because the pain is not real. These emotional responses may cause some healthcare providers to think that the pain is psychogenic or not of a physiologic nature. This further compounds the emotional frailty, which an acute hospitalization can create.

Each patient is best addressed on a case-by-case basis. Although a person may qualify as "obese," he or she may or may not be affected by weight. Staff must look for any adverse effect of weight on the patient in regards to psychosocial issues. Some events may be crying or verbalizing concern related to weight issues. Families may also need emotional support. Staff may offer support in the form of listening, discussing concerns, and offering suggestions. An emotionally stressed patient may be distrustful toward caregivers, this undermines cooperation and a sense of teamwork. The relational barriers can lead to inactivity and immobility.

CONSIDERING CAREGIVER INJURY

Many hospitals have formally or informally adopted the goal of a PT/OT assessment of the patient's functional status within 8 hours of admission. PT/OT staff can then provide caregiver orientation to include use of equipment. A regularly scheduled repeat/reinforcement of the same inservice/orientation should include use of equipment, and mobility techniques. All staff should be educated on proper techniques for moving the bariatric patient. A key point is to ensure that the number of personnel assisting the patient is adequate/appropriate for the task. Additionally, more helpers may not ensure a higher degree of safety. Adequate personnel coupled with adequate equipment may serve as the best combination for safety (see Table 10.6).

TABLE 10.6
Thoughts from the PT/OT Community

The following are thoughts from a group of physical and occupational therapists that were willing to share their views on safely mobilizing the very obese patient:

- For the first week of service, utilize the same therapist, this ensures trust and confidence in care.
- Always use a gait belt. If the belt is too small, strap two together.
- Always ask for help.
- Involve family members.
- Frightened patients will not move.
- Watching a "soap" on the television is not a reason to defer mobility activities. This usually means the patient is reluctant to accept care.
- Even two days in the ICU will decondition the bariatric patient. Begin treatment early!
- Set realistic goals.
- Provide monthly PT/OT on-unit inservice on mobility techniques.
- Recognize the benefits/burdens of a lift team.
- Plan a team conference within 24 hours, plan a PT evaluation within 8 hours!
- Recognize the value of an interdisciplinary team, including the patient.
- Plan for the actual needs of the patient.

TABLE 10.7
Resources for Gurneys and Stretchers

www.stretchchair.com
www.wyeastmed.com
www.gendroninc.com
www.convaquip.com
www.sizewiserentals.com

Transporting the bariatric patient to diagnostic areas may require special equipment and skill. Specially designed gurneys and stretchers will accommodate to 1000 pounds (see Table 10.7). The ability to maintain a semi-fowlers position in the stretcher is imperative, additionally adequate width while ensuring the stretcher will move freely through doorways is essential. Although doorway widths are mandated in patient care areas, sometimes doorways are not sufficiently wide in other areas of the hospital.

Tests themselves can pose concerns because of weight limits on the diagnostic or interventional equipment. Radiology and special procedures departments should be notified as soon as the request for treatment/diagnosis is made. A minimum of one hour should be anticipated for scheduled procedures so they are able to assign an appropriate room for the bariatric patient. Ensure that the patient's weight and girth are measured and recorded prior to the scheduled test in order to ensure accommodation.

Certain types of equipment can be helpful in activity and mobility. Several styles of lifts can be used to mobilize larger patients: portable, standing or lateral lift systems. Portable lifts can move easily throughout the hospital setting and staff is generally familiar with this type of technology. Standing lifts are valuable in mobilizing a patient with a sufficient amount of strength as to assist the caregiver in the standing process. The advantage of the lateral lift systems is that there is little to no tugging on the patient, and little physical strength is needed to move the patient. Of concern for the bariatric patient, is the risk for complications that can arise when the patient with OHS is place in the supine position. Some lateral lift systems cannot accommodate a semi-fowlers position. Respiratory considerations are important to recognize in that those patient who must maintain a semi-fowlers position may have respiratory problems when placed in a supine position. Some lifts are designed to lift and or transfer patients who weigh as much as 1000 pounds (see Table 10.8). Systems are available that, in the event of a patient fall, the patient can be lifted up off of the floor with the assistance of 1 to 2 people. The decision as to the appropriate lift system is best made by the physical therapist and other members of the healthcare team who understands the needs of the particular patient.

TABLE 10.8
Resources for Lift and/or Transfer Systems

www.hovermatt.com
www.arjo.com
www.sizewiserentals.com
www.kci1.com
www.liftaid.com
www.wyeast.com

MAKING CHANGES

A number of organizations have used a variety of strategies in hopes of reducing or preventing caregiver injury. Of these, transfer teams, appropriate equipment, and criteria-based protocols, among others have been useful in achieving this goal.

Transfer teams, sometimes called lift teams, are thought to reduce or prevent caregiver injury in some settings. For instance, in the early 1990s, during an organization-wide nursing shortage, one hospital discovered a significant increase in the number of nursing injuries. Coincidentally, there was a significant increase in workers' compensation and operation costs. Hospital staff members from a variety of disciplines collaborated to create a task force to address the nursing injury issue. The task force discovered the work of William Charney, who had developed the concept of a lift team that had been successful in other organizations that were addressing similar challenges. The task force decided to use Charney's lift team model to serve as part of the injury management team. Charney's research on the concept as a holistic approach to nursing injury prevention has proven successful when confronting lost time injuries related to nurses and lifting patients. The lift team is comprised of individuals who have been screened and found to be at low risk for injury. Members are carefully trained to perform patient lifting and transferring activities. Team members are provided with patient handling equipment and are skilled in using the equipment to perform strenuous patient-handling tasks. Lift team intervention has reduced the frequency and severity of nursing injuries that result from patient handling. The lift team has become an accepted adjunct to the patient care team. The lift team intervention is believed to have improved nursing morale and patient satisfaction with regard to patient handling tasks. Lift team intervention programs do prevent work-related injuries and should be considered by patient institutions.

Standard hospital equipment, such as chairs or bed frames, may pose safety risks for the obese patient and their caregivers. On the other hand, equipment specially designed for overweight patients can improve their quality of care, reduce the patient's length of stay, and make it easier and safer for caregivers to perform care measures. Heavy-duty walkers, which accommodate patients weighing from 350 to 1000 pounds, make it easier to safely assist in ambulating heavy, weak patients. Beds, support surfaces, and wheelchairs that support up to 1000 pounds are also available. And in addition, a number of lift designs are also available to mobilize the very large patient. Providing equipment specially designed for the bariatric patient is important for reducing work-related back injuries among caregivers and lowering the risk of related patient injuries. The challenge is the timely availability of equipment to patients and caregivers.

Healthcare facilities must have a plan in place to care for the special needs of the morbidly obese patient. Rather than attempting to make a standard size fit all, patients are best served when equipment and care is selected that is appropriate to the patient's size and needs. Preplanning with manufacturers and vendors to provide equipment for the morbidly obese patient is essential. Institutional policies and procedures to obtain transportation and transfer devices, bed frames and support surfaces, wheelchairs, walkers, and commodes or furniture must be available.

When selecting oversized equipment, it is essential to consider both the weight limit and width of the equipment. For example, a patient may not exceed weight limits for a standard bedside commode, but he or she may be unable to use a standard device because of the width of the patient's hips.

Most medical equipment suppliers rent or sell extra-wide wheelchairs, walkers or commodes that accommodate patients up to 1000 pounds (see Figure 10.1 to Figure 10.5). Some rental companies provide a number of oversized bariatric items as a bundle, providing a price break. Criteria-based protocols for use of specially designed equipment are designed to ensure more appropriate, timely, and cost sensitive use of equipment. Performance improvement teams offer a resource to develop and implement appropriate policies and resources for bariatric equipment needs.

Implementing changes to better manage the unique care issues for the morbidly obese patient can be challenging for caregivers. The initial cost of any change is often viewed as an obstacle.

TABLE 10.9
Bariatric Functional Status Protocol

Patient Care Systems has derived this functional status protocol, which uses categories for grouping bariatric equipment in acute care settings that will ease the ordering process and create a more consistent formula for safe, effective care.

Category 1
Rehab, Ready

This group is ready to do pivot transfers and bed-side standing. These patients help turn themselves and are mobile.

Rehab Ready — Under 400 lb. (specify height and weight)

Patient gets a 39 inch wide bed with foam mattress, fitted wheelchair, fitted commode chair, overhead trapeze, and walker.

Rehab Ready — Over 400 lb. (specify height and weight)

Patient gets a 48 inch wide bed with foam mattress, fitted wheelchair, fitted commode chair, overhead trapeze, and walker.

Category 2
Rehab, Nonambulatory

This group needs to be lifted into the chair or is uncertain of its weight-bearing skills. These patients are still considered high risk for skin breakdown and they are not much assistance in turning themselves due to immobility.

Rehab Lift Assistant — Under 400 lb. (specify height and weight)

Patient gets a 39 inch wide bed with low air loss mattress, fitted wheelchair, fitted commode chair, bariatric lift with trapeze handle, and walker.

Rehab Lift Assistant — Over 400 lb. (specify height and weight)

Patient gets a 48 inch wide bed with low air loss mattress, fitted wheelchair, fitted commode chair, bariatric lift with trapeze handle, and walker.

Category 3
Rehab, Complex

This group needs a lateral transfer to get into ortho-chair to monitor BP and tolerance to sitting for first couple times with uncertain trunk stability. These patients are still considered high risk for skin breakdown due to immobility.

Rehab Lateral Transfer — Under 400 lb. (specify height and weight)

Patient gets a 39 inch wide bed with low air loss mattress, fitted ortho-chair, Air Pal, bariatric lift, and walker.

Rehab Lateral Transfer — Over 400 lb. (specify height and weight)

Patient gets a 48 inch wide bed with low air loss mattress, fitted ortho-chair, bariatric Air Pal, bariatric lift, and walker.

Category 4
ICU, Immobilized

This group is not stable enough for out of bed orders yet and typically a very high risk with such complexities as ventilator dependent and preexisting wounds.

Immobilized — Under 400 lb. (specify height and weight)

Patient gets a 39 inch wide bed with lateral rotation therapy, and bariatric lift for patient handling.

Immobilized — Over 400 lb. (specify height and weight)

Patient gets a 48 inch wide bed with lateral rotation therapy, and bariatric lift for patient handling.

Source: With permission from Shawn Strahan, Patient Care Systems, Houston, TX.

Without a thorough understanding of the cost incurred in caregiver injury and the prolonged hospitalization of the patient, it may be difficult to economically justify introduction of specialized equipment, which very well may not be reimbursed by third party payers.

Most agree that in the healthcare delivery market, a successful organization must remain profitable while maintaining or improving the quality of its services. The Joint Commission on Accreditation of Healthcare Organizations recognizes the numerous barriers to change, and has adopted the philosophy of continuous quality improvement (CQI) as a cornerstone for change, and the management principles of CQI continue to gain acceptance in the healthcare industry. Performance Improvement is a recent innovation that utilizes these principles of CQI in making changes in health organizations.

TABLE 10.10
Challenges Across the Continuum of Care

Joe is a 560-pound, 42-year-old school teacher who took a serious fall in the upstairs bedroom of his two-story apartment. Challenges began the moment the 911 call was placed from his home. Knowing of Joe and the issues in transporting a very larger, injured patient, paramedics were reluctant to authorize the run, instead, initially reporting that all vehicles were out on calls. Paramedics were concerned with their own safety, and they were concerned about subsequent legal action in the event Joe was further injured in the attempt to move him out of his home. Additionally, Joe exceeded the weight limits for the transport gurney. Options for the transporting paramedic service were to: (1) place Joe directly on the gurney, unsecured, but in a rollover Joe and the attending paramedic would be seriously injured; (2) strap Joe onto the vehicle flooring, but skin ulcers were likely to develop from the trauma, and respiratory problems could occur by placing Joe in the supine position.

Once at the hospital, emergency department (ED) personnel had problems in moving Joe out of the vehicle and into the ED. Movement through the hospital was challenging throughout Joe's hospital stay. Administrators and department heads were reluctant to accept Joe into their respective departments, citing the inability to meet his needs. This lead to delays in care and a prolonged hospitalization. The failure to accommodate Joe's needs not only affected his clinical outcomes, but also it predisposed caregivers to injury as Joe became increasingly deconditioned, and his length of stay increased unnecessary.

Performance improvement (PI), based on the principles of CQI, seeks to make changes that improve the therapeutic, cost, and satisfaction outcomes associated with patient care. Decisions need to be made by those individuals closest to the patient, they are customer-focused, and change must continue to be on-going. A PI task force could include a physical therapist, occupational health coordinator, risk manager, safety specialist, ergonomist, front-line caregiver, administrator, among others. The task force might consider inviting a former patient who can actually provide input from the lived experience of being cared for as a patient who is overweight. Vendors are also valuable in this process because they are able to partner with hospitals that are looking for equipment that could be tailored to better meet the needs of caregivers.

These quality-based efforts can be unit-based or house-wide. Regardless, they are important to consider in that they can more accurately establish the actual needs of the organizations as they seek to reduce or prevent caregiver injury.

PREPARING FOR THE FUTURE

Although attempts to reduce body weight are common among Americans, the prevalence of obesity has continued to increase since the 1980s. Considering that more than 50% of U.S. adults are overweight, it is likely that issues of caring for the overweight patient will continue. In fact, of even more concern, not only has the percentage of adult Americans increased, but the number of overweight children has doubled. And even though some overweight people are able to lose some of their body weight, a majority regain that weight within 5 years.

Such increases will tremendously affect healthcare delivery since obesity is strongly associated with several chronic diseases. This may lead to hospitalization and the corresponding issues described earlier. Recent estimates suggest that obesity-related morbidity may account for 6.8% of U.S. healthcare costs. This increasing prevalence will impact acute care and may not only influence the frequency of admission, but will influence the severity of care that patients will require when hospitalized. Clinicians best serve the needs of the patient when policies and protocols are in place to care for the patient. Continued use of interdisciplinary teams is essential to more fully understand the interdepartmental impact of caring for overweight patients in the acute care setting. Furthermore, manufacturers and vendors need clinical input in order to more fully understand the unique equipment needs of the larger patient. Clinicians best serve their patients when they are able to partner with industry to creatively seek solutions to the challenges described. And further, outcome studies can provide the data necessary to sustain these efforts.

SUMMARY

Managing the complex needs of the bariatric patient can be time consuming and costly. Solutions to this industry-wide problem are not simple. The prevalence of overweight patients is increasing. Collaborative task forces are in the best position to understand the issues more fully, as each department is affected in a unique yet very important manner. A number of solutions may exist, however, transfer teams and use of appropriate equipment supported by criteria-based protocols are one strategy. Quality improvement theories may be helpful along with outcome studies to develop and sustain these efforts to reduce or prevent caregiver injury. The challenge to hospitals and caregivers is that all these changes must be done in a growing climate where profit and cost reductions outweigh safety. Preplanning care is designed to control costs by preventing caregiver injury and promoting patient safety. An interdisciplinary approach is likely to best serve the needs of the patient, caregivers, and the institution.

REFERENCES

Boernstein, D., Epidemiology, etiology, diagnostic evaluation, and treatment of low back pain, *Current Opinion in Rheumatology,* 8, 124–129, 1995.

Charney, W., Zimmerman, K,, and Walara, E., The lifting team: a design method to reduce lost-time back injury in nursing, *AAOHN J.,* 39(5), 231–234, 1991.

Charney, W., An epidemic of health care worker injury, in Charney, W. and Fragala, G., Eds., *The Epidemic of Health Care Worker Injury: An Epidemiology,* CRC Press, New York, 1999, p. 1.

Flegal, K.M., Carroll, M.D., Kuczmarski, R.J., and Johnson, C.L., Overweight and obesity in the United States: prevalence and trends. 1960–1994, *Int. J. Obes. Related Metabolic Disorders,* 22, 39–47, 1998.

Gallagher, S.M., Greenstein, R., and Parson, R., *The Bariatric Patient: a Therapeutic and Risk Management Perspective,* VHA Satellite Network, Dallas, TX, 1998.

Gallagher, S., Caring for the overweight patient in the acute care setting: addressing caregiver injury, *J. Healthcare Saf. Compliance Inf. Control,* 4(8), 379–382, 2000.

Gallagher, S.M., Ethical dilemmas in pain management, *Ostomy/Wound Management,* 44(9), 16–21, 1998.

Gallagher, S.M., Meeting the needs of the obese patient, *Am. J. Nursing,* 96(8), 1s–12s, 1996.

Gallagher, S.M., Panniculectomy: implications for care, *Perspectives,* 3(3), 1,4–8, 2001.

Gallagher, S.M., Restructuring the therapeutic environment to promote care and safety for obese patient, *JWOCN,* 26, 292–297, 1999.

Gallagher, S., Tailoring care for the obese patient, *RN,* 62(5), 43–50, 1999.

Jaxon et al., *Management of Cancer Pain.* Clinical Practice Guidelines. No. 9 AHCPR Pub No. 94–0592. Accessed 1/03 at: www.ahcpr.gov/gils/00000176.htm.

Joshi, M.S., Clinical performance improvement series. Classic CQI integrated with comprehensive disease management as a model for performance improvement, *Joint Commission J. Qual. Improv.,* 25(8), 383–395, 1999.

McCafferty, M. and Pasero, C., *Pain: Clinical Manual.* Mosby, St. Louis, 1999.

Mafoski, D., *Epidemiological Analysis of Occupational Injury in Health Care: Essential of Modern Hospital Safety,* CRC Press, New York, 1995.

Meittunen, E.J., McCormack, H., and Sobczak, S.C., Evaluation of patient transfer tasks using multiple data sources, *J. Health Care Saf., Compliance Inf. Control,* 4(1), 13–16, 1999.

NIDDK, Statistics related to overweight and obesity. Accessed 1/03 at: http://www.niddk.nih.gov.

Oria, H.E., Performance standard in bariatric surgery, *Eur. J. Gastroenterol. Hepatol.,* 11(2), 77–84, 1999.

Pasero, C., Making your pain committee effective, *Am. J. Nursing,* 97(3), 17–19, 1997.

Rogers, B., Health hazards in nursing and health care: an overview, in Charney, W. and Fragala, G., Eds. *The Epidemic of Health Care Worker Injury: An Epidemiology,* CRC Press, New York, 1999.

Weissman, N.W., Achievable benchmarks of care: the ABCs of benchmarking, *J. Eval. Clin. Pract.,* 5(3), 269–281, 1999.

Injured Nurse Story #14: Wake Up Call

by Out of There

I worked for several years on a medical/surgical oncology unit. One factor for leaving the hospital setting was related to a preventable back injury caused from understaffing. At the time, staffing ratios of nurses to patients was too low.

My assignment was four, high-acuity, immobile patients, each weighing approximately 250 pounds. As staffing was short, assistance was very limited for repositioning and changing patients when they were incontinent. The entire day, I was struggling to turn patients. I kept asking for help, but, as all the nurses were struggling to provide care, there was little help available.

The most frustrating situation for me was my inability to provide quality nursing care to one particular patient who was extremely weak and had multiple bouts of diarrhea stools. I weigh 130 pounds and am petite in size. She was too weak to roll over without help or to hold herself on her side. I was not strong enough to turn her onto her side and hold her up without help. We had no friction-reducing devices to help move and turn patients. We didn't even know products like that were available.

This patient lay in bed soiled for about 20 to 25 minutes before I could get somebody to help me. The family even offered to help me because they could see that staffing was a problem. I felt, over all, that this was a pretty sad situation, when families feel they have to help clean up a loved one when they are paying for care.

I felt terrible for the patient because the smell made it obvious to her family members she was incontinent of stool. I wanted to spare her dignity from lying in smelly stool and her skin from possible breakdown with being exposed to stool for that long. Putting an ill person through this type of situation is degrading and unacceptable.

The bottom line was there was nobody there to help me and I hurt my back. I made it very clear at the start of the shift and throughout the day about my concern with this assignment. I was angry. I was baffled. And, during the shift I hurt my back struggling to turn and move patients by myself. That was part of my reason for leaving.

Because I had an injury, I turned in an incident report. I reviewed the charts so I could report accurately. Then, I went to the employee health nurse to report the situation. I asked her to evaluate this situation and provide some training for supervisors and charge nurses on making assignments. I feel that if you're at the mercy of your charge nurse, they need more guidance on making appropriate assignments, that there is a responsibility for basic training in making assignments respective of physical limitations of patients. I had no feedback after turning in my report.

I was offered employment at a local clinic and accepted the offer. And that was it. I'm now very happy in my job.

The charge nurse and I have remained friends and I hope that she learned from that experience. Working hard five days a week, then getting a ridiculous assignment that puts me at risk, it's hard not to take personally.

The important thing is that my back is my livelihood. I have a lot of financial responsibility. If I would have had a permanent back injury, I would have had to deal with the consequences. This could have been devastating financially. And, I really do mean that. These are the years that I'm

1-56670-631-9/03/$0.00+$1.50

saving for my retirement and I have kids in college. It only takes one time to injure your back. To have an assignment that is completely overwhelming, there is really no excuse for that.

The next day, a different charge nurse made an adjustment in the assignment. This situation makes me believe you are at the mercy of the charge nurse, but if you complain, you will be labeled as a complainer. I have learned that we have to advocate for ourselves because you can be sure no one else will advocate for you. This was extremely unsettling because I wasn't sure if I could work the next day. I didn't know if I could continue working at all. You don't want to continue working with an injury. It was all a very negative experience.

Overall, I don't know what I did to my back that day. It hurt for the next couple of days and then went away. It seems fine now. I don't know whether I suffered some cumulative damage that day, but it scared me. Lying in bed with back pain that night, I realized what the outcome could be. I didn't know. When you think you're injured, that's when you start thinking what the ramifications could be. You think, "I've hurt myself. How bad is it? How will this affect me?" That's when the fear comes. I have a house payment and all these responsibilities. I can't afford not to work. That was the reality for me.

I, also, worked with another RN the night she hurt her back. I was recovering a surgical patient who weighed about 250 pounds. She insisted on getting up to the bedside commode after the doctor said she could get up. With much difficulty, I assisted her up to the commode but she slumped over. She was just too weak and couldn't support herself.

There was one mechanical lift somewhere in the hospital but we very rarely, almost never, used it. We never knew where to find the lift, the slings were kept somewhere else, and we just didn't have time to go looking for it. We never relied on the lift at all. We were expected to work fast and get our work done on time, so, we just picked the patients up and moved them ourselves. This is what all the nurses did.

The lady was ready to fall off the commode and we had to do a dead lift to put her back to bed. During that lift, the nurse assisting me, my friend, hurt her back. I feel terribly guilty about my friend hurting her back. I feel just terrible, absolutely horrible, because she hurt her back and it was my patient. My friend was severely injured and has never been able to return to floor nursing.

The nature of nursing needs to be addressed. I don't want to be a disposable commodity because of a back injury. This is my story. My injury was preventable and so was my friend's. She will never work as a nurse again. I've gone to work outside the hospital.

It was a terrible situation and I hope I never have to go through anything like that again. You don't realize how you feel about it until an injury almost takes you out of your career. You give 100 percent to your job, and, when you're hurt, you can't work, and, if you can't work, you can't pay your bills.

Until you've gone through something like this yourself, you just don't realize. You take it for granted, that you'll make it through the shift without being injured, until you're hurt or see a coworker disabled from lifting a patient. You really could lose everything you've worked for. And, once you're labeled as an injured nurse, you're not going to get hired anywhere else. No one is going to hire a nurse with a bad back.

11 Participatory Ergonomic Design in Health Care Facilities

Jocelyn Villeneuve

CONTENTS

INTRODUCTION

Many design projects are carried out without sufficient organized dialogue between the direct users, the designers, and the management. All too often, the fundamental decisions regarding design choices are made in a vacuum by a committee of technical professionals who do not give enough consideration to overall working conditions in which the users of the building — employees, clients, visitors and

1-56670-631-9/03/$0.00+$1.50

suppliers — will have to conduct their activities. That is why many dysfunctions in workplace design are identified, both in the patient care units themselves and in other related areas such as laboratories, food services, administrative offices and so on. Ergonomic interventions are required to correct these situations, when in fact the problems could easily have been avoided at the planning stage.

This paper describes an ergonomic user-focused approach ensuring that workplace designs provide the best possible health and safety conditions and satisfy the requirements of healthcare workers in terms of comfort, quality, and efficiency. Although focused on employees, it also has a positive impact for clients and other users.

We will first describe the method used to effectively structure internal project management using a participatory approach. The principal obstacles encountered in implementing this approach will then be discussed. A case study will briefly illustrate the positive impacts of such an approach.

Since one of the major problems in occupational health and safety is back injury caused by the manual handling of people or objects, it is a priority to propose the introduction of mechanical aids that will eliminate manual handling as much as possible, right from the design phase of the building.

Designing a hospital is extremely complex and the degree of uncertainty involved is sometimes quite high. This is a real challenge, particularly for the people involved who are not design professionals. Two methods that facilitate their participation are described: the dynamic simulation of future activities and visits to reference sites. A second case study illustrates the positive impact of well-organized visits.

This procedure can save time and money, while producing healthy, safe, and effective environments for everyone.

PARTICIPATORY PROCESS

The main deficiency in the management of a project is the poor structure of the internal design process. The roles of the various players are poorly defined. Often, there is no "Users Committee" to guide the professionals, meaning that choices are not tested sufficiently with users. The effects of this can be disastrous in some cases. An example is the newly redesigned radiology room in which it is impossible to manipulate a stretcher, even though the walls and doors were built at great expense.

In fact, the main concern of the ergonomist is to ensure that all the institutional players who have information relevant to the project are involved from the beginning of the design phase. The way people are involved is a key element in the success of the project. The approach proposed here is based on positive experiences in a number of institutions that truly took charge of their projects.

In a large-scale project, three types of committees are required. Their mandates are separate but complementary, as shown in Figure 11.1. The Steering Committee is composed of representatives of all the social players concerned, and is led by the general management. Its mandate is to define the project guidelines. The Technical Committee, composed of technical professionals and led by the project leader, is responsible for designing and carrying out the project in compliance with the Steering Committee's guidelines. Finally, the Users' Committees are formed by department or by theme, and are composed of representatives of the executives and workers concerned by the project. Their mandate is to help define requirements and then provide an informed opinion on the design proposals, based on their expertise in the field.

Close and constant contact should be maintained between the committees. In particular, each committee should delegate a representative to the Steering Committee, which is in fact the decision-making body. The project leader plays a key role in managing the contacts between the Steering Committee and Users Committees and the professionals on the Technical Committee.

OBSTACLES TO STAFF PARTICIPATION

A participatory approach to ergonomic design is not always easy to implement. A number of obstacles can arise, depending on the professional interests and prerogatives of the various parties

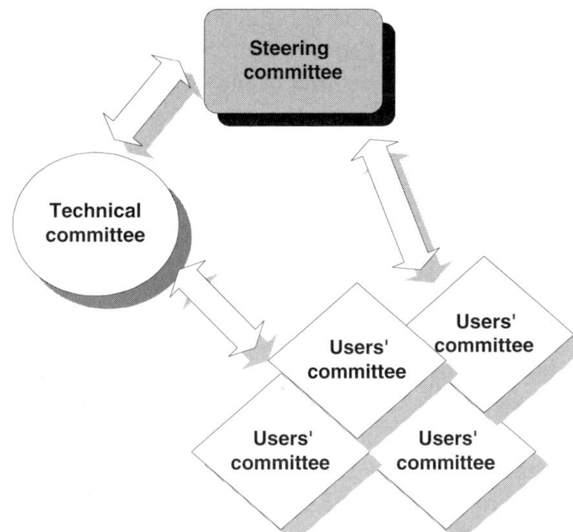

FIGURE 11.1 Project management structure. Close and constant contact must be maintained between the committees. The project leader plays a key role in managing these contacts.

involved in a project. Here are the principal obstacles encountered involving hospital management, design professionals (architects, engineers, etc.), as well as employees and their trade union representatives. These obstacles are easy to overcome if the hospital management really wants to involve employees and if the project manager is in favor of this approach and is sufficiently skilled in participatory management.

For hospital management, the primary obstacle is the sharing of information, because sharing information means sharing power. The participation of management and staff cannot be solicited without opting for transparency in terms of decisions relating to the project. However, some people do not really consider the consequences and tend to back out along the way. All they really want are assurances, without any debate by staff. Of course this is not possible. Undertaking a consultative process involves certain requirements on the part of management, which must then consider staff as a true partner in the decision-making process.

Some people claim that they are afraid of being confronted by unrealistic demands that cannot be met, and use this as a reason for lack of consultation. Staff satisfaction is assumed, saying that in any case, the new facilities will improve working conditions compared to the existing situation. This is not necessarily true, however, because the start-up of a new facility is not always smooth, particularly when work organization, technologies, and staffing have been changed. Start-up is much easier if staff has been involved right from the beginning of the project. Conclusive experiments have shown that demands by workers and department heads are realistic and do not necessarily increase the costs of a project. As one hospital manager pointed out, "The employees didn't make any frivolous demands. On the contrary, all of the proposals submitted were relevant and some were major."

Project managers are always afraid of budget overruns or scheduling delays, and put pressure on management to eliminate any interventions that risk disturbing the fragile balance between program delivery, scheduling, and budget. Yet, participatory management experience has shown that on the contrary, considerable savings can be achieved by significantly reducing modifications that must be made after the construction work has been completed.

For designers, architects, and engineers, corporatism is the main roadblock to staff participation. For the majority of them, any intervention by someone other than a technical professional is perceived as being a form of interference. The fact that an employee, nurse, maintenance worker, kitchen helper, etc. might have something relevant to say about a planning project or technical installation is inconceivable to them, if their proposals are not backed by an "expert." Ergonomists are often called on to play the role of intermediary between designers and people working on the floor.

Another major obstacle stems from the fact that designers have trouble managing incoherent or contradictory demands from employees, management, and doctors. Indeed, different groups sometimes make demands that are not consistent. Professionals' skill in interpreting these demands and negotiating creative compromises is therefore crucial. On the other hand, they are not always well informed of the impact the project will have on the social dynamics of the institution.

Professionals have trouble accepting last-minute interventions once they are at the final preliminary or detailed planning stage, and rightly so, because any significant change means revising their plans. Often, additional fees are also charged. It is therefore the responsibility of the institution to initiate the consultation process right from the start of the project, to avoid this type of inconvenience.

As for employees and trade union representatives, mistrust and skepticism are frequently observed. They wonder about management's actual motivations. Some trade union representatives sometimes have trouble subscribing to a logic other than that of confrontation. That is why it is very important to clearly define the mandate and powers of the joint committees (employees and management) from the outset. These committees usually have only recommending power, which can be real power if management takes the process seriously. Hence the importance of following up proposed recommendations and explaining why some of them were not chosen or were modified.

Giving staff time off with pay poses a problem. It is important to set aside the funds required to replace staff representatives, to avoid causing overwork for those who remain on duty.

ZERO LIFT APPROACH

The most promising approach for preventing back injuries in the healthcare industry as well as other industries would be to eliminate manual lifting by making maximum use of mechanical aids. Training programs on safe lifting techniques do not give good results without modifying the working environment and providing the proper equipment.

Dealing with this issue right from the building design phase represents an extraordinary opportunity for implementing a "zero lift" approach. Since the direct and indirect costs of occupational accidents are very high, investments made at the outset have a good chance of achieving rapid self-financing, while providing better working conditions for staff and greater efficacy in terms of operations.

The best way to proceed is with a step-by-step analysis of the flow of people and materials from the entrance to the exit of the building. During this process, it is important to identify the specific locations where handling is done and to make provisions for the most appropriate equipment for eliminating as much manual handling as possible. The equipment purchasing program should therefore be updated when the recommended equipment has not been planned for.

HANDLING OF PEOPLE

In healthcare institutions, the vast majority of back injuries are related to the transfer of dependent patients. It is important to clearly identify clienteles and departments that require heavy lifting of patients and to provide the appropriate equipment: patient lifts, adjustable-height stretchers, electric beds, gliders, etc. Staff should be trained in how to use this equipment and in safe patient transfer methods.

Recent studies have shown that the results of extensive use of ceiling lifts is very significant in terms of reducing the incidence and seriousness of accidents related to patient transfers. These studies also show that it is possible to completely self-finance these installations over a five-year period exclusively through savings in compensation to occupational accident victims (Villeneuve et al., 1999).

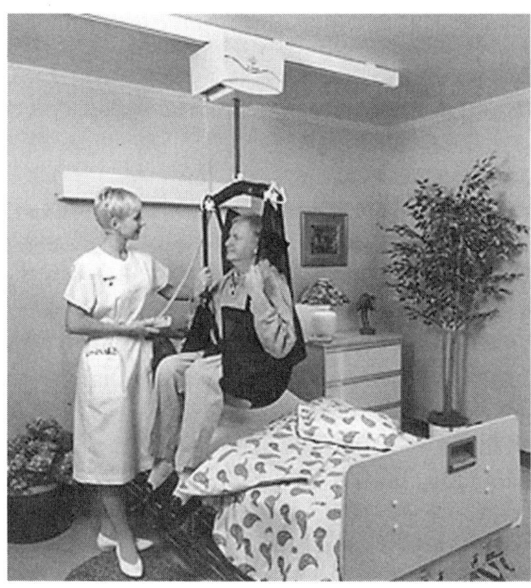

FIGURE 11.2 Ceiling lift system.

Handling of Objects

There are many auxiliary services involved in handling heavy materials. For instance, merchandise receiving or shipping platforms, warehouses, dietary services, laundry, housekeeping, and building maintenance departments.

In all of these departments, it is important to analyse the circulation of materials and to provide the best equipment possible to eliminate manual lifting of heavy loads: pallet trucks, hoists, trolleys, freight elevators, etc. When it is impossible to mechanize the handling of loads, it is important to reduce the weight of the objects, teach people safe lifting techniques, and ensure that there are enough employees on hand to do the work without injuring themselves.

Ceiling lifts will be installed in all rooms that are likely to accommodate dependent patients. In all, 45% of the beds will be covered by a ceiling lift. All therapeutic bathrooms will also be equipped with ceiling lifts. There will be adjustable-height electric beds in all rooms, except in the psychiatry department (see Figure 11.2).

CASE STUDY NO. 1: NEW HOSPITAL CONSTRUCTION

Here is an example of an ongoing ergonomics intervention in a new hospital construction project on the outskirts of Montreal (Canada) where a participatory ergonomics expertise was asked by the general management.

At the very beginning, users were grouped by service, to form a total of twenty groups in all. The groups were composed of the service head, employee representatives for each job title, and representatives of the service's professionals and physicians. The users' groups played a full and active role in the architectural design process. The architects submitted sketches, and the groups made comments or suggested modifications, until the proposed layout coincided with the service's clinical mission and anticipated operations. Simulations of future activities and visits to reference sites were organized, to allow the users' groups to visualize operating scenarios and proposed facilities.

The Impact of the Ergonomics Intervention on the Design

The project in question involved the construction of a new complex comprising three interlinking buildings. This paper focuses on one of the three buildings, namely the hospital block (Block

A), including the nursing and rehabilitation services on the upper floors, and the staff cloakrooms and food services on the ground floor. I will show how the ergonomic intervention triggered some major changes between the initial and final building designs. The changes in question involved four major factors.

Factor 1: The Rooms

Life-sized simulations of all the proposed standard room models were produced, and a number of specific design criteria emerged, including the position of the beds and technical panels, the space required around the beds, the shape and dimensions of the bathrooms, and others. In particular, the obstetrics room simulations showed that the plans as initially proposed were not functional, especially for the two-bed rooms, and as a result management decided to retain only the single rooms. This new factor meant that additional areas, not provided for in the initial program, would be needed. In the end, three room models were retained: LDRP (Labor, Delivery, Recovery, and Postpartum), LDR (Labor, Delivery, and Recovery) and P (Postpartum). The architects were given specific layout criteria for each type of room. The validation work with the Obstetrics Service, located on the second floor of Block A, had a significant impact on the overall design, and the architects' task of superimposing the four floors, each with different room models, was extremely complex.

Ceiling lifts were to be installed in all rooms that were likely to accommodate dependent patients. In all, 45% of the beds were covered by a ceiling lift. All therapeutic bathrooms were also equipped with ceiling lifts. There were adjustable-height electric beds in all rooms, except in the psychiatry department.

Factor 2: The Nursing Station

In the care units, the program provided for mini-nursing stations located in the middle of the building's wings. After consulting the service heads and users' groups, this configuration was rejected because the nursing resources on the evening and night shifts were insufficient to cover eight mini-stations in the four Block A wings. The choice of a centralized nursing station for two wings turned out to be more efficient in terms of both human resources and economy of movement for the staff, especially on the evening and night shifts.

Factor 3: The Psychiatric Courtyard

At the Psychiatric Service, an inner courtyard closed off by a walkway between two wings was proposed in the architectural design. After consulting the psychiatric users' group, this solution was rejected because the courtyard, with walls on all four sides, was reminiscent of a prison. In the new version, the walkway was eliminated, leaving the courtyard open on one side, with a decorative fence that could be used when necessary to close it off. The new courtyard was bigger and contained an additional games space as well as a landscaped rest area that could be accessed from the community room.

Factor 4: The Food Service

For the Food Service, the first version of the plans proposed a service zone separated from the production zone by the cafeteria entrance. This layout was not functional, because continual movements of clean and dirty equipment would often hinder circulation. Moreover, it complicated the work of the staff and made the service less efficient. In the final version of the plans, the service and production zones were contiguous.

The final plan of Block A was longer, allowing for more rooms around the perimeter. It offered a larger surface area in the centre of the wings, to house the nursing station and service offices in the central zone. The main entrance to the hospital was moved so that the cafeteria service zone could be located alongside.

The participatory approach enabled the design process to focus on direct users, by taking advantage of their knowledge of operations and tasks. As a result, the plans were developed coherently and significantly reduced arduous backtracking at more advanced stages of the planning process. The possibility of errors and omissions was also reduced, since the principal stakeholders were involved from the outset and approved the plans as they were developed.

THE CHALLENGE OF ANTICIPATING FUTURE ACTIVITY

The primary requirement of a design project is to produce a design that is consistent with the future activities of users. Consequently, the activities to be performed in the new facilities must be correctly anticipated.

This is a significant challenge, since the level of uncertainty is always high, and particularly so in a field as complex as the hospital sector, with its many different worker categories and evolving technology. It is an even greater challenge for users, who are not skilled at designing new facilities. Indeed, the mental representations of users are founded on past experience, derived mainly from the practices of the institution in which they are currently working. However, a design project involves major changes of practice, because the premises themselves change, as do the technologies, operating methods, care philosophy, and organization of work. Reference to what currently exists is therefore ineffective to some extent, and can even be an obstacle to the development of the project. In addition, if the participation of users is to be productive, it requires increased support, so that they are able to guide the professionals (architects, engineers, and designers) correctly in their design choices. Two methods are useful in providing this support, and the ergonomist plays a crucial role in both. They are *simulations of future activities* and *visits to reference sites*.

SIMULATION OF FUTURE ACTIVITIES

Simulation is undoubtedly one of the most effective ways of testing design concepts. Obviously, not all situations can be simulated, and priorities should be established on the basis of precise criteria, including the ease with which situations can be reproduced and the risks to human health and safety.

The proposed *dynamic simulation procedure* can be applied to all steps of the design phase — programming, design draft, preliminary plans and specifications, and detailed plans and specifications. It can also be reproduced at the three functional analysis levels — in other words, at the level of the building itself (macroscopic), the various departments (mesoscopic), and the individual workstations (microscopic).

The general logic of the simulation process is to develop priority future activity scenarios on the basis of the data used to define the design project. Placing real users in a layout representing the proposed design concept simulates the scenarios. The suitability of the concept for the predetermined future activity scenarios is then diagnosed. This gives rise to proposals for changes to the concept or the scenarios, and finally to a design that fully satisfies user expectations.

Simulation should reproduce, as faithfully as possible, an anticipated reality that exists only in the representation of the design players or with reference to a comparable situation.

It also provides an opportunity to confront different viewpoints in a positive way. Architects, engineers, departmental heads, employees and ergonomist all have very different views of the work. This diversity of viewpoints is not an obstacle to project development. On the contrary, it provides an overall vision of the projected situation without which the design exercise may be defective. Simulation is thus an excellent way of confronting viewpoints and reaching a creative compromise.

Simulation Props

The major props used to create dynamic future activity simulations are the enlarged plan, the three-dimensional representations, the full-scale simulation with mock-up and the prototype.

Any of these simulation props can be used, depending on the circumstances. All simulations begin with a plan. However, users may not be used to reading design plans, and their natural ability to do so may vary tremendously between individuals. Some people will find it easy to situate them in a given space and to project themselves into a two-dimensional plan. However, the vast majority find this task difficult, especially when the plan is on a small scale. For this reason, inexperienced users must receive guidance to help them read and interpret the plan, and it is better to work with larger-scale plans (1:50 rather than 1:100).

Three-dimensional simulation props are much better at providing users with a meaningful representation. The closer the simulation context is to the proposed reality, the better the results will be in terms of the reliability of the reference points used for the design.

Three-dimensional simulation software is now available at reasonable cost. On request, architectural firms can easily provide three-dimensional representations produced with Auto-Cad that are tremendously useful in visualizing the building as a whole or specific areas inside the building.

One excellent type of simulation is the full-scale simulation because it has the advantage of involving real users in action. It is relatively simple to organize, either in comparable existing facilities or in a room that is large enough to reproduce the situation with a mock-up of all fixed and movable equipment (see Figure 11.3).

Simulations such as this are appropriate when copies of the same layout model will be reproduced a number of times in the building. This is often the case in hospitals, where the different floors all follow the same model. They are also appropriate for testing operations of a critical nature, where errors may have serious human or financial consequences.

It is sometimes necessary to use prototypes to validate anthropomorphic dimensions and the functioning of certain expensive installations before making a final decision on design. This was the case when designing a clean linen-sorting belt in an industrial laundry where seated/standing stations had been introduced for workers (see Figure 11.4 and Figure 11.5).

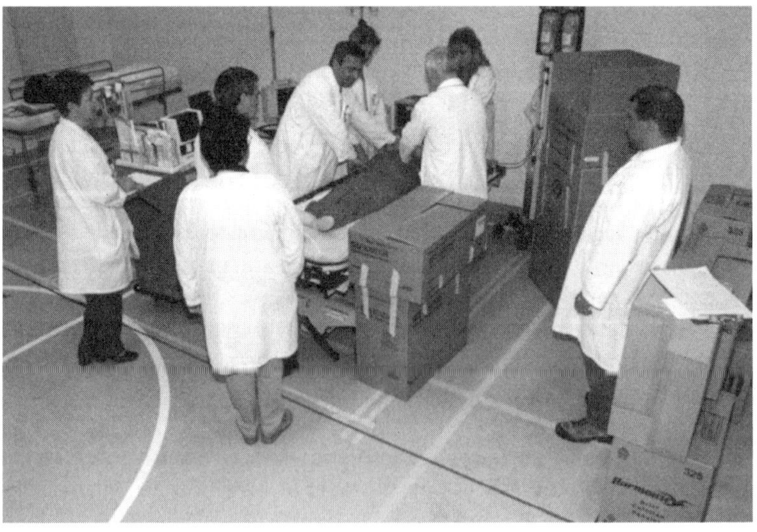

FIGURE 11.3 Simulation with mock-up in an emergency department.

FIGURE 11.4 Simulation with a plywood prototype for the design of a new linen-sorting conveyor.

FIGURE 11.5 Simulation with a plywood prototype for the design of a new linen-sorting conveyor.

Simulations Follow-Up

The simulation is aimed at testing the viability of a design concept in light of the predetermined future activity scenario. When it is complete, the people involved are in a position to decide whether or not the concept is appropriate for the work context as defined. This is known as the prognosis.

The prognosis can take three directions:

- The concept works well in the context defined by the scenario
- The concept does not work well and has to be reviewed in whole or in part (modified concept)
- The concept would work in a different context (modified scenario)

The design concept must be adapted to the future activity scenario, and not vice-versa. In design projects involving substantial sums of money, users have a right to expect that the proposed design will meet their expectations. The designer must satisfy users' needs, instead of trying to change working methods to suit the proposed design. This is not always easy, and creative compromise is often required, especially in renovation projects where additional constraints are imposed by the existing building.

The results of the simulation must be written up, and rigorous follow-up is required to define new measures until a satisfactory solution is found, in terms of operational functioning and the health, safety and comfort of users.

Successive simulations in different projects have made it possible to define the general design criteria for three typical care situations in a hospital setting: bed care, stretcher care, and care in a treatment chair. A chart containing information on the width of the doorways and hallways by type of activity has also proved to be extremely useful. These data are presented in the form of templates at the end of the chapter.

Visits to Reference Sites

Visits to reference sites are an excellent way of creating a more open attitude to new operating methods and new approaches that are often derived from new technology. Such visits encourage those involved to think about changes in practice and move away from what presently exists. They are therefore a vital component in the project definition process. They tend to be organized spontaneously, but all too often they are improvised and achieve only part of their goal.

Here are some basic conditions required to ensure that visits to reference sites are as useful as possible.

Forming the Project Group

Whatever the type of visit, the project group should include a decision-maker, the project manager, direct user representatives, a workplace health and safety representative, and the professionals involved. Each person has a different viewpoint and the questions he or she raises will help the group understand how the site works, how it is organized, and how all this is relevant to their own project.

Establishing Objectives

Wherever possible, the project group should meet beforehand to prepare the visit and establish precise objectives. What specific aspects does the group wish to consider? What information is needed on operations, care approaches, technologies, layout designs and ergonomic, health and safety questions?

The group members should draw up questions in advance, although there should of course be room for spontaneous questions that arise during the visit. There is nothing to prevent the questions from being sent to the site beforehand. This will allow the host team to prepare for the visit and obtain copies of any relevant documentation.

Selecting Sites

Preference should be given to sites that meet the project's general objectives. It is best to visit newly constructed or newly renovated buildings, because the technologies and layouts will reflect recent changes in the medical field. Other sites may be chosen because they are of more specific interest in that they contain interesting workstations, room layouts, equipment, or furniture.

The number of sites visited will depend on the complexity of the project and the expertise available within the group. Sometimes new functions may be added, with which users do not have much experience. In such cases, it is essential to make several site visits, so that users are able to give clear opinions on the department's future orientations and on the layout proposals submitted by the architects.

Informing the Host Team of the Visitors' Expectations

The host team should be told about the visiting group's expectations, so that it can ensure that the right people are available. Indeed, the choice of interlocutors is very important. "Tourist" type visits, led by the institution's public relations officer, should be avoided, since they will give a very superficial idea of the premises and the positive aspects of the building. It is just as important to know the negative aspects, and it is therefore important to meet with the people who are most familiar with operations. The local renovation or construction project manager should also be present.

During the Visit

An introduction to the department and the background to the project should normally precede the site visit. A classical way of conducting a visit is to follow the chronological order of the department's operations, beginning for example with the patient arriving in the waiting room and moving through the various stages up to discharge from the hospital. At each stage, the group should be free to ask questions.

The following are some examples of key questions that might be asked:

- What are the department's main goals and preferred approach?
- How many professional and managerial staff work on each shift?
- What is the client profile?
- What is the volume of activities, and are there any seasonal or other variations (evening, night, weekend)?
- What are the main paths — patients, visitors, staff, clean and dirty supplies?
- What are the main types of workplace accidents that occur?
- What technologies are used? Why were these particular technologies selected? What are their advantages and disadvantages, costs, and benefits? How reliable are they? How much maintenance do they need?
- What is the logic of the layout design selected? Why was this particular architectural design selected?
- What works well? What are they proudest of?
- What would they improve if they could start over?

It is essential to document the visit, by taking notes, photos, or a video, as needed. The premises and equipment should also be measured if necessary, and group members should ask questions about the suppliers, the quality of the after-sales service, and the advantages and disadvantages of the products, equipment and furniture.

It is best to hold a group meeting immediately after the visit, so that members can identify the elements worth keeping and those that should be rejected, while their memories are fresh.

Producing a Report

It is essential to produce a report of the visit. One person should be assigned specifically to this task, and his or her role is to gather the information collected by the group members. A preliminary report should be sent to members for comments and additions. A final version, illustrated with photos, can then be prepared, emphasizing the aspects to be used in the project. The report is then kept in the project file.

CASE STUDY NO. 2: MEDICAL IMAGING TEAM VISITS

This is an example of the visits organized for a group of medical imaging users involved in the construction of a hospital on the outskirts of Montreal. The project group was composed of the departmental head, the assistant head, the chief radiologist, technicians from each specialty area (ultrasound, x-ray, CAT scanning, etc.), the project manager, architects, an ergonomist, and (occasionally) some engineers.

In this case, it was particularly important to organize several site visits, because the service would be expanding considerably with the introduction of leading-edge technologies. In addition, several new functions were to be added, and the current players' knowledge of them was limited. The functions in question were nuclear medicine, magnetic resonance, mammography and angiography. There were also plans to introduce digital medical imaging, which would involve radical changes in the practices of all the players.

Three major objectives were identified for the visits:

- Become familiar with the operation of departments using technologies unknown in the present hospital organization.
- See if digital medical imaging would really work.
- Identify interesting layout designs that could be used in the project.

A total of ten visits took place. They helped enrich the expertise not only of the users, but also of the professionals who were not necessarily familiar with the installation conditions required for sophisticated equipment, or with the protective measures needed for certain hazardous materials such as radioactive products, x-rays, and electromagnetic waves. In fact, the existence of a protected environment (leaded divisions, leaded equipment, etc.), together with rigorous control of staff and patient exposure to hazardous materials and proper management of highly toxic waste, are the most important elements to be considered when designing a medical imaging department.

A visit to a fully digitized medical imaging department clearly showed that the method was greatly superior in terms of quality and efficiency to the traditional film-based method — a fact that nobody really believed at the outset.

At the suggestion of the architects, several other sites were visited to examine interesting design concepts that were subsequently used by the group as a basis for their own building design. The visits also allowed users to become familiar with the operation of larger departments similar to the one in which they would be working in the new building.

CONCLUSION

The participatory ergonomic method is not intended to be carried out in parallel to the normal procedure, but should form an integral part of it, and be conducted in close cooperation with everyone involved. To do this, it is essential that internal project mechanisms are well structured, and that provision be made for true cooperation between the social players in the institution, in particular the direct users of the future premises, i.e., management and employees, as well as clients and other users. They should be involved from the very beginning of the project definition process, because their expertise and knowledge are essential to the success of the project.

The design concepts proposed at the different design stages should be tested through appropriate simulations of future activities. Obviously, simulations will not be required in every situation. The functionality of many facilities can be established by professional expertise or reference to construction standards and comparable sites. The situations in which more elaborate simulations are required must be properly defined and carefully prepared.

Visits to reference sites are also essential in providing support to groups of users involved in a participatory design process. The visits allow group members to be creative, to think about inevitable changes to their practices, and to participate fully in the design process.

Construction professionals focus their attention on the building and technical aspects, and it is absolutely vital that they should do this. However, the approach proposed here focuses on users, and therefore provides an excellent complement to the work of the architect.

The challenge of this user-focused approach is to obtain more success with design projects, at a lower cost.

TYPICAL WORKING SITUATIONS

TEMPLATES

RIGHT-HANDED PERSONS
RELATION

Over 75% population

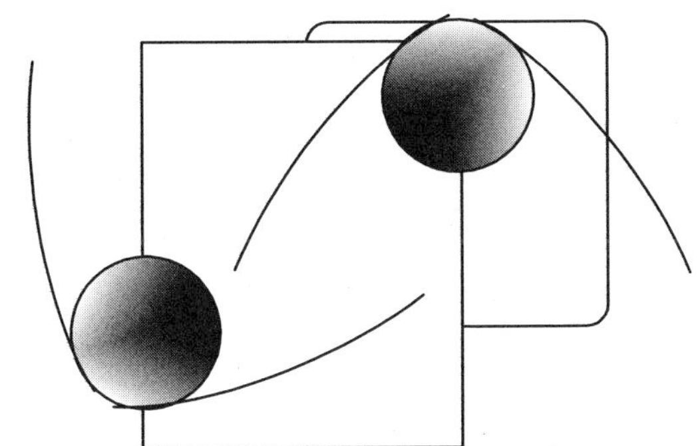

- Primary working space shall be at the right of the patient; also favorable for right-handed caregiver
- Provide enough working space at the left side for left-handed people (less then 25% population) and care giving

FIGURE 11.6 Right-handed persons relation.

BED- WORKING SPACE
Surface: 14 sq m

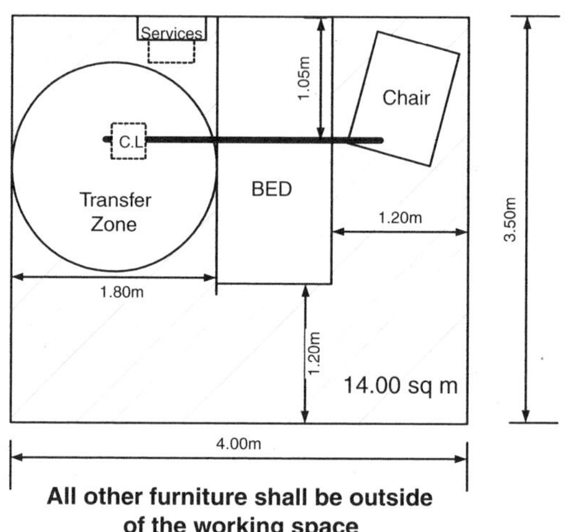

PRIMARY SIDE

- **Primary working space**
- **Patient transfer with wheelchair or stretcher**
- **Right side patient examination**
- **Bed side desk**
- **Medical panel**
- **Computer**
- **N.B. Ceiling-lift is very efficient for disabled patient transfer**

SECONDARY SIDE

- **Secondary working space**
- **Left side patient examination**
- **Rest chair**

FOOT

- **Wheelchair circulation**
- **Foot patient examination**

Scale: 1:50
Dimensions:
Bed: 100 x 230 cm
Chair: 75 x 100 cm
Side desk: 40 x 40 cm

FIGURE 11.7 Bed-working space.

STRETCHER- WORKING SPACE

Surface: 9 sq m

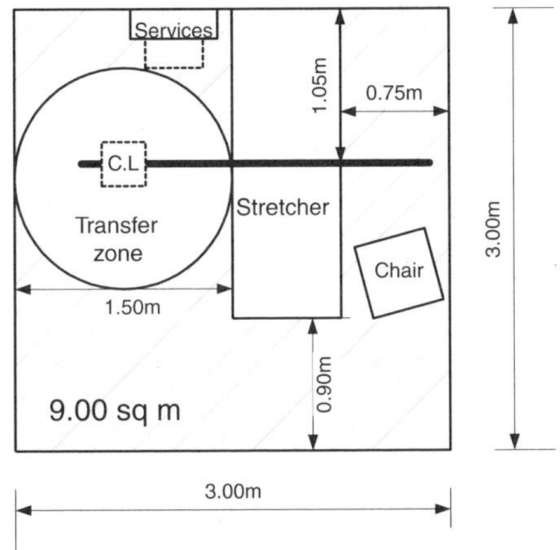

PRIMARY SIDE

- **Primary working space**
- **Patient transfer with wheelchair or stretcher**
- **Right side patient examination**
- **Bed side table (on wheels)**
- **Medical pannel**
- **Computer**
- **N.B. Ceiling-lift is very efficient for disabled patient transfer**

SECONDARY SIDE

- **Secondary working space**
- **Left side patient examination**
- **Folding chair**

FOOT

- **Walking circulation space**
- **Foot patient examination**

Scale: 1:50
Dimensions:
Stretcher: 75 x 210 cm
Chair: 50 x 50 cm
Side table: 40 x 40 cm

FIGURE 11.8 Stretcher-working space.

TREATMENT CHAIR-
WORKING SPACE
Surface: 7.5 sq. m

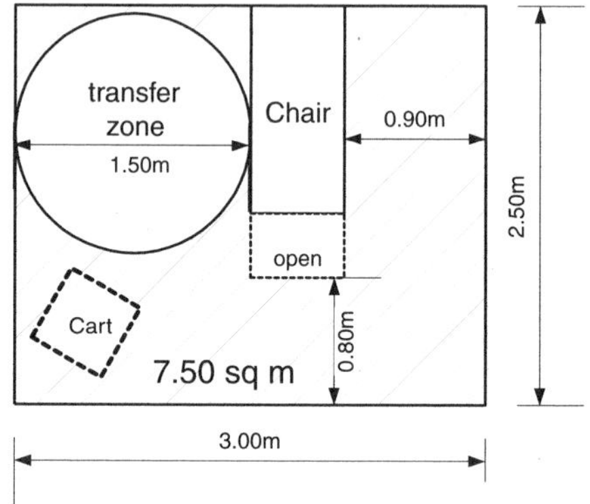

PRIMARY SIDE

- **Primary working space**
- **Patient transfer with wheelchair**
- **Right side patient care**
- **Service cart**

SECONDARY SIDE

- **Left side patient care**
- **Walking circulation space**

FOOT

- **Foot patient care**
- **Walking circulation space**

Scale: 1:50
Dimensions:
Chair : 60 x 130 cm
Open chair : 60 x 170 cm
Cart : 50 x 50 cm

FIGURE 11.9 Treatment chair-working space.

DOORS

DIMENSIONS (Inches)

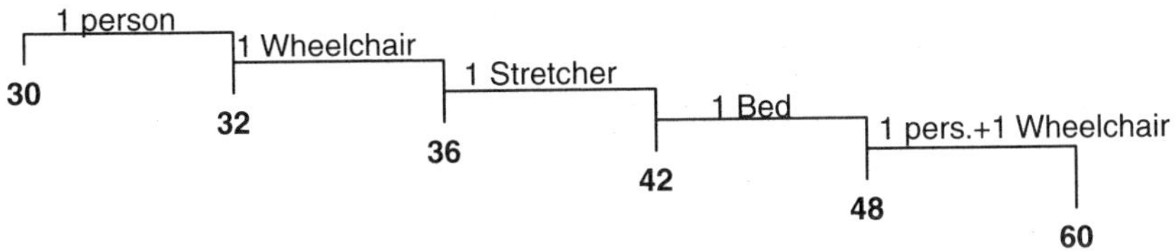

CORRIDORS

DIMENSIONS (feet)

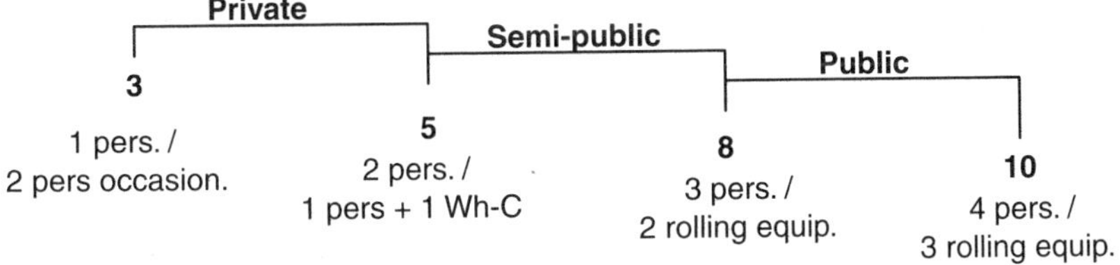

RAMPS

DIMENSIONS (feet)

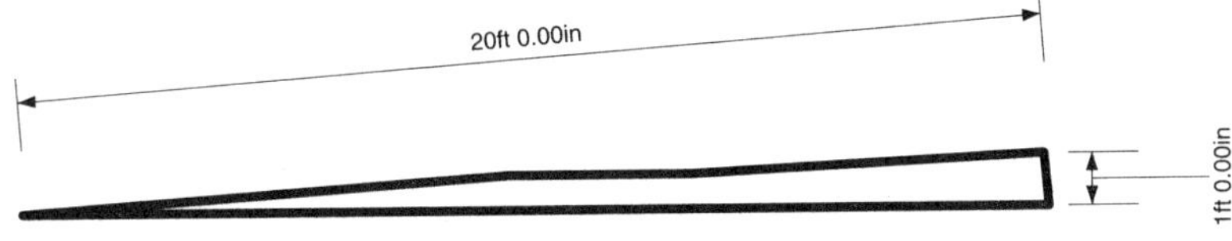

- Avoid ramps. If not possible, ramps ratio 1/20
- Ramps ratio 1/12 is too difficult for elderly people
- Provide a midway step rest on the ramps

FIGURE 11.10 Doors, corridors, and ramps.

RECOMMENDED READINGS

Bertrand, G. and Morissette, L., La chambre: milieu de vie et lieu de travail, Collection Parc, No. 2, ASSTSAS, 1996.

Daniellou, F., Le statut de la pratique et des connaissances dans l'intervention ergonomique de conception, Université de Toulouse - Le Mirail, 1992.

Estry-Béhart, M., *Ergonomie Hospitaliere: Theorie et Pratique*, ESTEM, 1996.

Garrigou, A., Bellemare, M., and Richard, J.-G., La simulation dynamique des activités futures, Intervention IRSST/SECAL, Final Report, IRSST, 1995.

Goumain, P. and Villeneuve, J., Étude d'ergonomie architecturale dans un centre d'hébergement pour personnes âgées, Proceedings of the Canadian Ergonomics Association, 1991.

Ledoux, E., Ergonomie et conception des espaces de travail: ... Travail et santé, 12(2), 34–37, 1995.

Maline, J., Simuler le travail. ANACT, Collection outils et méthodes, 1994.

Piché, B., Chéruet, R., and Goumain, P., Construire un nouvel hôpital : le Gatineau Memorial, *Objectif Prévention*, 21(1), 1998.

Villeneuve, J., et al. Objectif prévention, dossier thématique PARC, 17(5), 13–33, 1994.

Villeneuve, J., Goumain, P., and Elabidi, D, A comparative study of two types of patient-lifting devices for moving patients in long-term care, IEA Proceedings, Toronto, 1994.

Villeneuve, J., Le programme PARC: aide à la conduite des projets architecturaux, Collection PARC No. 1, ASSTSAS, 1996.

Villeneuve, J., Ergonomics in hospital design, 3rd International Congress, ICOH, Edinburg, U.K., 1997,

Villeneuve, J. and Thibault, B., Ing. chargé de projet, Les utilisateurs sur la planche à dessin au Centre de services ambulatoires du sud de Lanaudière, *Objectif Prévention*, 33(4), 2000.

Zimring, Craig., Site visits, in Sara O. Marberry, *Healthcare Design*, John Wiley & Sons, 1997, pp. 3–25.

REFERENCE

Villeneuve, J., The ceiling-lift : An efficient way of preventing injuries among nursing staff, in Charney, W., Ed., *Handbook of Modern Hospital Safety*, Part 3, Lewis, 1999, pp. 736–741.

12 Designing Workplaces for Safer Handling of Patients/Residents

Victorian WorkCover Authority

CONTENTS

1-56670-631-9/03/$0.00+$1.50

FOREWORD

There is a link between the layout and design of a workplace and the risk of musculoskeletal disorders. These guidelines were developed by industry for industry. They represent an important contribution to our knowledge about reducing risks through good design practices. Renovations or the building of new facilities provide an ideal time to incorporate occupational health and safety considerations into the planning process.

This practical material is specific to the health and aged care sector and focuses on the main areas where patient/resident handling occurs. The recommendations for the allocation of space are based on the space needed to perform the handling activities that occur in facilities. The guidelines consider both the needs of patients/residents and the occupational health and safety needs of direct care staff. An important feature of the guidelines is the inclusion of a consultation process with ultimate end users of the facility, during the planning phase.

I recommend these guidelines to all involved in the planning and design of health and aged care facilities.

ACKNOWLEDGMENTS

These guidelines have been prepared by the Workplace Design Working Party as part of the Victorian WorkCover Authority's Health and Aged Care Project, and are aimed at reducing injuries among staff who handle patients/residents. The Health and Aged Care project is linked to the Victorian WorkCover Authority's "Back Strategy," which aims to reduce the incidence and severity of back injuries in the workplace. The Health and Aged Care Project Consultative Committee, comprising key representatives from the Victorian WorkCover Authority and the health and aged care sector, provided valuable support and guidance throughout this project.

The working party was facilitated by Fiona Begg, an ergonomist with the Victorian WorkCover Authority, and assisted by an external consulting team comprising an architect and an ergonomist.

The Working Party gratefully acknowledges the contribution of all those people in the health and aged care sector who hosted site visits, participated in workshops or offered comments and advice at various stages of the project. The Austin and Repatriation Medical Centre kindly hosted the two workshops at the Royal Talbot Rehabilitation Centre, and staff of Royal Talbot assisted with the running of the workshops.

One particularly important source of information requiring acknowledgement by the working party is a report by the Swedish Institute for Hospital Planning and Rationalisation titled (in English) Hygiene Rooms function of space for personal hygiene in long term care. The project team gratefully acknowledges the valuable contribution of Tore Larsson and Kay Wilson for drawing the report to its attention and for translating it from Swedish into English. The learnings from this report have enabled us to develop our guidelines.

THE WORKING PARTY

The following organizations were represented on the working party:

Austin and Repatriation Medical Centre Australian Hospital Care Limited
Australian Nursing Homes and Extended Care Associations — Victoria
Department of Human Services
Injured Nurses Support Group
National Association of Nursing Homes and Private Hospitals
Peninsula Health Care Network
Royal Australian Institute of Architects
Victorian Association of Health and Extended Care Victorian WorkCover Authority

SUPPORTING ORGANIZATIONS

The following organizations have expressed support for these guidelines:

Australian Nursing Federation — Victorian Branch
Australian Nursing Homes and Extended Care Association — Victoria
Injured Nurses Support Group
Private Hospitals Association of Victoria
Victorian Association of Health and Extended Care Victorian Healthcare Association Limited

INTRODUCTION

WHY THESE GUIDELINES HAVE BEEN DEVELOPED

These guidelines are one step towards reducing the continuing high prevalence of musculoskeletal strain injuries among staff who handle patients or residents in Victorian acute and aged care facilities.

Key stakeholders of the health and aged care sectors and the Victorian WorkCover Authority are concerned about the high incidences of injuries, particularly back injuries, among staff who handle patients/residents.

In the health industry, 44% of all WorkCover claims are for back injuries, compared to 26% for the whole of the Victorian work force. Over half of all claims in the health industry are lodged by nurses.

Research shows that patient/resident handling is the most frequent cause of back pain and injury to nurses, and that poor work place design is a major contributing cause of these injuries, especially in patient rooms, toilets, bathing areas and corridors. Restricted space may lead to constrained and awkward postures during handling tasks, and poor workplace design may lead to unnecessary or double handling of patients/residents. There is, however, no single set of guidelines available to designers of acute or aged care facilities in Australia, which give adequate guidance in relation to design requirements for safe patient/resident handling.

The Building Code of Australia addresses questions of access for independent disabled people, but it does not consider the extra needs of access for disabled people who require assistance. It deals with some important safety issues such as fire safety, egress in emergencies, structural strength, and the design of stairs and balustrades. However, there are many more design factors with the potential to affect safe working conditions for the occupants, on which guidance is needed.

The Victorian WorkCover Authority has undertaken this strategic project in cooperation With the health and aged care sector to provide comprehensive guidance on how good workplace design can reduce manual handling injury risks.

AIM

These design guidelines are targeted at three prime groups: direct care workers who handle patients/residents, managers, and designers. The aim is to assist acute health and aged care service providers, staff, and designers to develop new, and existing, facilities in a way that will reduce risks to staff who handle patients/residents. The guidelines are intended to help a range of stakeholders, including funding agencies, owners, regulating agencies, accrediting agencies, planners, architects, project managers, business unit managers, and finally, the staff who do the actual day to day handling of patients/residents.

The guidelines may also be applied to existing workplaces when assessing patient/resident handling safety.

The guidelines provide best practice information for the design and layout of the main patient/resident handling areas within hospitals and aged care facilities, including:

- Bedrooms
- Bathrooms/ensuites
- Dining/lounge and recreational rooms (aged care)
- Corridors
- Other relevant areas, e.g., equipment/aids storage

Patient/resident handling requirements and practices vary enormously across the whole spectrum of health and aged care. It is essential that these requirements be clearly determined for the particular group of patients/residents in question before these guidelines can be usefully applied.

Tackling Injury Prevention during the Building Design Process

The prospect of building a new facility or refurbishing an existing building to accommodate patients or residents should be the trigger for an agency to make an all out attack on reducing staff injuries in the new/refurbished facility.

The organization should take the opportunity to:

- Review its safety polices as they affect patient/resident handling, and consider introducing policies to abolish manual lifting of patients/residents
- Completely review and revise its work practices in relation to patient/resident handling in the new/refurbished facility
- Decide what types and quantities of lifting equipment, fixed and mobile, will be provided, and determine the functional space and storage requirements for lifting equipment, including proximity to the point of use

These decisions should be taken in consultation with direct care staff and with business unit managers, in order to achieve the best solutions and a unity of commitment to the new work practices. Further information about consultation processes is given in Section 5 and Appendix 12A.

ISSUES IN SAFE HANDLING OF PATIENTS/RESIDENTS

The Size and Nature of the Problem of Injuries from Handling Patients/Residents

The Victorian WorkCover Authority report Workplace Injuries in the Health Industry 1996/1997 states that over half of all reported claims in the health industry are lodged by nurses, with average claims costs varying between $5000 and $7000.

A high proportion of claims within the health industry are related to back injury (44% in Victoria in 1996/1997). This is much higher than for Victoria as a whole where back injuries account for 26.4% of all WorkCover compensation claims. Body stressing and lifting/handling injuries accounted for 56.5% of all injuries in the health industry compared with 35% for all Victoria. Published research literature clearly shows that patient handling is the most frequent precipitating factor or cause of back pain and of overexertion back accidents to nursing staff.

The problem is not new. It was thoroughly documented 30 years ago by Ferguson (1968) in a landmark report titled *A Survey of Manual Handling in Repatriation Institutions*, commissioned by the Australian Commonwealth Department of Health. Until now, however, no comprehensive, strategic approach has been taken to deal with the problem. Ferguson (1968) reported that strain injuries were by far the greatest cause of lost time accidents in the Repatriation Department, and that manual handling, particularly of patients, was by far the most important source of such injuries. He concluded that the strain hazard can best be lessened by elimination, as far as possible, of the need for manual handling. He called for collaboration between various professions, including safety and engineering, from the design stage on, and extending to an examination of existing situations. He specifically called for systems of design and equipment to be considered on an ergonomic basis, in the light of their interaction with the human operator, and to be standardized within the department and reviewed continuously.

Workplace layout and design aspects that increase injury risk include working heights, lack of space or cluttered work-space, narrow doorways, and narrow passageways. Restricted workspace also undermines or prevents the proper use of lifting equipment for patient/resident handling.

According to Engkvist et al. (1992), the three spaces where most accidents occur amongst nurses are the patient's room, the toilet and the corridor. As a result of studying nursing assistants in nursing homes Garg et al. (1992) recommended that redesign is needed in patient rooms, toilets, bathing areas, beds, wheelchairs and transferring devices.

Detailed guidance on functional space for hygiene rooms in long-term care was developed through a major research project conducted for the Swedish Institute for Hospital Planning and Rationalization (1979). This work included site inspections at 20 different hospitals and nursing homes built or renovated in the previous 10 years, and interviews with patients and nursing staff. Trials — 280 in all — were then conducted in full scale mock ups in a laboratory, increasing the space by 100 mm at a time, until the task could be performed satisfactorily without bumping into the walls. Ten rooms were then constructed in the laboratory and another ten (total) in two hospitals, and the designs assessed. Based on this research, a number of design examples were developed and published in the report. This report of the Swedish Institute for Hospital Planning and Rationalization (1979) provided an important foundation for the guidelines presented here.

Common Problems Identified in New Buildings

The consultants visited eight relatively new or renovated acute or aged care facilities in order to detect whether, under current regulations, standards, and design practices, any design problems were still occurring which adversely affect the safety of patient/resident handling.

It was quite clear from these observations that the current Building Code of Australia and current standards do not provide adequate guidance for designing buildings for safe handling of patients/residents.

The patient/resident handling problems related to building design, which were identified by consultation with staff and by inspection of these eight facilities, included the following:

- Storage for patient handling equipment was inadequate and inconveniently located in almost every facility visited. Bathrooms were often used as storerooms, sometimes precluding their use as bathrooms. It was noted that in some aged care facilities, space is needed to temporarily store personal furniture after a resident dies.
- One recently built nursing home had bedroom doors too narrow to get beds through. If double rooms are included, beds may need to be moved around between rooms at times to assist gender management.
- Some bedrooms were too small to use a lifting machine. It was suggested that consideration should be given to having at least one bedroom big enough for a mobile shower/bath to be used (using a pat slide to transfer the patient).
- In toilets, some nurse call buttons were located out of reach of the patient/resident — typically on the wall behind the toilet.
- Some toilets were too close to the side wall to allow a staff member on each side to assist the patient/resident. It was suggested that large shared ensuites are better than smaller individual ensuites in many situations.
- Carpet squares were reported to be better for cleaning and hygiene control. Direct stick carpets allowed easier movement of mobile equipment. Some carpets were reported to be inappropriate for use with incontinent patients, because they retain odors. This was partly due to concrete floors. To overcome this problem some facilities sealed the concrete prior to laying carpet.
- Small-wheeled equipment was reported to be difficult to move on some carpets, but generally large wheeled equipment was easy to move.
- In one instance the plinth under the bath was observed to be too wide for a lifting machine.
- It was reported that some wash basins need to be set at the correct height to allow wheelchair access, including adequate clear space under the basin for the seated person's legs.
- Furniture was reported to be an impediment to resident movement in some dining rooms and activity rooms. In one case, four-legged tables had been replaced by round pedestal

tables high enough to allow wheelchair access. Frames and wheelchairs were reported to be a problem in some dining rooms.

- It was suggested that facilities need to be designed to accommodate the range of patients/residents likely to occupy them in the foreseeable future, taking into account policies such as Ageing in Place. One facility, which had been designed as a hostel (low dependency), was being used as a nursing home (high dependency), giving rise to safety problems.
- It was suggested that more use should be made of automatic doors, and that consideration should be given to changing regulations that restrict their use in some facilities.
- It was reported that, in one case, security doors restricted movement of beds, and in another case, fire exit doors opened onto a grassed area, creating potential patient handling problems in emergencies.
- Nurse call systems were reported to cause a number of difficulties affecting staff safety. In several cases nurses in ensuites could not hear call signals, so that other staff could not summon help. In closed-in designs, nurses "disappear" from view and can be difficult to locate if needed.
- Space beside beds and/or at the foot of beds was observed to be inadequate in some cases. Location of privacy curtains restricted staff movement in some cases, even when room size was adequate. In many cases furniture or fittings obstructed staff access to suitable positions for manually assisting patients/residents.

Selecting Patient/Resident Handling Equipment in Relation to Building Design

These are some issues, from the building design perspective, for consideration when selecting equipment for handling patients/residents.

- Does the layout of the facility accommodate policies relevant to safe patient/resident handling, such as a "no lifting" policy aimed at eliminating all lifting of patients/residents? In particular, is there adequate space for safe use and storage of equipment, and safe access?
- Do the agency's purchasing procedures include a requirement for health and safety assessment and approval of all items prior to purchase?
- Do the agency's purchasing procedures include provision for consultation with direct care staff in the selection and trial of patient/resident handling equipment?
- Is there sufficient information available about the proposed design of the building to ensure that the equipment will be compatible with the building and the new work practices?
- Is the patient handling equipment and furniture compatible with the design of the building?
- What is the footprint (i.e., floor area) covered by the equipment?
- What additional space is required by the occupant (e.g., arms and legs)?
- What space is required by the person operating the equipment?
- Will the equipment fit into or through all the spaces where it needs to go?
  - Adjacent to beds (three sides)
  - Adjacent to toilets
  - Adjacent to baths
  - Through doorways (bedroom, bathroom, toilet, ensuite)
  - Along the corridors and around the corners
  - Under height adjustable beds
- Is the plinth under the bath narrow enough to allow the base of the patient lifter to be maneuvered into the correct position in relation to the bath?

- Is there sufficient and safe storage space for each item of equipment, and can it be stored close to the point of use?
- Are the floor surfaces on the routes over which the equipment will be transported compatible with the design of the equipment (consider: ramps, carpets, steps, lift doorways, size of wheels, steering characteristics)?
- Will the equipment itself constrain staff movement, and therefore contribute to a risk of strain injury?
- Does the equipment have any special anchorages, supports, etc, and do the floors and walls require any special protection from damage by equipment?
- Is the equipment suited to a range of sizes of staff and patients/residents?
- Can you be sure that the position of fixed equipment for handling patients/residents is correctly determined before it is installed?
- How maneuverable is equipment, including force required to push/pull?

GENERIC FUNCTIONAL REQUIREMENTS AND OPERATIONAL ISSUES

General Influences on Handling of Patients/Residents

Two main influences on safe handling of patients/residents are:

- The building environment, including room layout, types and location of equipment, furniture, fixtures and floor finishes
- The level of dependence of the patient/resident

These can affect safe patient/resident handling as follows:

Staff Work Practices

- Common work practices include patient/resident handling being performed mainly during a couple of intense morning hours and evening hours. Rooms and equipment are therefore extremely busy during a short period of time. Some facilities, however, do spread tasks requiring patient/resident handling, such as showering and washing, over the whole day.
- Habits and traditional ways of working influence the way personal assistance is given. For instance, staff may rely on manual lifting and transfer, rather than using lifting devices.
- Training and education in safe patient/resident handling is often limited to manual lifting techniques rather than risk assessment and control, and is often not updated.

The Type of Facility

- Aged care residential units and wards for acute care accommodate patients/residents with varying degrees of functional disabilities.
- Some patients/residents might be completely bed-bound and in great need of personal assistance. Duration of care and rehabilitation varies between types of patients/residents.
- There is often a mixture of single and two-bed rooms, and in acute health facilities some four-bed rooms.
- The layout and design of some buildings makes reorientation of equipment difficult.
- Newly acquired aids and equipment do not always fit existing buildings and therefore cannot be used.

The Functional Capacity of the Patient/Resident

- The degree of functional and cognitive capacity of the patients/residents, and their need for assistance and aids, determines how the personal hygiene tasks of the patient/resident can be performed.
- Residents in aged care residential units are often frail and can have behavioral problems.
- The posture that the patient/resident is able to assume during the handling procedure affects the way handling is performed, and therefore has important implications for the design of the room.
- The type of aids used for patient/resident handling is also important to the design of the room.

Organizational Culture and Resources

- Manual handling safety policies, such as a policy to eliminate all lifting of patients/residents, should be the starting point for designing work practices and work places.
- Safety practices, management, and supervision will affect handling practices.
- The availability of sufficient, suitable, workable handling aids, including fixed and mobile equipment, determines whether optimum lifting practices can be employed.

Classifying Patients/Residents According to Functional Capacity

The functional capacities of patients/residents are important determinants of staff handling procedures and essential considerations when developing the best functional layout and design of rooms. This section of the guidelines classifies patients/residents into groups based on mobility and dependency. It is also important to consider future occupants of the buildings, as well as possible changes over time in the functional capacity of individual patients/residents. This is particularly relevant for aged care where residents who are now being admitted are frailer and more dependent compared to those admitted in the past.

When planning health facility and aged care rooms, two factors are critical: the patient's/resident's way of ambulating, and his/her need for assistance. Depending on mobility, there are four main patient/resident categories, namely:

- Those who move about without any form of aid
- Those who need some form of walking aid such as a stick, four-point stick, frame or support from staff
- Those who move about in a wheelchair
- Those who are bed-bound

In the last two categories, it is likely that there will be a need for a lifting device for transfer to and from toilet seat or shower-trolley. It should also be noted that, even in the second category, a patient/resident may need staff assistance for on-bed movements.

Within these categories there are great variations. This includes the need for staff assistance, where three main dependency levels can be identified, namely:

- Those who manage without assistance
- Those who need assistance from one staff member
- Those who need assistance from two or more staff

These categories can be combined in 12 different ways:

Alternative	Ambulatory Aid	Dependency Levels
1	None	1
2	None	2
3	None	3
4	Walking aid	1
5	Walking aid	2
6	Walking aid	3
7	Wheelchair	1
8	Wheelchair, can stand	2
9	Wheelchair, cannot stand	2
10	Wheelchair, can stand	3
11	Wheelchair, cannot stand	3
12	Full assistance	3

Despite this large range of functional ability, the space requirements and design demands for health and aged care facility rooms for several of the categories are very similar. The number of categories can therefore be reduced to four main groups according to different dependency needs. All patients/residents can be allocated to one of these groups.

In all four groups, the patients/residents have some form of functional disability, which impedes their mobility and impacts on their level of required assistance. The disability can be related to physical and/or mental disorders. The patients/residents might need different types of aids and different levels of staff assistance to move around and attend to their personal hygiene. In addition, some people from all categories may need assistance with on-bed movements.

Group A: Fully Independent

Patients/residents who walk with or without walking aids. They do not need assistance from staff to move about.

Walking aids that might be required include sticks, four-point sticks, wheeled frames or other supporting equipment. The patients/residents in this group might suffer from stiffness and restricted mobility of shoulders and arms, hips and legs, diminished functional ability due to partial paralysis and disturbed sense of balance. The patients/residents might rely on some staff assistance but generally can move about independently.

Staff must be aware of the potential for these patients/residents to need assistance in a one-off situation or emergency. These patients/residents generally need some support aids in toilets, bathrooms, ensuites, etc. This is particularly relevant for aged persons who may not have any disabilities but due to their frailty do not have the strength to assist themselves, particularly in toilets where grabrails are provided. The patients/residents who usually fall into this group are: acute medical patients, acute surgical patients who have recovered from surgery, or low level aged care residents, or high level (wandering, dementia) aged care residents.

Group B: Partially Dependent

Patients/residents who walk with or without walking aids, yet need some or full assistance from one or two staff to move about.

Walking aids that might be required are sticks, fourpoint sticks, wheeled frames or other supporting equipment. The patients/residents in this group might suffer from restricted mobility of shoulders and arms, hips, and legs, and thus have difficulties in sitting down and getting up, and dressing and undressing. Diminished functional ability of one half of the body due to partial paralysis, disturbed sense of balance, and reduced vision is common in this group.

Staff must often provide partial assistance to these patients/residents in getting in and out of bed, on and off toilets, and showering/bathing. This group of patients/residents need support aids in toilets, bathrooms, ensuites, etc. They generally comprise acute medical patients, acute surgical patient with some mobility, rehabilitation patients, low and high level aged care residents or patients/residents who require assistance with on-bed movement.

Group C: Dependent

Patients/residents who need a wheelchair to move around. They still manage some tasks on their own, or they need some or full assistance from one staff member to cope.

Those who manage on their own have good functional capabilities of arms and hands, but might suffer from reduced leg and trunk function due to paralysis or amputation.

Those who need assistance for transfers in relation to personal hygiene tasks are able to stand on their legs. Their disability might be reduced functional capacity of trunk and/or one or all extremities, or one half of the body due to partial paralysis, disturbed sense of balance and/or cognitive impairment. The staff assistance may entail using a lifting device.

Although these patients/residents can help with transferring their weight they often need full assistance from staff in getting in and out of bed, on and off toilets, and showering/bathing. Support aids in toilets, bathrooms and ensuites are essential.

They are acute medical patients, acute surgical patients with minimal mobility, rehabilitation patients in the early stages of their program, or high level aged care residents.

Group D: Fully Dependent

Patients/residents who need a wheelchair to move around or are bed-bound. During transfers some of these patients/residents are able to stand with support. For those who cannot stand, the staff must apply different forms of transferring techniques or use lifting devices in order to transfer patients/residents. In addition, the use of bath trolleys may be required. The patients/residents in this group might suffer from reduced functional capacity of the trunk and/or one or all extremities due to paralysis or amputation, or reduced functional capacity of one half of the body due to partial paralysis, and/or cognitive impairment.

In all tasks, staff must assist these patients/residents in their movements. Support aids in toilets, bathrooms, and ensuites are essential. These patients/residents are severe medical/surgical acute patients or high level aged care residents.

These groupings fall into two main functional capacity categories impacting on design.

1. Patients/residents requiring minimal staff assistance (groups A and B)
2. Patients/residents requiring either significant or full staff assistance (groups C and D plus some group B who need assistance with on-bed movement)

As a consequence, these two categories will have a significant impact on the functional layout and design of the rooms accessed by patients/residents. For instance facilities for category 1 patients/residents could be designed around Australian Standard AS 1428-r Design for Access and Mobility. By contrast, facilities for category 2 patients/residents need greater space to allow staff to assist patients/residents in a safe working environment.

It is important that careful consideration be given to future occupants of the building, as those who may be category 1 today may be category 2 tomorrow. If there is an intention to cater for ageing in place, the design should cater for the most dependent. In any case, a clear statement of likely future uses of the facility should be included in the master planning documents and in the facility design documents.

General Design Considerations

General Planning Considerations

An issue that arises due to the various purposes of the acute and aged care facilities is the actual function of each room, particularly in the case of bedrooms.

Acute health facilities

In acute health facilities the patients' stay is often short term, and as soon as they have recovered or are capable of being cared for in their own home, they are discharged. As a consequence the rooms do not have to cater for personal furnishings and may be more clinical. This can often result in specific rooms (e.g., intensive care, birthing), shared rooms (two- and four- bed), central bathroom facilities, and shared ensuites.

Depending on the patient's ailment, the number and size of mobile and fixed items of medical equipment that may have to be accommodated in the room can vary.

Rehabilitation facilities

Given their very nature, these facilities accommodate patients who are undertaking rehabilitation programs prior to being discharged home. They often stay medium term (3 to 8 weeks) and have a medical condition relating to stroke, amputation, motor accident, or similar conditions. Therefore the facilities need to resemble a homelike environment while catering for their intensive rehabilitation program, which can often see them progress from highly dependent to independent patients.

Aged care facilities

In accordance with government policy, aged care facilities need to be as residential as possible, while providing for a safe environment for both residents and staff.

These facilities are the residents' permanent home and as such, they often bring their own personal belongings and have a higher need for privacy. This results in single bedrooms, with space for personal furnishing and a number of visitors, single ensuites with storage space and greater emphasis on privacy.

Although these guidelines will not be providing detailed information on total room sizes, the above items need to be taken into consideration in association with the specific design requirements.

Patient/Resident Handling Equipment

Staff should be provided with, and encouraged to use, suitable aids to assist them in handling patients/residents. There are different types of fixed and mobile equipment, including electric and hydraulic machines, which vary in weight capacity (typically up to 250 kg) and size (typically up to 1150 mm in length and 650 mm in width). Figure 12.1(a) to Figure 12.1(e) show typical lifting machines.

It is important to select machines that are easy to push (including on carpet if necessary) and easy to maneuver, with handles well positioned to give good mechanical advantage when turning, and with wheels that swivel easily. Also check that machines are stable and cannot overbalance when loaded.

Adequate storage space should be provided for lifting machines close to the point of use. In aged care facilities, residents are not normally transported through corridors in such machines. One staff member may push the resident in a wheelchair followed by a staff member pushing the empty lifting machine.

Fixed overhead lifting devices have great potential to make patient/resident handling safer and more efficient, and their use should be strongly encouraged wherever practicable. They are becoming more popular, particularly in bedrooms, bathrooms and ensuites where they have the potential to save space. However, they can restrict room layouts. For example, in a bedroom, the bed cannot be moved to a new position as the overhead lifting machine is fixed. This is particularly relevant

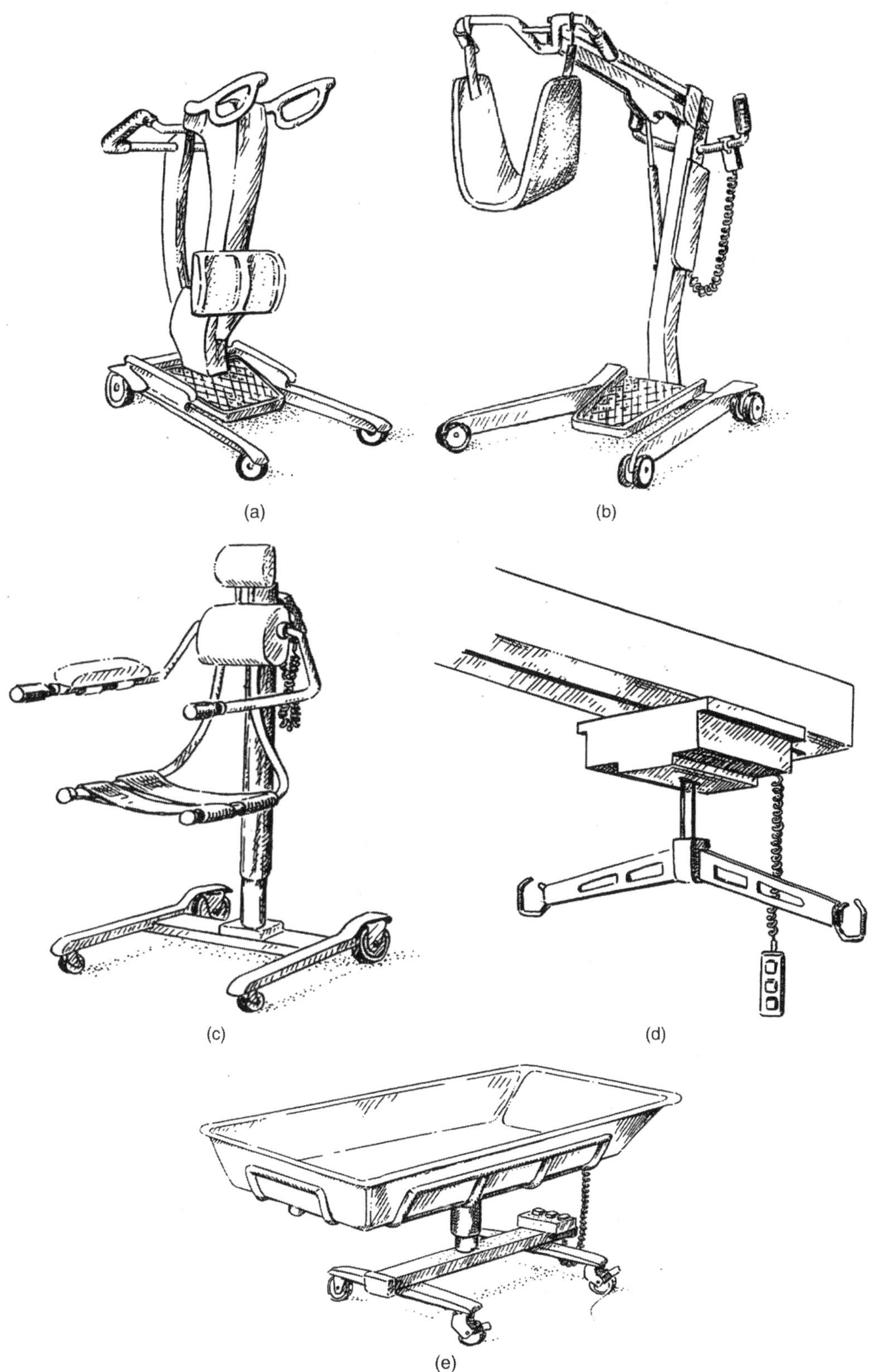

FIGURE 12.1 (a) Standing lifter. (b) Sling lifter. (c) Bath hoist. (d) Fixed overhead lifter. (e) Bath trolley.

for aged care facilities. The ceiling supports may require additional structural members to take the load. Some people believe that such devices make an aged care facility look less homelike. Clever design can overcome this problem, for example, by installing recessed tracks, and providing small cupboards in the wall for the device to be stored out of sight.

Even if a decision is taken not to include overhead tracking in the original construction of a new facility, ceiling structures should be designed to allow the installation of ceiling tracks at a later date. It is better to consider this at the building design stage, to avoid conflicts with services in the ceiling space and structural ramifications.

For patients/residents who can't sit up, a more recently introduced patient/resident support aid is the bath trolley. The patient/resident can be transferred directly onto the bath trolley in their bedroom using a lifting machine or patient slide, and then wheeled to the bathroom where they can be bathed and dried. They can then be returned to their bedroom where they can be transferred into bed. However, such procedures should not breach the privacy and dignity of patients and residents.

Another welcome trend is the purchase of adjustable beds, which enable staff to raise or lower the height of the bed to assist in the handling of patients/residents. These beds are often larger in size than standard beds.

There are many other lifting and manual support devices. One source of information is the Independent Living Centre. The center is part of Yooralla, and can provide information on the selection of equipment and aids that can assist people to be more independent in the home and the workplace.

Furniture and Fittings

There are a number of items of furniture and fittings found in acute and aged care facilities, some of which are necessary in the delivery of medical services (e.g., oxygen, gas, suction, IV, and blood pressure machines) and other items (e.g., bedside tables, fixed shower screens, privacy curtains, etc.). These fixtures can impede staff when handling patients/residents. Choosing appropriate movable fittings, such as bedside tables, can contribute to better access and therefore to safer handling procedures. This also includes appropriate adjustable beds (electronic/hydraulic) and special lounge chairs that can assist in reducing staff injuries.

There should be enough electrical outlets for all personal and agency equipment, and they need to be located so as to minimize cords trailing across circulation areas, which can cause tripping hazards or impede movement of wheeled equipment.

Doors and Door Openings

Doors and doorways can either assist or hinder staff when handling patients/residents. Even simple things such as types of door handles and their positioning can be critical. For example, lever type handles at the appropriate height enable staff to easily open doors while supporting or maneuvering patients/residents. Consideration should also be given to the weight of the door to ensure that it is easy to open and close. Full height doors can be relatively heavy. There are a number of different door types on the market including solid, semi-solid, and hollow core. Door openings need to be high enough to allow access for equipment likely to be used such as intravenous drip poles, fracture frames, and electric beds. Generally 2040 mm high (standard door opening) will suffice. In special circumstances this may have to be increased to 2400 mm high.

All doorways through which patients/residents maneuver must have at least 900 mm clear width when the door is fully opened (for both hinged and sliding doors) (Figure 12.2(a) and Figure 12.2(b)). Sliding doorways must be wide enough and have stops to avoid damage to fingers when holding the handle or hand grip during opening. This allows for staff assistance, large equipment, wheelchairs, and walking frames. In addition, to enable uninhibited access for disabled persons, Australian Standard AS 1428 - Design for Access and Mobility (Section 7.3.1) requires a 460 mm minimum clear space of wall on the latch side of the door.

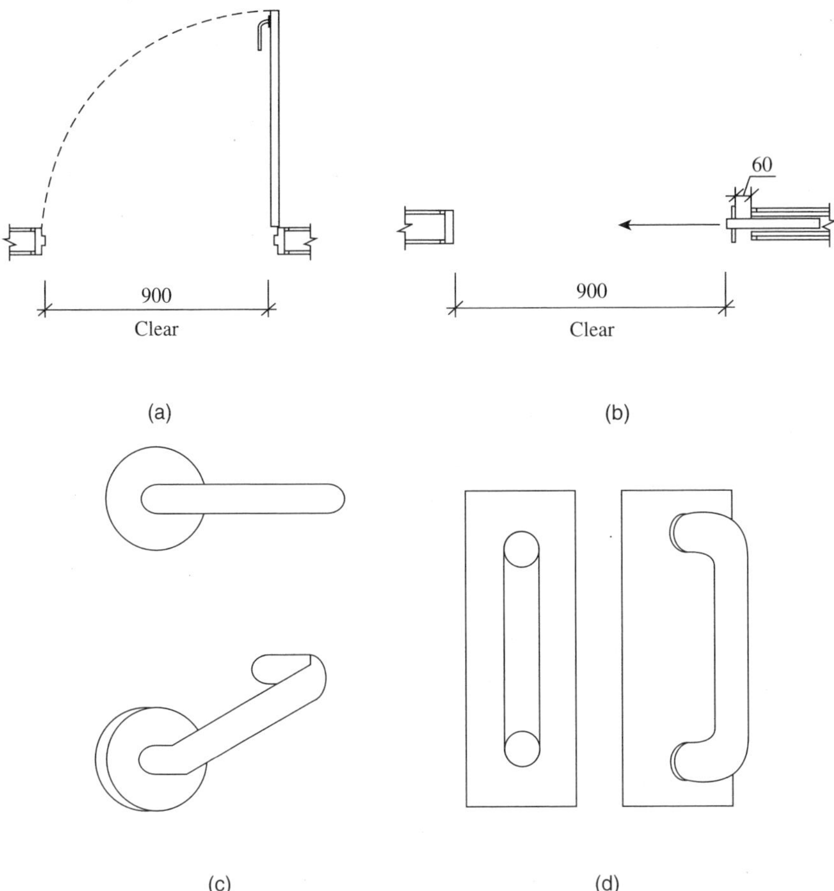

FIGURE 12.2 (a) Minimum clear width of opened hinged door. (b) Sliding door. (c) Lever type door handles (hinged doors). (d) "D" Pull door handles (sliding doors).

For doorways through which beds or large patient/resident trolleys are transferred, a minimum clear width of no less than 1070 mm (for a 1 meter wide bed, with a wide enough corridor to allow turning, i.e., 1800 mm) and preferably 1350 mm is recommended. This can be achieved by providing two door leafs: one at 950 mm (which allows for 900 mm clear space during general usage) and the other 400 mm.

Generally, aged care bedroom doors are 1070 mm minimum and acute bedroom doors are 1350 mm to cater to medical equipment that is fixed to the side of the bed. Door openings to bedrooms must be wide enough so that beds can be maneuvered through without being tilted onto their sides.

A key element that needs to be considered when designing acute and aged care facilities is door swings. Doors must not open into a zone that impedes maneuvering of patients/residents, nor swing out into a circulation area. The movement of staff and patients/residents needs to be incorporated into the design process. Automatic sliding doors have great advantages in certain situations, especially where space is restricted. Fire doors linked to hold open devices controlled by smoke detectors reduce impediments to safe patient/resident handling and should be installed wherever possible.

Another issue associated with doors is door furniture. For handles, lever type is best for hinged doors and "D" pulls for sliding doors (Figure 12.2(c) and Figure 12.2(d)). The positioning of such door furniture is equally important in providung ease of use and reduce injuries.

Door closers on swing doors can be a hazard, and should generally not be used on doors to patient/resident rooms, ensuites, toilets and bathrooms, or in other doorways frequented by patients/residents. Door closers can contribute to the risk of injury from patient/resident handling because the staff member often has to hold the door open against a closing force, while trying to assist the patient/resident. It can be very awkward trying to reach the door while supporting the patient/resident, then stepping back to allow space to pull the door open, holding it open while moving through the doorway, and then releasing it without allowing the door to swing back against the patient/resident. Door closers with hold open devices may reduce some of the risk, but they should be used only where there is a specific need. It is recommended that door closers be used only where absolutely essential, and generally not in patient/resident traffic areas.

Floors and Floor Coverings

Floor coverings can impact on staff work practices and on occupational health and safety in five ways:

1. Cleaning/maintenance procedures
2. Maneuverability of equipment
3. Risk of slipping or tripping
4. Spread of flame, and the density of smoke produced
5. Fatigue on feet and legs (the types of shoes staff are wearing should also be considered)

In order to reduce the risk of slips and falls, floor surfaces should comply with Australian/New Zealand Standard AS/NZS 3661 - 1993 Slip resistance of pedestrian surfaces. A hazard can exist at the junction of different floor finishes (e.g., where vinyl meets carpet). At such points careful consideration needs to be given to low profile junction or diminishing strips. Having different types of floor finishes in the one room (e.g., carpet and vinyl) should be avoided as it often results in varying floor levels (diminishing strips) and can create a feeling of unsure footing. Unexpected changes in floor friction create a risk of slipping.

Any carpet that is used should be low profile and securely attached to the floor structure allowing for easy movement of wheeled equipment and wheelchairs. However, this provides a hard surface to stand on and may contribute to fatigue and aches and pains for staff, who walk or stand on the surface for long periods. Careful consideration needs to be given to reducing such impact while not impeding staff pushing/pulling equipment. The use of cushioned back vinyl may overcome this without increasing noise or vibration.

Shock absorbent underlays may reduce stress on staff, provided that they do not make equipment difficult to move.

In practice, rooms generally have the following floor finishes:

Room	Facility	Floor Finish
Bedrooms	Acute	Cushioned vinyl or carpet
Bathrooms/Ensuites	Aged care	Carpet
Corridors	Rehabilitation	Cushioned vinyl or carpet
Dining rooms	All	Nonslip vinyl or epoxy
Lounge rooms	Acute	Vinyl or carpet
	Aged care/Rehabilitation	Carpet
	All	Vinyl or carpet
	All	Carpet

Where carpet is used on concrete floors, it is recommended that the concrete floor be sealed prior to laying the carpet, or that vinyl backed carpet, which can be seal welded, be used. This assists in preventing urine and other body fluids penetrating the concrete slab and then generating unpleasant odorss through bacterial action. The use of appropriate cleaning procedures and equipment is also important.

Nurse Call Systems

Nurse call systems play an important part in staff handling of patients/residents, particularly in emergency situations. If the system is inadequate, staff often don't wait for assistance which can lead to inappropriate manual handling of patients/residents resulting in work related injuries. The major problems that can occur with a simple/traditional nurse call system include:

- Staff not being aware how critical the situation is when the nurse call system is activated
- Staff not being able to relay messages to other staff for back-up
- Staff not being aware that the nurse call system has been activated (e.g., staff member is in ensuite and does not hear the nurse call system being activated in another room)

To help with these problems, consideration should be given to installing a radio paging nurse call system. This should include personal pagers for staff and could incorporate pendant nurse call buttons for the most dependent patients/residents. Such systems can be integrated with the fire detection system and can provide for quicker response in emergency situations. The strategic location of nurse call buttons in rooms should be carefully planned. It should be noted that in toilets, a nurse call button located on the rear wall will be out of reach of many patients/residents. Where drop down grab rails are installed on both sides of a toilet, a call button and toilet roll may need to be attached to the drop down grab rail. An additional call button could be provided within reach, when the drop down grab rail is folded away.

Electrical and Mechanical Fittings

The appropriate location of electrical and mechanical fittings (e.g., air conditioners, ducted vacuum systems) can also have an impact on staff handling procedures and prevention of staff injuries. Design considerations include:

- Type of equipment to be plugged in
- Frequency of location
- Head clearance
- Ease of use and access
- Restrictions on patient/resident usage
- Elimination of cords across walkways

Lifts

Lifts can have an impact on patient/resident handling procedures. The key elements to be considered include:

- Door openings (width and height)
- Internal dimensions (allow for staff to stand on either side of bed/trolley)
- Position of controls
- Door hold-open times
- Accuracy of leveling between lift floor and external floor
- The horizontal width of the gap between the lift floor and the external floors, relative to the diameter of the wheels of mobile patient handling equipment including lifting machines and beds

Fixing of Grab Rails and Handrails

The design sizing and location of grab rails and handrails is nominated in AS 1428 - Design for Access and Mobility. It is highlighted that the fixing of such supports "shall be able to withstand a force of 1100 N applied at any position and in any direction without showing visible signs of deformation or loosening of the fastenings."

SPECIFIC SPATIAL REQUIREMENTS

The factors that have the most impact on designing rooms for appropriate patient/resident handling procedures are the patients'/residents' functional capacity, their dependence on staff assistance, and the use of equipment. The staff work patterns are also important to the overall design of the facility.

The patients/residents have been divided up into two main categories with four subgroups (Section 3). The patients/residents in these groups have some degree of decreased functional capacity.

The care team should develop a handling plan for each patient/resident. A range of equipment should be available to cater to different handling needs.

The demands that patients/residents with functional disability make on interior design and equipment are often very different to those of able bodied people. When designing and positioning the furniture and equipment, it is important to consider the different needs. Staff may need to have the equipment and furniture placed on the right or the left dependent on whether that person is right or left handed. The patients'/residents' requirements and preferences may be in conflict with those of staff. It may be that general solutions, which satisfy all wishes, cannot be made. In some situations, the different demands can be met by duplicating the equipment or by having equipment that is easily moved. In other cases, the demands can be met by having special alternative areas.

These guidelines have focussed on the following main areas in which manual handling of patients/residents occur:

- Bedrooms
- Ensuites/bathrooms
- Lounge/dining rooms (aged care only)
- Corridors
- Equipment storage

This section provides recommendations on the spatial requirements in individual rooms for manual handling tasks, taking into consideration staff work patterns and lifting equipment. It does not necessarily provide for total room layouts or specify minimum room sizes, as these factors will be influenced by other activities which occur in the room.

BEDROOMS

There is a range of practices for staff handling of patients/residents that occur in bedrooms, including transfers in and out of bed and on-bed handling. These can vary from minimal staff assistance to total dependence on staff and lifting equipment. Hence spatial requirements can vary significantly.

However, in most acute and aged care facilities at any given point in time during the life span of the building, a bed could accommodate the most dependent patient/resident, being category D, as detailed in Section 3. As these patients/residents place the greatest demand on spatial requirements, this section on bedrooms will focus on their handling needs. All bedroom layouts will depend on the size of the bed, which can vary greatly.

Another influence on designing bedrooms is work practices. For example, in acute (surgical) facilities the practice of transferring patients from their beds onto trolleys to be taken to surgery has recently changed, with the whole bed now being transported to the surgery unit.

A minimum space of 650 mm needs to be allowed on both sides of the bed for making the bed. Beds that are less than 500 mm high may require more space for making the bed.

Another key element in the spatial design of such rooms is the location of other loose furniture and equipment such as a bedside table, visitor chair, and medical equipment. Allowance for space to accommodate such furniture and equipment must be taken into consideration.

When designing bedrooms, consideration also has to be given to items that will be located out of the clear zone. This includes storage of wheelchairs, walking frames, location of television, wardrobe, shelving for flowers, nurse call buttons, medical support equipment, and access to ensuite.

Single-Bed Rooms

Figure 12.3(a) is based on a lifting machine that does not exceed 1200 mm in length and 700 mm in width. It assumes that the bed has enough clearance underneath for the lifting machine base.

Figure 12.3(b) is planned using a fixed overhead lifting device. Fixed overhead lifting machines may also be used to transport patients/residents directly from their bed to the ensuites and back. It should be also noted that the spatial area required for this procedure could also be adopted for instances where staff provide physical support to less dependent patients/residents in transferring to and from bed. The wheelchair in this example is based on a maximum size of 700 mm (w) × 900 mm (l). It assumes that the staff member will push the wheelchair in backwards.

Figure 12.3(c) shows space required for patients/residents being transferred onto a trolley by staff using a "Patslide."

Multiple-Bed Rooms

Often acute health and aged care facilities have more than one bed in a room. This can include two-bed and four-bed rooms in acute health, and 2-bed rooms in aged care. As a consequence there maybe times when a number of staff are in the one room assisting two or more patients/residents; with beds being configured differently (e.g., footend to footend or side by side). This section provides for those situations.

Figure 12.4(a) shows maneuvering a mobile lifting machine in a side by side situation. To optimize space it is based on positioning the mobile lifting machine between the two beds if access is required from both sides (refer to Figure 12.3(a)). Since only a curtain separates the two beds, slight intrusion is possible into the other bed zone when actually maneuvering the lifting machine into position, providing it does not create other safety problems, such as knocking into objects on the other side of the curtains.

Figure 12.4(b) shows spatial requirements in a two-bed room with a fixed overhead lifting device.

Figures 12.4(c) and 12.4(d). show spatial requirements when beds are located end to end. Note that the dimensions shown are not the actual total bedroom floor area.

If using a "Patslide" to transfer patients/residents onto a trolley then refer to Figure 12.3(c) for spatial requirements of trolley and staff.

Consideration must also be given to using beds on wheels, which will enable staff to move the bed on the occasions that they need additional circulation space.

Another key element in the spatial design of such rooms is the location of other loose furniture and equipment such as a bedside table, visitor chair, and medical equipment. Allowance must be made for space to accommodate such furniture and equipment.

Ensuites and Assisted Toilets

Ensuites and assisted toilets are often planned for the movement of independent patients/residents with disabilities. Detailed consideration in allowing for staff assistance is often overlooked however, for example the situation of two staff assisting in the transfer of a patient/resident and the use of mobile lifting machines.

Figure 12.5(a) to Figure 12.5(g) show the spatial requirements for manual handling around individual fixtures.

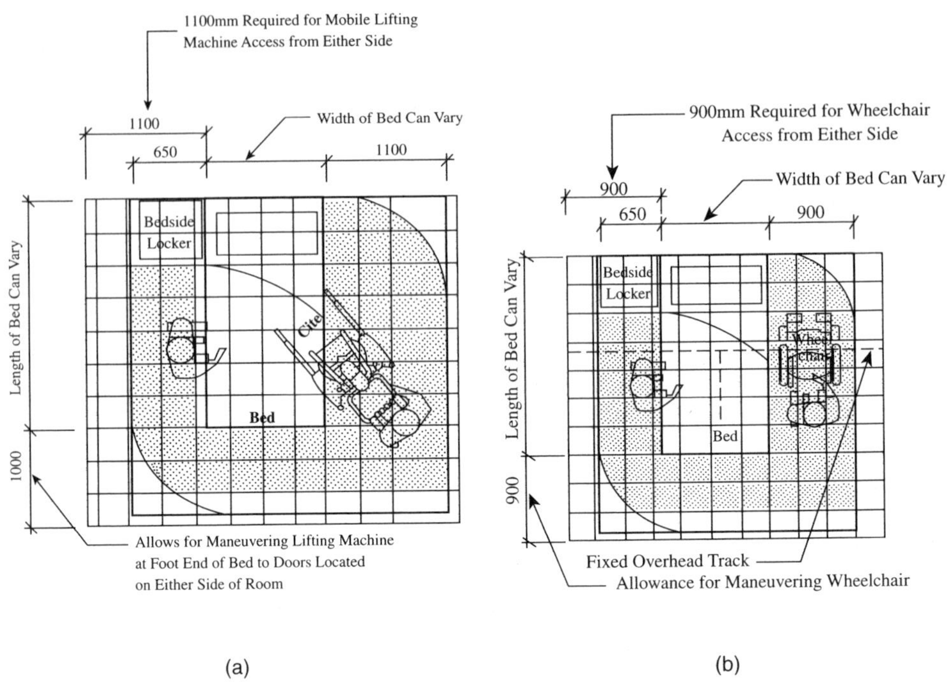

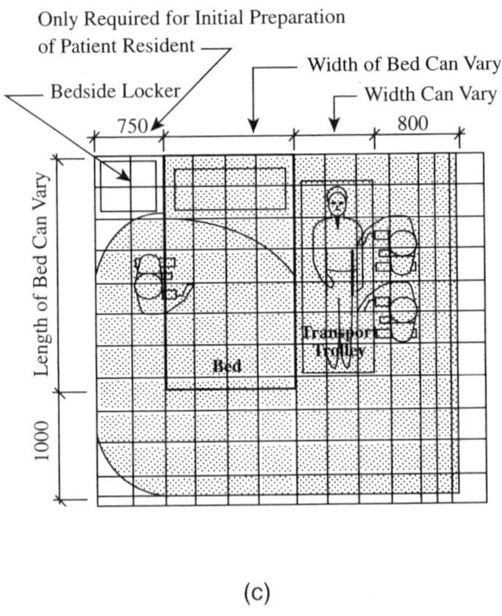

FIGURE 12.3 (a) Single-bed room mobile lifting machine. (b) Single-bed fixed overhead lifting machine. (c) Single-bed room space required for "Patslide" transfer to trolley. The dimensions shown in the diagrams are not the total bedroom floor area (i.e., the perimeter walls of the room are not shown). The measurements shown are clear space requirements and any furniture or equipment within this zone should be easily movable (i.e., no fixed or large/heavy furnishings within this area). If wheelchairs or lifting equipment are larger than shown, spatial dimensions may have to be increased. Spatial requirements are based on adjustable beds, so that staff do not have to bend when attending the patient/resident or making the bed.

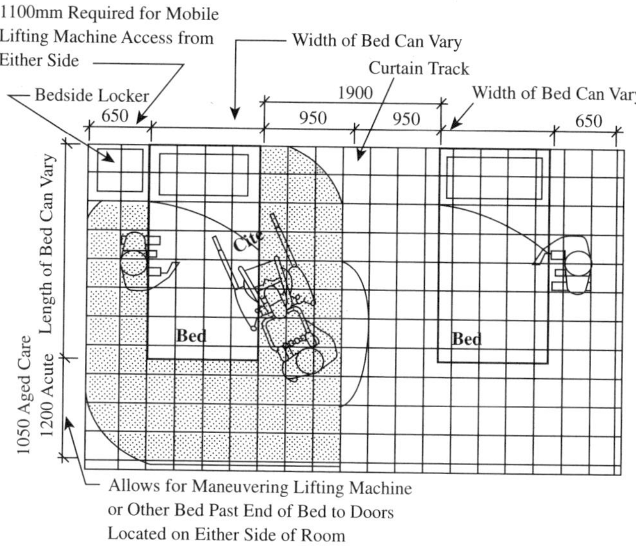

(a)

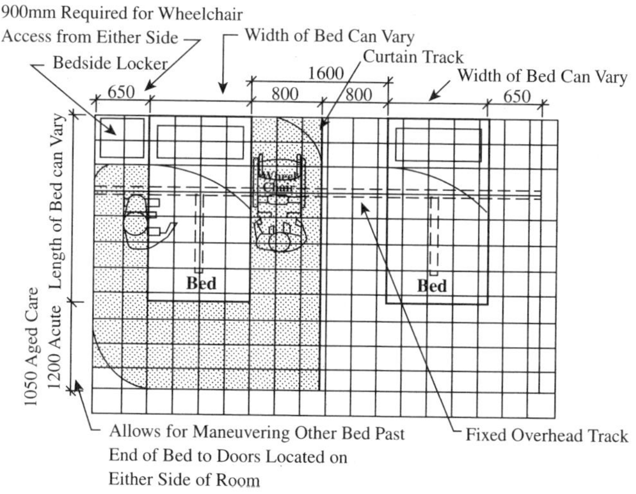

(b)

FIGURE 12.4 (a) Multiple-bed rooms, side by side mobile lifting machine. (b) Multiple-bed rooms, side by side fixed lifting machine.

(continued)

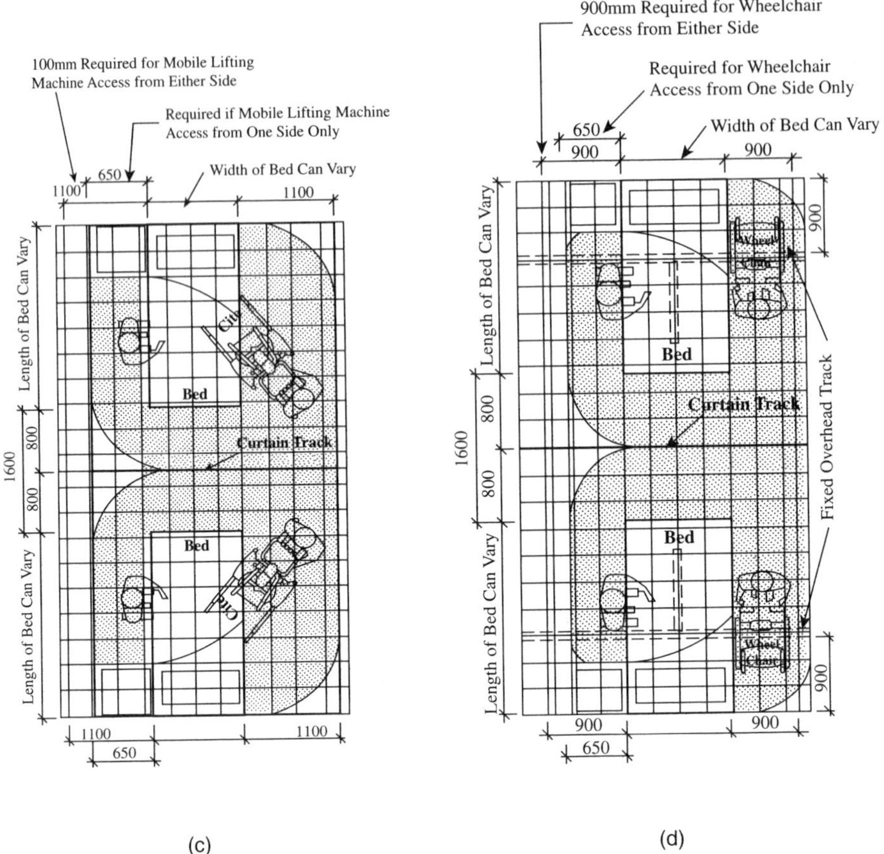

FIGURE 12.4 (CONTINUED) (c) Multiple-bed rooms, footend to footend mobile lifting machine. (d) Multiple-bed rooms, footend to footend fixed overhead lifting device. The dimensions shown in the diagrams are not the total bedroom floor area (i.e., the perimeter walls of the room are not shown). The measurements shown are clear space requirements and any furniture or equipment within this zone should be easily movable (i.e., no fixed or large/heavy furnishings within this area). If wheelchairs or lifting equipment are larger than shown, spatial dimensions may have to be increased. Spatial requirements are based on adjustable beds, so that staff do not have to bend when attending the patient/resident or making the bed.

Toilet Transfers

Note that these diagrams exclude hand basins and showers, as they will be provided outside the minimum spatial zone required for toilet transfers.

Another consideration which should be incorporated into the design is enhancing the ability of patients/residents to contribute to their rehabilitation by assisting themselves or staff when they are transferring. The positioning and type of grab rails is critical in these situations.

For design category 1 patients/residents, the Australian Standard AS 1428 - Design for Access and Mobility is usually most appropriate. But for design category 2 patients/residents, where you are providing for up to two staff to be involved, it is recommended that the use of drop down rails on either side of the toilet be included. These provide flexibility, allowing enough space for staff to assist on either or both sides, and they can be easily folded away for patients who do an independent transfer from a wheelchair located parallel to the toilet.

The grab rails need to be sufficiently strong and sufficiently well mounted to withstand a force of 110 kg applied at any point in both downwards and sideways directions. It is likely that a vertical support from the floor to the end of the rail will be needed. If the rails are installed too far away from the toilet (to the side) the horizontal component of the force exerted by the patient/resident will need to be greater.

Many commercially available drop-down grab rails are not strong enough, and many are provided with inadequate fixing mechanisms. During building design, the structural consultant should specify the anchoring required.

Ensuites

For ensuites there can be a range of design layouts depending on where the individual fixtures are positioned, and whether they are single or shared ensuites. Shared ensuites can provide a safer handling environment in a cost effective way, providing that privacy is not compromized. Figure 12.6(a) to Figure 12.6(c) are examples of some layouts. The critical element is the spatial dimensions.

Wheelchair access to basins and tables

When installing wash basins for use by independent wheelchair users, the underside of the basin should be 660 mm above the floor, with no obstructions under the basin at that height for a distance of 254 mm back from the leading edge of the basin. The sink should project at least 457 mm from the wall or from any structures or fittings at foot level, that is from the floor up to 165 mm above floor level. Exposed drain pipes and hot water pipes must be insulated.

Assisted Bathrooms

The most common types of assisted bathrooms use:

- Fixed peninsula type bath (attached to a wall at one end, with clear space on three sides)
- Mobile shower trolleys

Mobile shower trolleys have become more popular recently, particularly in facilities where patients/residents are immobile and totally reliant on staff assistance. The practice of incorporating showers within assisted bathrooms is becoming obsolete due to newer facilities providing ensuites to all bedrooms. However, to satisfy the requirements of the Building Code of Australia, assisted bathrooms without assisted showers must have showers that comply with Australian Standard AS 1428 - Design for Access and Mobility. The provision of toilets in an assisted bathroom is optional, although there is an advantage in being able to accommodate patients/residents who are incontinent and may need to access toilets quickly when in the bathroom.

A key feature in an assisted bathroom (with peninsula type bath) is the ability to maneuver the lifting machine so that the base can easily fit under the bath (Figure 12.7(a) and Figure 12.7(b)). Designers may overlook the difficulty in maneuvering such equipment when a person is in the carry seat.

Another consideration, particularly for aged care facilities, is that patients/residents should not be transported from their bedrooms to the bathroom in a lifting machine, as this contravenes the Commonwealth dignity requirements. Hence the person will need to be lifted from a wheelchair into the bath via a lifting machine.

Where the bath and toilet are fixed items, the use of a fixed overhead lifting machine can reduce the amount of circulation space required. This has the potential to save space and reduce building cost.

To ensure staff are not putting undue stress on their backs, the height of baths is also important. The options include a fixed bath, which is installed at a higher level, or the provision of a hydraulic bath where the height can be adjusted, the latter being preferable because they are more suitable for staff of varying heights.

The use of nonslip floor finishes in wet areas is essential.

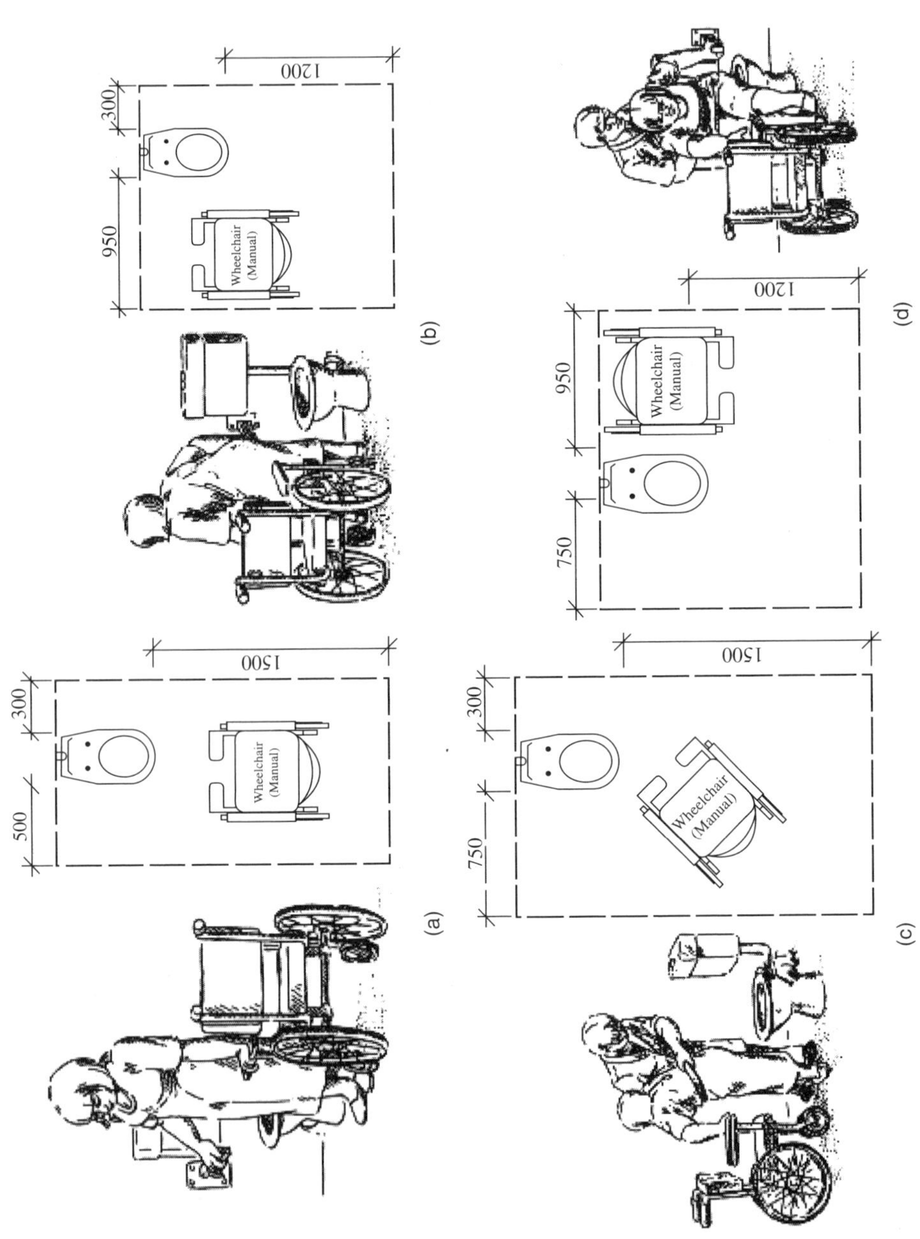

500
300
1500
Wheelchair (Manual)
(a)
950
300
1200
Wheelchair (Manual)
(b)
750
300
1500
Wheelchair (Manual)
(c)
750
950
1200
Wheelchair (Manual)
(d)

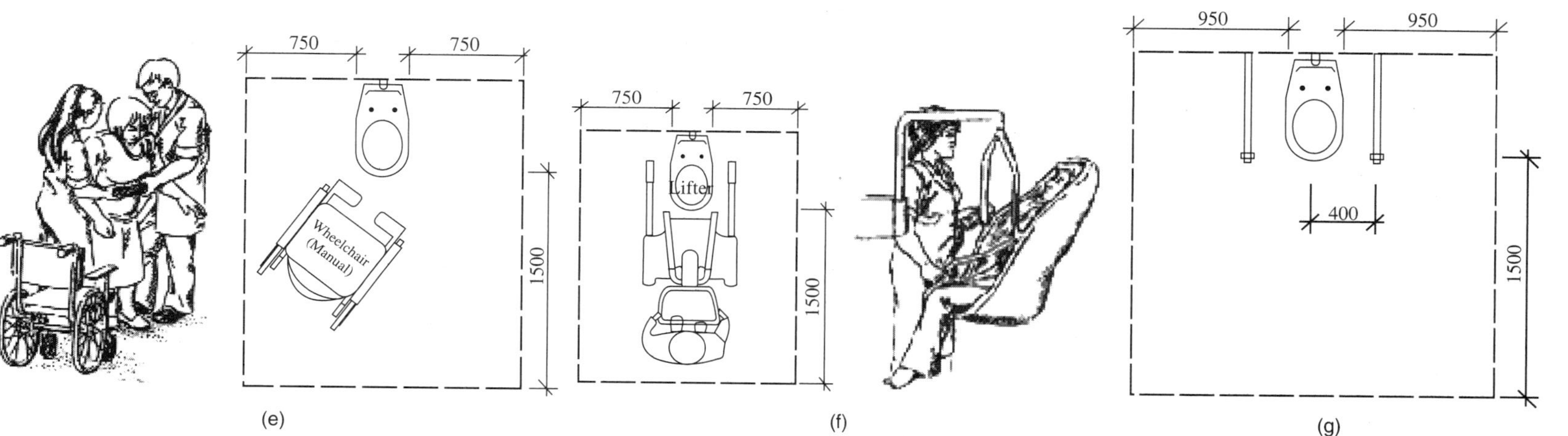

FIGURE 12.5 (a) The functional area for a wheelchair-bound patient/resident that can stand on one leg. Requires assistance by one staff member with frontal transfer. The wheelchair is placed as indicated in the figure. (b) The functional area for a wheelchair-bound patient/resident that requires assistance in side transfer. (c) The functional area for a patient that can stand on both legs. This transfer, requires the option of two staff members assisting by standing on either side. The wheelchair is placed as shown in the diagram. This is also sufficient space for a patient to transfer with the help of a swivel board/turntable. (d) The functional area needed by an independent wheelchair person — frontal transfer. (e) The functional area needed for a patient in a lifting machine and with the assistance of one or two staff members on either side. The transfer to the lifting machine has been done outside this area. (f) The functional area neede by an independent wheelchair person — side transfer. (g) Drop-down grabrails on either side of toilet. This allows for the option of full staff assistance, or sideways transfer from either side. Notes: Doors swings and sliding doors are shown as options. If swinging inwards, they need to be provided with removable hinges or an option to allow them to swing outwards in emergency situations. The location of the shower floor gradient is important in relation to toilet transfers to and from wheelchairs. If too close, then the wheelchair will tend to roll away from toilet. Water should drain to the centre of the shower area. If there is a shared ensuite between two rooms, then the location of the doors is even more critical and will result in increased circulation space within the ensuite. The shell construction of a prototype ensuite would assist staff in understanding and endorsing the design layout. It is recommended that all ensuites within a facility should be of similar layout and design. This will assist staff in developing standard handling procedures. The use of nonslip floor finishes is essential.

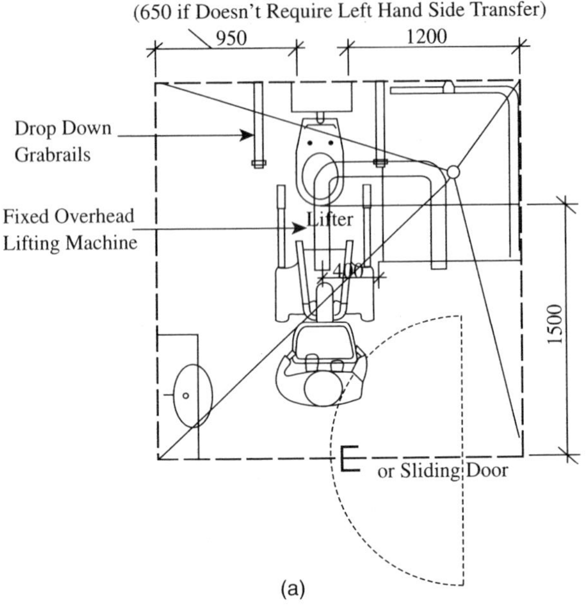

(a)

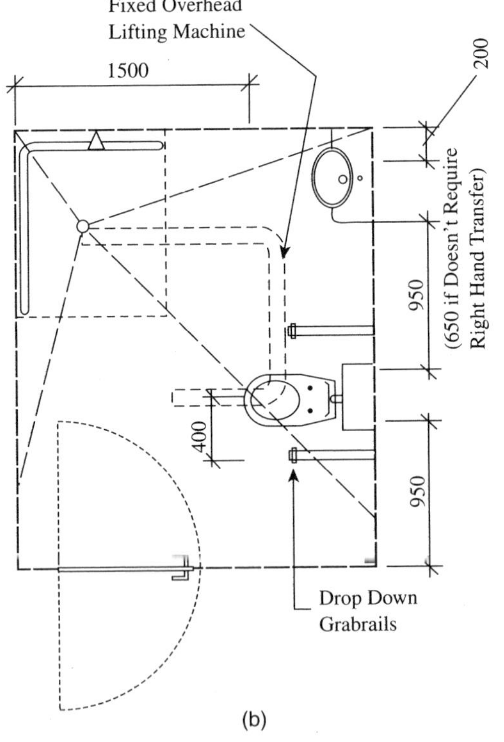

(b)

FIGURE 12.6 (a) Single ensuite. (b) Single ensuite. (continued)

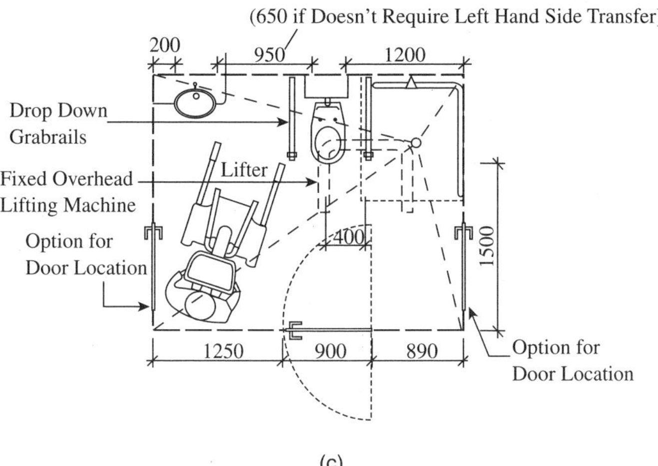

(c)

FIGURE 12.6 (CONTINUED) (c) Single or shared ensuite. Door location is optional and will vary depending on relationship to bedroom layout. If the ensuite is shared between bedrooms it will require an additional door, and the depth of the room may need to be increased. The position of the hand basin is optional, but needs to take into consideration movement of wheelchairs (i.e., Australian Standard AS 1428 - Design for Access and Mobility).

Lounge/Dining Rooms (Aged Care and Rehabilitation Facilities)

The spatial requirement for handling procedures in lounge rooms where residents are being transferred from wheelchair to a seat/chair is very similar to the Figures 12.5a to 12.5e (toilet transfers).

The main issue in lounge rooms is often the proximity of other furniture. This is particularly relevant with the increased usage of large recliners, which are heavy to move and more difficult for residents to get in and out of. There is also a trend to provide a number of small lounge rooms spread throughout the facility rather than one large central lounge room. As a consequence each small lounge room still provides for the television, stereo, and ornaments, which has resulted in proportionally reduced circulation space. It is recommended that a floor area allowance of 3 m^2 per person be provided for lounge rooms (Reference - Department of Human Services Aged Care Residential Services Design Guidelines).

In dining rooms, a person in a wheelchair will most likely remain in his/her wheelchair to sit at the table. It is usually only those residents using a walking stick or frame who are likely to require some staff assistance in sitting down and getting up. Often the main issue is getting to the table through the maze of seats in restricted circulation space. Careful design consideration has to be given to residents in wheelchairs and to those using walking aids maneuvering around in a dining room. This should include allowance for storage of such equipment away from the tables. As a general design consideration, an allowance of 2 m^2 per person should allow for tables, seating (including staff assisting residents), and circulation.

For wheelchair access to tables, a pedestal table is preferred with edge mounted legs. The underside of the table should be 660 mm above floor level or higher, and the table height should be approximately 787 mm. The table length, from edge to support pedestal, should be at least 533 mm and preferably 600 mm.

Corridors

Although not a great number of direct patient/resident handling procedures take place within corridors, the main issue is maneuverability, particularly for people in wheelchairs, people using walking aids, and staff pushing equipment. In addition, the building regulations stipulate minimum widths of corridors for emergency egress purposes.

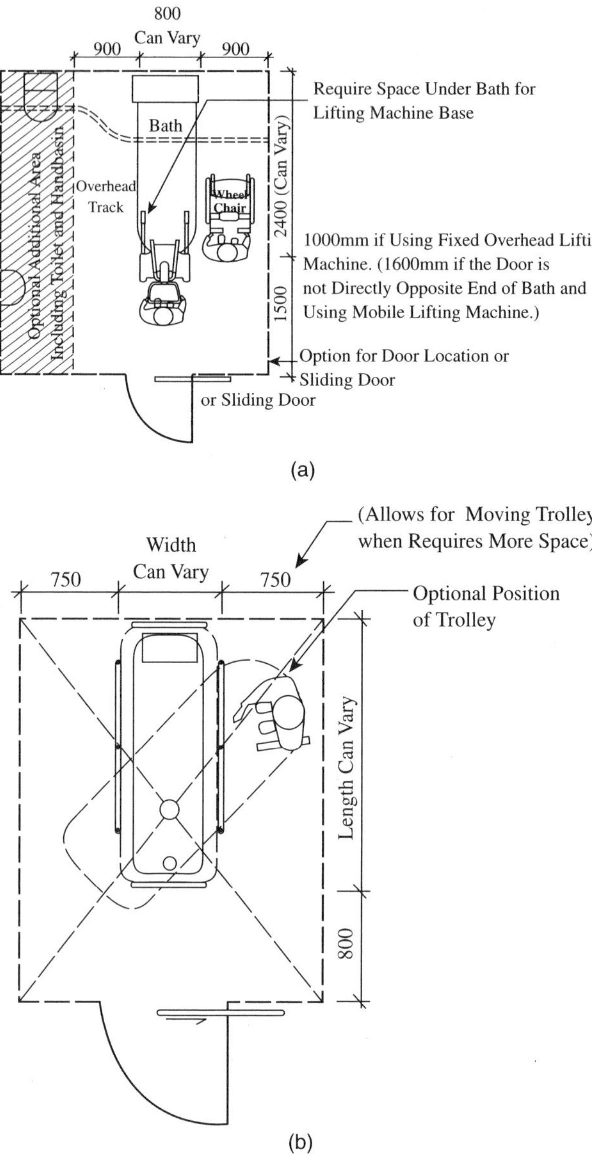

FIGURE 12.7 (a) Peninsula/island bath. (b) Mobile shower trolley.

Generally, acute care buildings require a minimum clear width of 1800 mm (i.e., handrail to handrail), (note that main access corridors to areas such as theatre, x-ray, should be 2400 mm clear). Corridors in aged care facilities maybe reduced to 1500 mm, however door openings may need to be wider or it maybe advantageous to widen corridors to 1800 mm clear at points where bedroom doors open onto the corridor (note that the Building Code of Australia currently requires "1800 mm in nursing homes where patients are transported in beds").

Specific reference is made to the Building Code of Australia clause D1.6 (f) which relates to corridor widths and door opening sizes.

It is recommended that in aged care facilities the corridor width be 1500 mm clear (subject to door opening size), but be increased to at least 1800 mm clear width to allow for larger pieces of equipment to pass and where there are bedroom doors opening opposite one another (Figure 12.8(a) to Figure 12.8(c)).

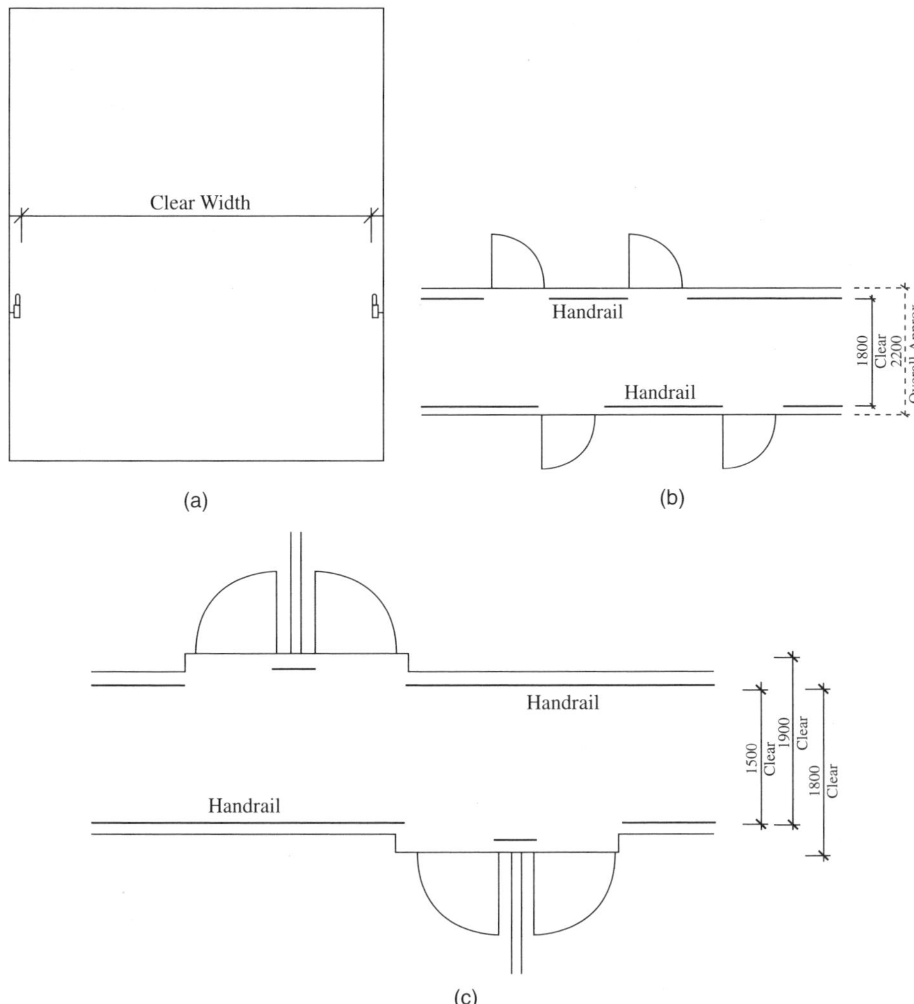

FIGURE 12.8 (a, b, c) Aged care facilities.

It is essential that clear width for corridors not be used for the storage of equipment or protruding hand basins. Consideration should therefore be given to providing recessed hand basins and storage areas for equipment such as lifting machines.

Equipment Storage

The spatial requirements for the storage of handling equipment such as mobile lifting machines, wheelchairs, commodes and shower chairs are often overlooked or ignored. As a consequence, such equipment is then left in corridors, bedrooms, or other bathrooms, which, in turn, creates safety problems for staff and patients/residents. It is important that the total storage requirements be ascertained during the design phase.

Mobile Lifting Machines

The strategic location of storage for mobile lifting machines is important with respect to accessibility by staff. If they are stored too far away, staff may be reluctant to use them, and staff travel is increased, which may put patients/residents at risk due to the delay in time.

It is recommended that mobile lifting machines be available to staff within 20 meters. Their actual storage should allow for easy access. A preferred option is directly off a main corridor and in a recessed alcove. They should not be stored in circulation areas where they inhibit movement.

Wheelchairs

The storage of wheelchairs often presents a dilemma: do you store them in one central location or do you provide space within the actual bedrooms? This will depend on the usage. For example, in aged care facilities that residents may occupy for long periods, once they have commenced using a wheelchair, they will probably continue to use it for the rest of their life. As a consequence, it is preferable to store their wheelchairs in their bedrooms, where they will need to use them first and last thing of the day. Provision of a specific storage area within the bedroom is preferred so that while residents are in bed, the wheelchairs can be stored out of circulation space.

However, in acute facilities where the patients' stay is often short and their individual requirements can vary significantly from patient to patient, the storage of wheelchairs may be better in small strategically located store rooms, which may also accommodate other equipment.

Suitable parking places need to be provided for charging of battery operated lifting machines and wheelchairs.

Commodes and Shower Chairs

Commodes and shower chairs can be stored in the shower area of the ensuite, as they are used whenever patients/residents use the ensuite. Shared ensuites only require one commode/shower chair per ensuite.

EFFECTIVE CONSULTATION WITH DIRECT CARE STAFF IN THE DESIGN OF FACILITIES

Why Consult Direct Care Staff?

The term "direct care staff" refers to the workers who perform the everyday work of the business unit, as opposed to senior managers of the organization. Direct care staff in this case means workers who routinely handle patients/residents.

Consultation with direct care staff should be seen as an important business opportunity to add value to a project by improving the quality of the finished product, and by contributing to long term harmony within the new facilities.

Consideration could also be given to consulting patients/residents, especially in the case of facilities for longer term patients/residents, but the importance of the health and safety of staff needs to be given high priority in design decisions.

Most health or aged care design projects necessarily have some degree of staff involvement in various stages of the design process. This process is normally adequate for obtaining the views of senior managers, but often it may fail to elicit much of the very useful knowledge, experience and insights that the staff who work directly with patients/residents have developed over time. The front line workers are a valuable source of detailed information, because they know the work practices and the workplace more intimately than anyone else, thus helping to get the design right in the first place, and avoiding the need for costly alterations to newly occupied facilities. If they have been properly consulted, workers are likely to have a greater sense of ownership of the end result, regardless of any shortcomings.

Under Victorian law, elected health and safety representatives must be consulted regarding any changes to the workplace that may affect the health and safety of workers (Occupational Health and Safety Act 1985). The Victorian Occupational Health and Safety (Manual Handling)

Regulations 1999 require that an employer must take all practicable steps to make sure that both the equipment used and the work practices carried out are designed to be safe from a manual handling point of view. Similar provisions exist in other Australian juristictions. The regulations require assessment of manual handling risks, where practicable, in consultation with any health and safety representative for the designated work group of which those employees are members.

When to Consult Direct Care Staff

Staff consultation consumes time for all the parties involved and therefore has the potential to add to the overall financial cost of the design process. A time efficient schedule of consultation is needed.

Effective consultation may reduce the overall costs associated with a project, especially if calculated over the lifetime of the facility.

Since much of the valuable knowledge of the direct care staff is in the fine detail of the work processes and workplace layout, it is essential that they have considerable input at the design development stages. Most of their consulting time should be invested there, refining the details.

However, some of the most important decisions affecting safe patient/resident handling are made at the earlier design stages, and some of these decisions may benefit from direct care staff input. For example, the relative levels of the floors in adjacent buildings will affect the slopes of any adjoining ramps where patients/residents may have to be moved between the two buildings. In a psychogeriatric setting, for example, the absence of a natural indoor walking circuit for patients may involve staff in a lot of unnecessary supervision and handling of wandering patients/residents.

As a third example, easy access between certain departments may reduce patient/resident handling, but to ascertain this may require direct care staff input at the master plan stage.

Examples of specific issues requiring consultation at different stages of the design process are set out in Table 12.1.

Preparing the Ground for Effective Consultation

The first essential prerequisite for an effective consultation process is mutual respect and trust between the direct care staff and the leaders in the design process: the agency's project team, and the external consultants, especially the architects. Ideally a productive working partnership can be developed between the direct care staff and the designers. Like any other human relationship, this partnership needs to be soundly established, and maintained in a spirit of good faith.

From the organization's point of view, this relationship is likely to be typical of its normal human resources management practices as carried out at line manager level. Managers may have to invest considerable effort to personally accept the need for — then to establish and maintain — a process that genuinely seeks out and considers the views of direct care staff. It is important that managers regularly inform and provide feedback to staff, including final outcomes and reasons why their ideas are or are not being incorporated into the design.

Therefore a consultation process needs to be established, which:

- Is clearly documented and circulated to participating staff and their constituents beforehand, including the purpose and scope of the consultations, the time lines, and the names and roles of all participants
- Includes a written record of the proceedings of the consultations, in the form of minutes of meetings or a brief memo, file note or report, with copies circulated to all participants and to the project committee
- Includes appropriate feedback to the participants regarding issues raised in the consultation, especially any issues not resolved in discussions, or any staff requests that are not included in subsequently approved stages of the design

TABLE 12.1
Agenda for Consultations throughout the Design Process

Stage of design process	Agenda for direct care staff consultation
Master plan	Patient/resident handling load and safety- between buildings, between different levels and between departments.
	Agency develops patient/resident handling safety policy, such as a "no lift" policy, and program including workplace design processes
Feasibility study	Establishment of formal consultative process for the project, budgeting for staff time
	Selection and induction of staff representatives into the process, including adequate documentation, briefing by design team and skills training
	Identification of the functional capacities of patients/residents likely to occupy the facility in its life time
	Staff to be requested to start to think about their needs and patient/resident handling practices
Schematic design	Test design philosophy and assumptions
	Define patient/resident handling safety objectives, specifications, and performance indicators
	Users inspect other good examples of similar existing facilities
	Relationships between other components: meals, therapy, support services, and their location
	Room functions and relationships identified
	Overall layout, room sizing, work practices, space required to use patient/resident handling equipment, storage practices, access, visibility
	Preparation of room data sheets
	Feedback on schematic design approval
Design development	Major consultation re 1:50 fit-out plan
	Suitability and location of furniture, fixtures, finishes, services
	Mock-ups (prototypes) if required
	Specialist ergonomics review if required
	Feedback on plan with explanation to staff by architect approval
Contract documentation	Material selection
	Inspection of samples
Construction	What services will be interrupted during construction and relocation?
	Logistics of relocation and occupancy
	Site inspection as final visual check and to assist smooth transition to new facilities
Post-occupancy	Participation in evaluation of the functional and safety outcomes of the design (including manual handling safety)

- Clearly specifies the decision making procedures and the powers and responsibilities of the different parties and the different committees involved in the design process
- Is conducted in language that all participants can easily understand

Helping Direct Care Staff to Contribute Fully to the Consultation Process

To get the best out of the investment of staff time the following procedures are recommended.

Staff need to be allocated sufficient paid time to participate. The work involved should be recognized as important core business of their work unit, and they should not be expected to do extra work in their own time. Their positions should be back-filled to release them to prepare for and attend meetings. This means that the resources have to be paid for out of someone's budget — either that of the relevant business unit, or charged against the project budget.

Staff may need training to give them sufficient technical knowledge and skills to actually communicate properly with the designers. Some of the skills required include:

The language of architects (i.e., most lay people do not understand a "GPO" to mean a power point)

The communication tools used by architects, especially two-dimensional, black on white, line drawings including plan views and elevations, and the symbols used on these drawings; use of more visual media such as 3-dimensional representations and color may help workers to better visualize the design

Direct care staff also need practical help in visualizing the design. In particular, they need help in translating the scale of the drawing to real life. Unlike a map, architects' drawings do not have scale bars — only a scale ratio. To translate this to actual size, it helps to have a scaled ("engineer's") ruler to read actual dimensions straight off the drawing, and a tape measure to demonstrate the actual dimensions proposed. The layout of full scaled room plans using marking tape on the floor, with actual sizes of beds, furniture, equipment, can greatly assist in this process. Another useful technique is to make scaled cut outs (on graph paper) of furniture and equipment including, if possible, human space envelopes, and place them on the architectural drawings to identify the optimum layout of the workplace. An ergonomist can be consulted regarding the human space envelopes.

Staff may need to be encouraged to participate fully in discussions. This may require some explanation of, and possibly coaching in, formal meeting procedures where these are used as part of the consultation process. Chairpersons of meetings need empathy and skill to ensure that the views of all participants are heard and fully considered. The incorporation of workshops with small working groups can often encourage participation.

Practical steps are also needed to compensate for a possible lack of assertiveness by staff in an unequal power relationship. Many hospitals have a long established and well defined hierarchy of power, with medical practice at the top, nursing subservient to medical practice, and other employees having a lower status again. The direct care staff that need to be consulted may be relatively low in the hierarchy within their own business unit. While they may have good ideas about their work, if they feel intimidated by the unequal power relationship with their employer or the architect, they may not articulate these ideas and valuable information will be lost. This is especially the case for staff that have limited English language communication skills.

Some staff may fear ridicule if the design does not work out well in practice. They need to be supported, but also need frank and open discussion of the implications of their ideas in order to help tease out the positive and negative aspects of their suggestions. Such frank discussion should be seen as a mark of respect, of taking their suggestions seriously, and is preferable to polite dismissal of their ideas, or worse still, silent dismissal of a suggestion without discussion.

Apathy by staff is a potential barrier to consultation, especially if they are not convinced of the worth of their involvement. Encouragement, and a clear explanation of the importance that will be placed on their valuable input, may help overcome such reticence.

Who to Consult

Both the method and the outcome of choosing who to consult is important to the success of the design process. The aim is to obtain informed opinion from those who do the everyday work in the unit. Normally, a business unit manager is already one step removed from the everyday work routine. However, it is important that people responsible for making decisions about work practices, namely the business unit managers, are also present and actively involved in the discussions along with the direct care staff representatives.

The selection process should be such that all the staff feel that they are represented and that they all have a chance to put ideas forward or ask questions. A call for nominations and a democratic election by workers may be an optimum method of selecting representatives.

The following attributes may help a worker participate fully in the consultation process:

- Good communications with and respect by fellow workers
- Awareness of and commitment to the duties and responsibilities that go with the consultation process
- Availability for the duration of the process
- Good knowledge and experience of working in the unit
- Ability and willingness to communicate ideas and ask questions
- An interest in improving staff safety, well-being and efficiency

When the facility being designed is a new enterprise and the direct care staff have not yet been selected or employed to work in the facility, and therefore cannot be consulted, efforts should be made to consult direct care staff in a similar facility.

Challenges for Architects and Designers

These guidelines should help managers to properly brief their design consultants, and should help everyone involved in the design process to be more confident in challenging the solutions put forward by the designers.

Architects and designers need to work with the project management team at the beginning of the project to ensure that a timely, thorough, and cost effective process is established for getting the most out of direct care staff consultations.

It can be potentially frustrating for architects and designers, with many years of experience and many successful health or aged care building projects to their credit, to have to spend valuable time waiting for a user group to come up to speed, or to have to go over the same old ground with each new user group. Project fees are minimized by competitive tendering and therefore budgets — and designers' time — are limited.

Experience has indicated that architects and designers involved in health and aged care facilities design in Victoria vary widely in their methods of user consultation and in their degree of empathy with direct care staff. This variation may extend between firms, as well as between individual architects and designers within the same firm.

Architects and designers need a high level of interpersonal skills, tact and humility when seeking to resolve differences of opinion between themselves and direct care staff. Until trust has been established, some staff may display strong feelings about the design process because of historical or organizational factors, which have nothing to do with the architect. Workers' emotions can be heightened by a concern that a new building project is a rare opportunity to solve a lot of their workplace problems — perhaps once in their working lifetime — and they place very high stakes on getting it right.

Architects can be the victim of inappropriate internal consultative mechanisms, which result in more work for the architect because the user consultations were not as productive as they needed to be. Internal organizational politics and unresolved power struggles can preclude some of the right people from being consulted, or can render the consultation process less effective because of oppressive time schedules or inadequate information. Failure to delegate for fear of relinquishing deeision making power may be a temptation for some senior managers.

Architects need to tune into the direct care staff members' concerns and needs, take them seriously, and develop mutual respect and trust with the direct care staff. The direct care staff consultation process should be viewed as a business opportunity to improve the design and the acceptance of the design, rather than as a process necessary merely to appease the users and the client.

Architects need to communicate in language that all participants in the consultations can easily understand.

Specialist Ergonomics Input

It is advisable to engage specialist ergonomics input to assist the design process.

Workplace ergonomics seeks to optimize the working environment to best suit people's capabilities and limitations. In regard to the design of health and aged care facilities, specialist ergonomics input can:

- Help the staff to participate more effectively in the consultation process
- Provide quantitative information about the sizes, space requirements and capabilities of people, and the implications of this information for the design

It is preferable to use an ergonomist with specialist experience in the design of health or aged care facilities, and who either has a specialist qualification in ergonomics, or who is a Professional Member of the Ergonomics Society of Australia (Certified Professional Ergonomist). The staff of the facility, and/or their elected occupational health and safety representative, should be consulted when selecting the consultant.

The ergonomist can coach staff in the accurate interpretation of drawings, using a variety of techniques including tape measure, full-scale markings on the floor, and scaled templates of equipment and functional work spaces. The ergonomic investigation can start with a walk through survey of the existing work place and work practices to identify existing hazards and opportunities to remove hazards at the design stage. This also helps to establish a rapport between the ergonomist and the direct care staff so that they can quickly begin to understand each other's point of view and language.

The ergonomist can identify crucial aspects of the work practices or of the building design that have a bearing on patient/resident handling risk factors, and can help the users to develop improvements.

Whilst it is possible for ergonomists to review drawings in their own office, with minimal or no time spent on site, it is difficult for the ergonomist to make certain judgements without knowing the details of, and the rationale for, the proposed work practices. Therefore a site visit by the ergonomist and consultations with direct care staff are recommended as important steps in the ergonomics input to the design process.

REGULATORY AND FINANCIAL CONTEXT

Building Regulations

The design and construction of acute health and aged care residential facilities must comply with the relevant building regulations. These include the Building Code of Australia (BCA), Australian Standards, and those of relevant authorities such as water and power regulatory authorities.

The Building Control Act requires building projects to obtain building permits prior to commencement of any building works, and usually a certificate of occupancy at the completion of the building works and prior to the occupancy of the new works. This is achieved through compliance with the Building Code of Australia. This is a performance based document allowing for some flexibility in the adoption of its contents, providing such interpretation does not put people at risk.

The Building Code of Australia provides for a number of building classifications that in turn impacts on specific building requirements. Acute health and aged care residential facilities fall under two classifications:

- Hospitals and Nursing Homes — Class 9A buildings
- Hostels — Class 3 buildings

Due to the recent restructure of aged care residential services by the Commonwealth Government to provide for one Residential Classification Scale (RCS) and "Aging in Place" the requirement to have separate building classifications for nursing homes and hostels may become obsolete. As a consequence the Australian Building Codes Board is currently reviewing the building requirements for nursing homes and hostels. However, the outcome of this review is unlikely to be implemented for some time.

Another key element of building regulations is the Building Code of Australia referral to relevant Australian Standards. One that has a major impact on the design of acute health and aged care residential facilities is Australian Standard AS 1428 - Design for Access and Mobility. However, this Australian Standard primarily focuses on patients/residents who can assist themselves with the appropriate physical support aids and space for maneuvering. As a consequence the spaces allocated in this Australian Standard do not allow space for staff assistance or mobile lifting machines. However, as these guidelines address the needs of patients/residents who require staff assistance they often exceed the minimum requirements of Australian Standard AS 1428 - Design for Access and Mobility.

Occupational Health and Safety

The Victorian Occupational Health and Safety Act 1985 requires employers to provide safe workplaces and safe systems of work. In regard specifically to patient/resident handling, the Victorian Occupational Health and Sarsty (Manual Handling) Regulations 1999 require, among other things, that an employer must take all practicable steps to make sure that both the equipment used, and the work practices carried out, are designed to be safe from a manual handling point of view.

The Cost

The cost of implementing the recommendations of these Design Guidelines should be seen in terms of return on capital investment. Any additional capital expenditure of increased floor space and or special lifting equipment can be offset by the reduction in WorkCover claims and other human resource related expenditure, leading to reduced recurrent expenditure.

However, each project should consider its specific needs and evaluate these against the design guidelines. For example, a particular accommodation unit may have 50% of its occupants totally independent and the other 50% relying on staff assistance to varying degrees. As a result, only half the facility may need to conform to the spatial and lifting procedures as set out in these design guidelines. The remaining part of the facility need only comply with Australian Standard AS 1428 - Design for Access and Mobility. However, as the life of the building is likely to be 25 years, one needs to be careful in considering the patient/resident types it is providing for now and throughout the future life of the building.

REFERENCES

Barrett, R. and Milburn, R., Lumbar loads in occupational bed making — a static planar analysis, *J. Occup. Health Saf. Australia NZ*, 13(1), 35–46, 1997.

Borys, D., Health and Safety in Hospital Building Design, Dissertation, graduate diploma in Occupational Hazard Management, Victorian Institute of Occupational Safety & Health, 1993.

Commonwealth Department of Health and Family Services — Aged and Community Care Division, Guidelines for use in ascertaining the quality of the built fabric of aged care facilities, Date unknown.

Department of Human Services — Victoria, Aged Care Division. Residential Care Guidelines, 1995.

Drury, C., Influence of restricted space on manual materials handling, — Industrial back pain in Europe, *Ergonomics*, 28(1), 167–175, 1985.

Engkvist, I. et al. Over-exertion back accidents among nurses aides in Sweden, *Safety Science*, 15, 97–108, 1992.

Engkvist, I. et al., Interview protocols and ergonomics checklist for analysing over-exertion back accidents among nursing personnel, *Applied Ergonomics*, 26(3), 213–220, 1995.

Garg, A., Owen, B., and Carlson, B., An ergonomic evaluation of nursing assistants' jobs in a nursing home, *Ergonomics*, 35, 979 – 995, 1992.

Keyserling, W.M., Punnett, L., and Fine, L.J., Trunk posture and back pain: Identification and control of occupational risk factors, *Appl. Ind. Hygiene*, 3, 87–92, 1988.

Kuorinka, I., Lortie, M., and Gautreau, M., Manual handling in warehouses: the illusion of correct working postures, *Ergonomics*, 37(4), 655–661, 1994.

Mital, A., Nicholson, A.S., and Ayoub, M.M., *A Guide to Manual Materials Handling*, Taylor and Francis, Washington, DC, 1993.

Pressalit Multi System Technical Manual.

Pryor, R., *Back Injury — A Resource Manual*, Australian Nursing Federation (Vic. Branch) and Victorian Healthcare Association, (1997.

Swedish Institute for Hospital Planning and Rationalization, Hygiene rooms — functional space for personal hygiene in long term care, SPRI Report 21, 1979.

Takala, E. and Kukkonen, R., The handling of patients on geriatric wards, *Applied Ergonomics*, 18(1) 17–22, 1987.

Victorian WorkCover Authority. Workplace injuries in the health industry 1995/96 and 1996/97, 1998.

Appendix 12A: Safe Handling of Patients/Residents — Workplace Design Process Checklist

This checklist is designed to be used by business unit managers and direct care staff representatives at all critical stages in the design process for new acute or aged care facilities. Different questions will be relevant at different stages as the design process proceeds.

Occupational Health and Safety Policies and Practices

1. Does the agency/facility have any occupational health and safety policies that have implications for the way buildings are designed to minimize the risk of injury to staff who handle patients/residents?
2. Does the organization have any policies relevant to the safe handling of patients/residents, such as a "no lift" policy? If so, how do these affect the way buildings should be designed to minimize the risk of injury to staff who handle patients/residents?
3. Has the organization carried out risk assessments of patient/resident handling in accordance with the requirements of the Victorian Occupational Health and Safety (Manual Handling) Regulations 1997?
4. Are there opportunities within the design process to ensure compliance with the Victorian Occupational Health and Safety (Manual Handling) Regulations 1997?

Master Plan

1. What patient/resident movements will take place between different departments, or between different floors, or between different buildings?
2. What work practices are proposed for these transfers, and has a risk assessment been carried out on the proposed work practices?
3. Are the proposed work practices sufficiently safe, or should changes to the master plan be considered (slopes of ramps, minimizing changes of floor level, minimizing patient/resident handling distances)?

Consultation Process

1. Has a process for staff consultation been developed and documented including the purpose and scope of the consultations, the time schedule, and the names and roles of all the participants?
2. Has the process been communicated to staff?
3. Have the staff expressed acceptance of the proposed process?

4. Has budgetary allowance been made to pay for staff time spent on the consultation process?
5. Are all members of the consultation team able to attend all consultation meeting?
6. Have the latest drawings been circulated to all participants?
7. Do any participants need assistance with understanding the drawings?
8. Do any participants need help with visualizing what the design will be like at full scale?
9. Are the outcomes of consultation meetings documented and circulated to participants?
10. Are the outcomes of any unresolved issues communicated to all participants, with reasons as to the resolution?

Feasibility Study

1. What type of patients/residents are likely to occupy the facility in the foreseeable future (i.e., 10 years)?
2. What type of assistance are these patients/residents likely to need in their bedroom, bathroom, toilet, and the corridor?
3. What method of assistance will be used in each location — no lifting, team lifting, mobile lifting machines, or fixed lifting equipment?
4. Will the patients/residents need to be moved regularly to receive various therapies (e.g., electroconvulsive therapy)? What handling practices will apply for movements between therapies?

Schematic Design

1. How much space is needed in each critical location to safely perform each of these transfers (beside the bed, at the toilet, at the shower, at the bath, though doorways, in corridors)?
2. What fittings are needed for the patients/residents to assist themselves, and where should these be located (e.g., grab rails, fold down shower seats)?
3. Will there be patients/residents who wander, and are there any special features in the layout of the building that need to be considered to minimize staff handling of these patients/residents (e.g., a natural walking circuit, places where they might congregate)?
4. Will beds ever need to be moved through doors (even on rare occasions or on an emergency basis)?
5. What width of beds will be used?
6. How wide do the doorways and corridors need to be to allow beds to be wheeled in and out?
7. Considering the answers to the previous questions in this section, are there any remaining patient/resident lifting tasks that have not been considered, or for which safe handling procedures have not been established?
8. In the existing set up, or in similar set ups in other places, are there any fittings or items of furniture that obstruct staff from optimum handling practices (e.g., bed side tables, fixed lifting equipment)?
9. How many mobile patient/resident transfer aids (wheelchairs, lifting machines, bath chairs, commodes, etc.) will be needed in the new unit, and where should they be stored so that they can easily be used and will not obstruct patient/resident rooms, corridors, or bathrooms?
10. Have all the patient/resident handling and space issues identified in this section been considered by the project team and fed into the design process?

Design Development — Checking the Plan for Patient/Resident Handling Risks

1. On a copy of the 1:50 fit-out plan, for one patient/resident in each type of room, mark with a pencil the spot locations or routes where the patient/resident may be transported assisted by staff. Include within-bed transfers, within-room transfers, and movements about the unit as a whole.
2. On the edge of the plan, indicate the type of handling procedure proposed for each transfer (type of lift, if any, number of staff, type of equipment).
3. Mark with a cross any places where, in your opinion, patient/resident handling could be difficult or risky.
4. Highlight on the plans the location of any furniture or fittings that may make handling difficult or unsafe.
5. For each of the potential risks identified, check for compliance with your unit's policies and practices for safe handling of patients/residents.
6. For each of the potential risks identified, refer to the relevant part of Section 4 of the design guidelines for safe patient/resident handling and check whether sufficient space has been allowed.
7. For each of the potential risks identified, check what aids are needed (e.g., grab rails, patient/resident handling equipment); check where these are located, and where mobile equipment will be stored.
8. Is any further information needed to help assess risks (e.g., mock-ups, specialist ergonomics advice, visits to other units)?

Contract Documentation — Material Selection

1. If carpet is to be used, will it be suitable for pushing mobile equipment on?
2. Do all pedestrian surfaces, both wet and dry, have suitable slip resistance characteristics?

Construction

1. Will any services or spaces that staff currently use be changed during the period of construction?
2. Will these changes have an impact on staff safety?
3. Will there be risks to staff during relocation to, and initial occupancy of the new unit (e.g., bed pushing, other unusual manual handling tasks)?
4. If possible, visit the site as construction proceeds, and "walk through" as many patient/resident handling tasks as possible as a check of the space, equipment, furniture, and fittings needed to perform each task safely.
5. Are any changes to the design required as a result of the site visit?
6. Have changes to work practices, furniture, or equipment been prepared for, prior to moving in?

Post Occupancy Evaluation by End Users

1. Identify, through review of incident reports and consultation with all staff, any minor or major shortcomings of the design, with special reference to strain injury prevention.
2. Conduct a formal inspection audit of the unit using the Workplace Design - Safety Audit Checklist (Appendix 12B).

3. Recommend any changes needed to the unit's policies and procedures to control any risks identified.
4. Summarize any changes needed to future designs of similar facilities.
5. Submit the summary report to the agency project management team, to the relevant safety committee and to any relevant external bodies (e.g., the Department of Human Services for public hospitals, the Board of Directors for private hospitals).

Appendix 12B: Safe Handling of Patients/Residents Workplace Design Safety Audit Checklist

This checklist is designed to:

- Be used in conjunction with the guidelines for Designing Workplaces for Safer Handling of Patients/Residents
- Help unit managers and direct care staff to audit the design shortly after occupying newly built acute or aged care facilities
- Help evaluate how safe the work practices are for staff when handling and moving patients/residents
- Help reduce any remaining risks related to handling and moving patients/residents within the unit
- Facilitate feedback to people responsible for the design of future facilities, in order to learn from the experience with this unit
- Assist in auditing patient/resident handling safety aspects at various stages throughout the design process
- Help assess patient/resident handling safety in existing facilities

Name of facility or business unit ____________________
Location ____________________
Owner or Health Care Network ____________________
Inspection carried out by (list names) ____________________
Checklist completed by ____________________
Position ____________________
Signed ____________________
Date ____________________
Signed (Business Unit Manager) ____________________
Signed (OH&S Manager) ____________________

Question	Yes/No N/A	Comments	Recommendations for action or for future designs
Is there enough space on both sides of all beds to allow safe on-bed movement of patients/residents?			
Is there enough space on both sides of all beds to allow transfers to and from bed?			
Is there enough clear space at the foot of all beds to allow safe handling and movement of patients/residents?			
Do furniture or fittings near beds impede safe patient/resident handling?			

Question	Yes/No N/A	Comments	Recommendations for action or for future designs
Are the privacy curtains located so as to enable unimpeded access to, and movement of, patients/residents?			
Can the beds be easily moved within the bedrooms when required?			
Can the beds be easily moved into and out of the bedrooms when required?			
Can the patient handling equipment be easily moved around within bedrooms when required?			
Can the patient handling equipment be easily moved into and out of the bedrooms when required?			
Can the patient/resident handling equipment be easily moved in and out of ensuites and assisted toilets when required?			
Can the patient handling equipment be easily moved around within ensuites and assisted toilets when required?			
Is there enough space on both sides of all toilet bowls to allow staff to safely assist patients/residents as required?			
Are appropriate patient/resident aids, such as grab rails, provided in optimum positions (so as to assist but not impede access) within ensuites and assisted toilets?			
Can the patients/residents easily reach the nurse call buttons in toilets and ensuites?			
Is the floor surface in ensuites suitable for safe movement and handling of patients/residents, i.e., nonslip, free from steps or steep gradients, and with adequate drainage?			
Is the floor gradient for the shower located far enough away from the toilet to avoid wheelchairs rolling away down the slope?			
In the event of an emergency in a toilet or ensuite, is suitable access available from outside (privacy latches operable from outside, inward swinging doors or removable doors)?			
Can the patient/resident handling equipment be easily moved around within assisted bathrooms when required, and can the lifting machine go close enough to the bath?			
Are appropriate patient/resident aids, such as grab rails, provided in optimum positions (so as to assist but not impede access) within assisted bathrooms and strongly constructed and mounted?			
Can the patients/residents easily reach the nurse call buttons in assisted bathrooms?			
Is the floor surface in assisted bathrooms suitable for safe movement and handling of patients/residents, i.e., nonslip, free from steps or steep gradients, and with adequate drainage?			

Question	Yes/No N/A	Comments	Recommendations for action or for future designs
If overhead lifting equipment has been installed anywhere in the unit, is it easy to use and does it serve its intended purpose?			
Do the patient/resident lounges, activity rooms and dining rooms provide a safe environment for patient/resident movement handling?			
Are the corridors safe for patient/resident movement and handling (handrails, floor surfaces, slopes, obstructions, smoke doors)?			
Is there enough dedicated storage space within the unit for patient/resident handling equipment?			
Is the patient/resident handling equipment stored sufficiently close to where it needs to be used?			
Is there adequate access to and from the unit as a whole, including emergency exit routes, to enable safe patient/resident handling?			
Is there sufficient appropriate patient/resident handling equipment available within the unit?			
Is the design of the facility compatible with implementation of a policy to eliminate all lifting of patients/residents?			
Can help be summoned quickly and without fail via the nurse call system?			
Are there any other issues regarding safe handling of patients/residents in this unit that warrant attention?			

Appendix 12C: How the Guidelines Were Developed

The Victorian WorkCover Authority, as part of the Health and Aged Care project established a Workplace Design Working Party to develop workplace design guideline layouts to assist staff in safe handling of patients/residents. The working party comprised representatives from:

- Victorian WorkCover Authority
- Private and public service providers
- Peak employer and employee organizations
- Department of Human Services
- Royal Australian Institute of Architects

Consultants were appointed to develop the Design Guidelines in consultation with, and under the direction of, the working party.

The Design Guidelines were developed in seven phases.

Phase 1: Review of Existing Information

The first step was to identify and review available information relevant to safe patient/resident handling, the design of acute health and aged care residential facilities, and relevant building regulations and Australian Standards.

One particularly valuable source of information was a report by the Swedish Institute for Hospital Planning and Rationalisation titled Hygiene Rooms — functional space for personal hygiene in long-term care. The project team gratefully acknowledges the valuable contribution of Professor Tore Lamson and Kay Wilson for drawing the report to our attention and for translating it from Swedish into English.

Phase 2: Evaluation of Existing Facilities

Phase 2 of the process was to visit and evaluate eight hospitals/aged care residential facilities that had been either recently upgraded or newly constructed, and evaluate their performance in terms of staff handling of patients/residents. This was followed by collating and analyzing all the information gathered in phase 2 with all relevant information being consolidated for usage in the development of the design guidelines.

Phase 3: Consultative Workshop

To achieve consultation with a broader spectrum of people involved with manual handling of patients/residents a workshop was held attended by approximately 60 people. This was a proactive approach to obtain constructive input from key stakeholders regarding suitable room layouts for patient/resident handling. The workshop program was structured to maximize both participation and consultation. It focused on three key issues: room layout, consultation processes, and the Building Code of Australia and Australian Standards. This material was fed into the guidelines.

Phase 4: Drafting the Guidelines

The consultants prepared a first draft, which was reviewed by the working party and refined in preparation for wider distribution and comment.

Phase 5: Consultation and Feedback

A second draft was distributed to key stakeholders in the health and aged care industry for comment. A second workshop was held to assist in industry consultation and feedback. About 20 written comments were submitted in response to the second draft.

Phase 6: Mock-Up

A field trial using a full scale mock-up of selected patient/resident handling activities was conducted at the Austin Hospital following release of the second draft, in order to test the guidelines in a simulated trial.

Phase 7: Finalization of the Guidelines

The consultants refined the guidelines in consultation with the working party on the basis of feedback received in writing and via the second workshop.

FURTHER INFORMATION

Victorian WorkCover Authority Ergonomics Unit
Level 24
222 Exhibition Street
Melbourne VIC 3000
GPO Box 4306
Melbourne VIC 3001
Telephone: (03) 9641 1744
Facsimile: (03) 9641 1573
Email: manual_handling@workcover.vic.gov.au
Website: wvvw.workcover.vic.gov.au

For specific guidance regarding workplace layouts and designs in your facility, you may need to engage an architect and/or ergonomist. These professionals can be found under "architects" and "ergonomists" or "occupational health and safety professionals" in the Yellow Pages. Ensure that any consultants engaged have appropriate qualifications and experience and provide contact details of clients that they have previously worked with.

Injured Nurse Story #15: A Nurse's Story

by Nurse in Pain

PART I

I'd like to tell you a story. Once there was a girl who had wanted to become a nurse ever since she was in the seventh grade. Growing up, she was very shy; but, as she got older, she came out of her shell somewhat. She wasn't the best student. In fact, she actually failed algebra and had to repeat it. One of the reasons she didn't turn in homework at times was her parents arguing.

When she applied for nursing school, her father refused to allow her to go there. She enrolled at the local hospital school instead. She was not able to stay in the dorm, so, she commuted on a daily basis, sometimes not making it because she had no transportation. Her parents again.....

The next year, her mother told her no more school unless she went to business school "to make a living." So, she enrolled and graduated with an AD in office sciences. Finally, she applied for University College of Nursing and was accepted. Again, as a day student, her attendance and grades were marginal.

After two years, she left and married, mainly to get out of the home environment. Fortunately, her spouse was, and still is, her best friend. They moved out of state following his army enlistment. At time of discharge, they moved back to the state.

Soon there were two children. So, the girl, needing to work, found employment with the local general practitioner. He was just like a "grandfather" to her children. This doctor encouraged her to return to nursing school, which she did. During school, she commuted 100 miles a day and worked three jobs. Twelve years after starting Nursing School, she graduated with honors.

Her first job as an RN was with the same doctor. She then became a hospital nurse and worked at the hospital for 22 years. She still commuted 100 miles a day. During those years, she was on many committees and reached the top of the clinical ladder. She earned national certification in her nursing specialty and was an officer in the local chapter of nurses within that nursing specialty. She worked for many years as Charge Nurse and was a preceptor for many new nurses.

Her children were both honor students and both have bachelor's degrees. They are both close to her and her husband.

Then, one day at work, she had a patient who required extensive orthopedic surgery. This patient was from out of town and, due to her financial status, was assigned to a semiprivate room. The nurses arranged with the hospital to allow the patient a private room so that her husband or her mom could stay with her.

Then, later that night, the nurse was told to move the patient immediately, in the middle of the night. That night was the last one the nurse ever worked. You see, she damaged her spine. She developed a severe neuropathy and now needs a cane or crutch to walk. She can no longer walk alone. She no longer drives due to lack of sensation in either foot and inability to lift her left leg at all.

So, she went from a highly paid RN at a major hospital to nothing. She lost her health insurance. Her RN license is still in effect; but, she will not be able to use it. She needed to refinance her home and her car in order to pay medical bills. Two doctors and a hospital turned her over to collection. She paid an attorney $33,000 to appeal her worker's compensation denial.

1-56670-631-9/03/$0.00+$1.50

It took 2 years to come to hearing. She also applied for Social Security Disability. That took 2 years, also, and she has yet to receive a monthly payment from SSA.

One day, she was sick and needed to go to the ER 50 miles away. A relative took her as she can no longer drive. The relative couldn't wait for her but told her if she could make it across town to the store where the relative was employed, she could take her home that evening after work.

In the ER, this nurse was made to wait. She was "triaged" as nonemergent, and, when she told them all she had was Medicare, she was told by the triage nurse, "Yeah, they all say that."

Four hours after arriving at the ER, she was escorted to a room by this same triage nurse who told the doctor, "She SAYS she hurt her back and cannot work."

The doctor said to her, "Well, what do you expect me to do for you, drugs?" The nurse said, "No, I have a UTI, it's Sunday, and I cannot find a doctor. I have had a fever since late Saturday and have been trying to get by until I could call someone Monday."

The doctor said, "Well, you could have taken Tylenol but I guess you want me to order that, too."

The nurse said, "No, all I would like is a prescription for an antibiotic."

The clerk said, "We don't file for Medicare when it is secondary insurance." The nurse then paid the ER bill using money she had borrowed from the relative. She went to the telephone to try to get a ride over to the relative's store.

At this time, a supervisor came by and stopped to talk. This supervisor had been a staff nurse on the unit where the nurse worked. The nurse had been his preceptor.

He told the nurse, "We can arrange a cab." The nurse said, "Thank you." They stood and talked for a while and then the supervisor promised to call her soon. This was the first human and "humane" contact she had had this entire day.

The cab went to the front of the hospital instead of the back ER entrance, so another was dispatched. The nurse recalled that she had never felt so low when she read the story of another back-injured nurse on an online forum.

PART II

When I was injured, I was moving an occupied bed with a 200+ lb. patient and an overhead traction setup. That's something we were required to do nearly every day, move patients to receive the next day's postoperative patients. In addition, each time a postoperative patient arrived in the second bed position, we needed to move the first bed aside. I was pushing; I had a nursing tech at the foot to guide the bed. I have done this perhaps 200 times over the years and have done it the same way each time.

I had 22 years experience with this facility, twenty of which were on Adult Orthopedics. I am a certified orthopaedic nurse (ONC). We had skills labs once a year to be checked off on lifting and pushing, etc. In fact, I have taught the back/lifting courses myself.

The number of lifts a day varied, usually from maybe two to about six a day. These were usually from bed to chair, maximum assist with two, with the potential of a patient being so weak that they would need to be helped to the floor. Then, of course, it was a total lift back to the bed or chair. This was with more help, though.

Postoperative patients were moved from the transport stretcher to the bed every time. We did team nursing and, as the team leader, I (the RN) was required to be there to receive. Each 10-patient team usually had three to five postoperative admissions each day. We handled 1800 surgeries, more or less, each year.

The only equipment available was one "smooth mover," a plastic sliding board, for a unit of 32 patients. We had, at one time, a Hoyer-type scale, but it was no longer on the unit.

We also were required to lift from, and place onto, beds the large type of continuous passive motion machines (CPM) for total knee surgeries. This would be possibly four to six times per patient. My load would be, with an LPN, 10 patients. Half of these would ordinarily be total hips, maybe three would be total knees, and the rest a mix of med/surg/trauma (fractures, etc.)

My injury was to the L3–L4 disc on the right and the L5–S1 disc on the left. There is an annular tear on the L5–S1 and a protrusion on the L3–L4. Following my injury, I have peripheral neuropathy in my left leg, with foot drop on the left. From my lower tibial area down, there is diminished sensation. The top and side of my left foot are numb. In addition, I have constant pressure across the sacral area. By choice, I do not take narcotic pain meds. I take ibuprofen for pain. I do take neurontin, baclofen, and amitriptyline for the neuropathy.

I have sciatica-type pain in both my hips. I have weakened reflexes in both knees and in the left ankle; my right ankle has good reflexes. I can't dorsiflex the left foot. I can't do a straight leg raise on the left. An assisted SLR causes severe pain in the right flank. In addition, I have, of course, the concomitant depression and anxiety.

I have been seen by a neurosurgeon, a neurologist, and a pain management specialist. I had a second opinion series at a back institute in another city. I have had four back procedures. In addition, I have had three epidural steroid injections and a discogram — all of which were quite painful and have resulted in no improvement. I walk with a cane. Being injured has impacted greatly on my physical relations with my husband.

I no longer drive due to diminished reflexes and lack of sensation in my foot. I do not grocery shop or do much shopping at all. My family does most of the household "chores" such as laundry and yard work. I used to enjoy computers and the internet. I now am limited to standing at the keyboard to type. This was not necessary before my injury.

The hospital has done NOTHING to support me. They fought me through the workers' compensation process and the nurse manager actually came to the hearing to refute my claim. She also required me to call in EVERY day until my lawyer put a stop to that. When workers' compensation accepted my claim immediately at the end of the hearing, the hospital entered a dispute and another hearing was scheduled for 6 months later. I have NEVER been contacted by the hospital, their lawyer, or anyone else other than my lawyer.

We refinanced my car, an old van, and took a second mortgage on the house, which had been paid off, just in order to pay bills.

I now receive Social Security disability. I went from making $55,000 a year to $13,000 on SSDI. This is not a permanent thing. They will review periodically and can revoke it at any time. Right now, I am rated 100% disabled by their physicians. (Since this writing, I have been approved for my retirement pension.)

I will never be able to work as a nurse again. For a while, I tried applying for various nonhospital positions such as doctors' offices, but have not even received a call back. I applied for a Quality Improvement position in the hospital I worked in, but the interviewer called attention to my cane and said I would probably not be physically able to handle calling people on the phone to obtain insurance preapproval. I knew I didn't get the job as soon as he said that. I considered an ADA suit against the hospital and still may file one.

We had had several nurses out with back injuries. I myself had just recovered from a torn rotator cuff related to a job injury. One nurse was out from a slip and fall, and another from lifting equipment (CPM). I believe I was the first to lose my job, though.

To someone entering nursing, my first inclination is, "NO, NO, NO! PLEASE, DON'T GO!" But, seriously, I believe my advice would be, "Look out for yourself, as no one else will."

13 Worker Control: The Best Means to Reduce Musculoskeletal Disorders (MSDs)

Bernadette Stringer

CONTENTS

The purpose of this chapter is to make what might seem an outrageous claim — that based on a reading of the scientific literature, one of the most cost-effective interventions to reduce musculoskeletal disorders among healthcare personnel and its resulting economic burden[1] may be a fundamental reorganization of the workplace to empower those who work in it. Further, this chapter will argue that this is a practical, realistic approach, advocated by a variety of established experts,[2,3] especially for public, not-for-profit institutions, and that this approach could provide ancillary benefits such as improved productivity, quality of patient care, and recruitment advantages.

Of course, this may sound especially radical given the present type of hierarchy and de-skilling that exists in U.S. hospitals, especially under managed care.

While on one level, this is a radical proposal, it will be further argued that in fact the changes that would be necessary to test this hypothesis are relatively modest for the majority of healthcare personnel. On the other hand, innovations in the exercise and control of management functions would mean a significant upheaval in the working lives of those who currently run hospitals and other healthcare institutions. The growing body of evidence to support the hypothesis that having lack of job control and exposure to other psychosocial risk factors leads to higher rates of work-related illness, including musculoskeletal disorders (MSDs) of the back and neck,[4] will be outlined. Theories to explain the underlying processes will be presented, followed by an argument

1-56670-631-9/03/$0.00+$1.50

about why this type of workplace reorganization, along with reduction of physical loads, is required in order to meaningfully reduce MSDs and how it could be achieved.

BACKGROUND

Since the 1960s studies have found a link between work organization factors, stress, and increased work related disease, especially coronary vascular disease (CVD),[5] and more recently, musculoskeletal disorders (MSDs).[6]

> Work organization factors:[7] the way work processes are structured and managed. For example, complexity of the job, workload, workplace hierarchy and culture such as level of support from managers and between coworkers, amount of overtime, inadequate staffing, job status, job security and reimbursement, and hours of work such as rotating shift work.
>
> Interventions: the new program or equipment that a research project wishes to evaluate. Examples are a re-design of the work processes to increase workers' decision making capacity, the introduction of additional staffing ratios or new support personnel and new work schedules.
>
> Outcomes: endpoints that are measured in a research project to assess if an intervention is having a desired effect. They could be at the level of the worker, the patient, or the hospital. Examples are a reduction in the number of number of sick days overall or per worker, employee turnover rates, patient satisfaction or compliance with treatment, as well as many others suggested by professional organizations.[8]

Work Organization Factors, Coronary Vascular Disease, and Musculoskeletal Disorders

Direct and indirect evidence of the health-related effects of poor psychosocial work environments has been described although the characteristics of environments are not defined in a standard manner.[9] While most studies reviewed for this chapter measured "demand," "decision latitude (job control)" and "social support," other characteristics such as those related to job status and downsizing, were also measured and linked to high levels of stress characterized as job dissatisfaction and burnout or poor health and diseases. A link between poor work environments (e.g., low job control, high demand) and back injuries (MSDs) specifically, has been reported in a number of studies[10–12] and recognized by the European Union's Agency for Health and Safety at Work and the U.S. National Institute of Occupational Safety Health (NIOSH).[13,14] Although studies in which a link was not found (negative studies) between poor work organization characteristic and MSDs, many more positive studies can be found. The following more detailed reviews of healthcare worker studies illustrates the type of evidence that exists on the link between psychosocial and physical load/biomechanical factors and MSDs.

A study in which 3769 hospital workers (43% RNs and nurses' aids) were followed prospectively to identify work organization factors associated with an increased risk of sustaining compensable (workers' compensation) musculoskeletal injuries, found risk associated with demand and job control equal or slightly greater than risk associated with patient lifting.[15] After accounting for workers' individual factors, risk of sustaining *upper-body* injuries was four times greater for workers with low job control (decision latitude), almost three times greater for workers with low supervisor support, and over two times greater if working during high sickness absence (high workload) periods. Workers reporting high control, high support, and low sickness absence rates were the comparison groups. The risk of *lower-body* injuries was less than for upper-body injuries; over one and one half times greater for workers with low job control and over twice as great if working during periods of high sick time. Workers in medium and high level repetitive arm and awkward shoulder posture groups, had one and one half and two times the risk of sustaining upper body injuries and those in medium and high patient lifting/manual handling categories had two to four times the risk of sustaining lower-body injuries. While these risks were somewhat higher than expected, they were consistent with previous findings.[16,17] It is noteworthy that an interaction effect between physical and psychosocial factors was not found in this study.

In another prospective study that followed healthcare workers from an institution that was downsizing,[18] symptoms of distress such as emotional exhaustion and time off with neck and back pain increased over time. Researchers reported that the psychological demands and hours of work, as well as work's interference with family, were the most important predictors of psychological distress and musculoskeletal pain, not individual, personal factors such as willingness to accept change. Unlike the previous study, the aim of this study was to identify aspects of work during downsizing, which could influence employee health.

Studies employing a variety of epidemiological designs, have also shown two to three times more coronary heart disease (CHD) in workers reporting high demand/low control work [19,20–22] and in high cost/low gain environments.[23,24] Sickness absence has also been linked to poor work climates in different industrial sectors,[25–27] including healthcare. A large, recent U.S. study of over 21,000 nurses that not only found poorer health at baseline to be associated with high demand, low control, and social support, but nurses experiencing all three had the greatest health declines in physical health four years later.[28] This may help explain the high sickness absence of 7.2% among full time, Canadian healthcare workers, especially nurses, when compared to 4.8% for all other workers in 2000.[29,30] A link between mental illness and simultaneously working in poor quality psychosocial environments has been established to some extent[31,32] including in healthcare. A 1992 case-control study of Québec nurses' insurance claims (certified sick time of more than 3 days) found that nurses working in high demand climates had about four times higher rates of sickness absence for all reasons.[33] A 1998 study, based on another cohort of Québec nurses, first reported an association between burnout and working in high demand/low decision latitude environments, then, after following the cohort for 20 months, reported an association between low social support and certified leave for all reasons, especially mental illness.[34,35] Although nurses appear to have an increased risk of suicide when compared to other occupational groups,[36,37] further study of the U.S. nurses' cohort, found that nurses reporting a stressful work environment, over time, had a significantly increased risk of suicide.[38] Poor environments are linked to a variety of other outcomes in workers and patients, as well. Outcomes such as: increased blood pressure[39] and blood serum cholesterol levels,[40] being less physically active,[41–43] smoking more,[44,45] increased drug use,[46,47] other types of damaging behavior,[48] elevated levels of burnout,[49,50] job dissatisfaction,[51,53] and patient satisfaction/dissatisfaction,[53,54] and patient death.[55,56] The association between nurse job dissatisfaction and high turnover[57] has been confirmed again recently in a study that included nurses from the U.S., Canada, England, Scotland, and Germany. A separate analysis of U.S. nurse and patient data reported that nurses' job dissatisfaction, four times that of other workers,[58] was associated with high "intention to leave" rates and that each additional patient per nurse, produced a 23% increase in nurse burnout.[59]

REASONS

Epidemiological evidence suggests a link between lack of control over one's work environment (alone or in combination with other factors) and an increased risk of MSDs and other ill-effects although the reasons why are not yet completely understood. Possible explanations are available from everyday experience, from cognitive scientists[60] and other experts who have offered explanatory models.[61,62] This section will offer possible explanations of why work organization factors appear, independent of physical factors, to cause MSDs, as well as, cause heart disease. Two models have been outlined by other researchers while two are my own speculations, which incorporate additional, more recent research on the social interactions and responses to stress of females.

Stress, Stressors, and Strain

The human body possesses "stress" responses to certain stimuli that evolved over tens of thousands of years of our ancestors surviving often harsh and dangerous situations.[63] The primary purpose of this stress response was to overcome, deal with, or escape danger. Physiological processes take place inside our bodies to give us the extra energy, the stamina, calmness, the powers of concentration, and other forces needed to cope with the situation.

The most commonly acknowledged stress response is usually referred to a "fight or flight" but another "female" response has recently been posited. It has been named "tend and befriend."

It has long been recognized that problems arise when the "fight or flight" stress response becomes chronic.[64,65] There seems to be a price to pay for a too-often-on response also termed a strained response. Common everyday experience has long produced a language to express what has taken place inside an over-stressed body. Who among healthcare personnel have not heard or spoken phrases like: "I feel completely worn out," or "My body is aching," or "When I get home I just need to vegetate for a couple of hours in front of the TV." A *stressor* is defined as "an environmental demand or threat that taxes or exceeds a person's ability to meet the challenge," while a *strain* is defined as "a person's response to a stressor in psychological and physiological terms."[66]

Nursing Statistics

As the stress models are discussed it is important to keep in mind the demographics of the healthcare workforce; factors such as composition, age,[67] and gender. As well, that while unionization in the U.S. healthcare sector is approximately 20%, in Canada and most of Europe, it is over 85%.[68] These statistics have implications for all attempts at work reorganization. To illustrate their importance, numbers describing the population, primarily RNs, are provided. Most arise from the U.S. National Sample Survey of Registered Nurses (NSSRN)[69] conducted every four years and annual estimates by Statistics Canada's.[70]

In 2000, nursing including RNs, licensed practical nurses (LPNs), and registered psychiatric nurses (RPNs) were 35% of all healthcare workers in Canada[71] and in the U.S. in 1998, RNs and LPNs made up 21% of all healthcare workers (RPN data not available).[72] In 2000, there were 2,201,439 U.S. RNs (population 300 million) and in 2001, there were 231,512 Canadian RNs (population 30 million). On average, 95% of RNs were women aged 45 and 44 years, although 12.1% of RNs were under 30 in the U.S. and 39% were under 30, in Canada. About 59% of RNs in the U.S. and 64% in Canada, worked in hospitals and 59% and 54%, worked full time. Of all practicing U.S. RNs, 25% had diplomas while in Canada, 74% practiced with diplomas alone. Of the remainder, most had undergraduate university degrees.[73,74]

Stress Models

Cognitive psychology has argued that the "mind" is modular with many parts that cooperate or even compete to make sense of and deal with the environment around us. Perhaps in the interplay of mental systems lies the answer to how a sense of "control" can mediate (moderate) the effects of stress. Say, for example, a nurse on a busy emergency ward faces many demands that her "stress response" system reacts to by producing the well-known physiological fight-or-flight effects. Unless another of her mental systems overrides or mitigates the effects produced by that stress response system these physiological fight-or-flight effects will continue to be generated until the stressor(s) is removed.[75] On a Friday emergency ward night shift this will not likely happen until she leaves the hospital. But say there is another mental system that is capable of overriding the stress response system. Cognitive psychologists have described something called the "supervisory attention system," which can do precisely that. Let us hypothesize that a feeling of control causes the supervisory attention system to send instructions to the stress response system that says something like "don't worry too much, I am in control, this is my territory, and if things get too bad, I can always bail out." This causes a "settling down" of the fight or flight response. But our nurse, working in a contemporary environment, is more likely to feel that she has no control over her work environment, that someone else makes all the important decisions, that work just keeps on piling on and there is no end in site, except the end of her shift. Even that is not guaranteed if she has been asked (or is forced) to work overtime (a 2002 business report estimated that Canadian RNs worked a quarter million hours of overtime each week, approximately the equivalent of 7000 full-time jobs).[76] Plus, she desperately needs her paycheck, which she believes doesn't adequately reimburse her anyway, since she is the family's primary wage earner and supports two children. What instructions will her supervisory attention system likely send in such circumstances? It is a message of hopelessness or despair, exacerbating (aggravating) the stress response, resulting in strain.

This process of "control mitigating stress" can also be illustrated by the everyday experience of those of us lucky enough to have contact with the rural world. Work on family farms can be tough and physically and mentally demanding, with frequent long hours of work to get the crop in and yet when farmers talk about stress what is the typical subject? The price they will be paid for their produce, the weather, pests, etc. What do these have in common with the nurse scenario? All are things the farmers have no control over. The hard work is typically not what farmers feel is stressful, but rather the things that are beyond their control.

Karasek and Theorell have proposed and tested the job strain or high demand/low control, theoretical model, which is based on work organization factors, to explain occupational stress.[77] (see Definitions Box.) Through their evaluations and those of many others, they have established the damaging effects on workers' health, although earlier research focus on male dominated work primarily, of a lack of job control or limited autonomy and discretion (decision latitude) in the application of skills, while simultaneously being required to work hard, physically and/or mentally and fast, with deadlines, for long periods. They state that, "a lack of control over how to meet the job's demands and how one can use one's skills defines a state of arousal that inhibits learning; strain-induced inhibition of learning, in turn, further increases arousal by impairing confidence and self-esteem."[78] The idea is that if workers have the power to decide how to best tackle work related problems they will have the self confidence to continue to tackle problems until they can resolve them since they will be in an atmosphere that allows them to learn from each attempt. Exploring the difference between North American physicians' and nurses' levels of autonomy helps illuminate this. Physicians work hard, often meeting challenges and life and death deadlines, but they do not suffer the same high levels of stress-induced disease as nurses or nurses' aids. This model posits that this is so because physicians have much higher levels of autonomy to apply and further develop their skills.

Another theoretical model, which is complimentary to the high demand/low control model is Siegrist et al. effort-reward imbalance or cost/gain model. It adds a labor market dimension.[79] The effects of broader social impacts such as inadequate remuneration, little job security, or poor prospects for promotion while having a job requiring a large effort (high demand) also results in disease. Although less well evaluated than the high demand/low control model, there is growing evidence that these work-related factors result in two to three times as much coronary heart disease. While job insecurity for example, might be expected to lead to workers disengaging and not working to an acceptable quality and level, this depends on the worker's opportunities. If workers' ability to change jobs is limited, the effort reward imbalance will be sustained over a long time and "high cost/low gain conditions are likely to elicit recurrent feelings of threat, anger, and depression and demoralization, which in turn evoke sustained autonomic arousal."[80]

I would also suggest that the previous models are enhanced by Taylor et al. "tend and befriend" stress model.[81] Although reports on this model have focussed on the reasons why women live longer, it is based on biological evidence demonstrating that, through evolution, men and women have been programmed to respond very differently to stressors.[82] Women, because they were the ones to become pregnant, give birth, breastfeed, and nurture children, were often not able to "fight" or take "flight." Because of this, and other distinctive biocultural traits, women tended to cooperate in order to survive.

Authoritarian, almost militaristic management structures are descended from the ways in which males hunted or conducted battle. Instead of the more relationship-based, let's-talk-about-it, more cooperative structures that made sense to females, hierarchies in hospitals could be considered culturally "male." Although these structures have been considered counterintuitive to many nurses and other healthcare workers who are primarily women, perhaps the recent technological changes and downsizing have generated even more strain.

Given that the vast majority of front line workers in healthcare are women, perhaps something in the organization of work is fundamentally in conflict with us. There is anecdotal evidence in a Canadian government report that appears to support this,[83] that nurses feel the new "professional management" running hospitals has led to increased alienation. Nurse managers talk about wanting to return to "the old days" when supervisors were mentors or colleagues and came up through the ranks. Perhaps another way of explaining this desire is that women are "hardwired" and culturally attuned to prefer "organic hierarchies" in which status is acquired more democratically, from group interactions.

A BRAVE NEW WORLD

Of course, even if none of the above explanations proves sufficient (or if they are all partly true) the fact remains there seems to be a link between MSDs and the characteristics of work organization, more specifically, low control/high demand work and a lack of social support. The next section will explore how control might be increased and what a healthcare work environment, which would inspire feelings of control, could look like.

Prior to proceeding though, results from a study on factors consistently associated with low injury rates and another on the health impact of lack of input into institutional level decision-making, should also be highlighted. These studies further substantiate the need for empowerment to be at the individual, job control level, and embedded in the organization as a whole.

"Empowerment of the workforce" was the factor that emerged first from a comprehensive literature review to identify factors consistently linked to lower injury rates across industries. In this review, conducted by Dr. Harry Shannon and colleagues from McMaster University's Occupational Health Program, delegation of safety activities and an active role in health and safety of top management, were also reported to be important factors associated with consistently lower injury rates.[84]

In another study, consisting of a survey to almost 4500 Finnish healthcare providers from seven hospitals throughout the country, employees response scores to questions on "procedural" and "relational" justice was linked with medical certificates for sickness absences lasting three days or more, provided by employers. The following are examples of questions used to assess relational and procedural justice: Your supervisor considered your viewpoint, your supervisor was able to suppress personal biases, your supervisor took steps to deal with you in a truthful manner, procedures (in your hospital) are designed to collect accurate information necessary for making decisions, and procedures are designed to provide opportunities to appeal or challenge the decision. Workers with low "relational justice" and "procedural justice" scores were up to twice as likely to have medically certified sick leave, compared to workers with higher scores, even after adjusting for individual levels of job control, workload, and support from coworkers. The health effect was independent.

CAN HEALTHCARE WORKERS TAKE CONTROL?

This is a question that nurses and other healthcare workers are unlikely to ask since the answer seems obvious. Nurses are responsible for the patients found throughout facilities, they run wards on all shifts although they rely on the advice of many others on days and to some extent, evening shifts. During the night shift, in many institutions, front line nurses rotate the role of charge nurse, and primarily rely on each other and aids and orderlies to provide the care that keep patients alive and recuperating. As well, nurses and healthcare workers have successfully developed and maintained effective essential services 24 hours a day over weeks and months during a growing number of strikes. It could be argued that unionized nurses and other healthcare workers are more likely to have the organizational experience necessary to carry out successful job control interventions but not necessarily. As highlighted in a recent Canadian government document, unions and management should be actively involved in work reorganization discussions.[85] It would be preferable if negotiated collective agreements supported interventions and their evaluation through legally enforceable clauses outlining the employer's commitment to worker's job control and the conditions of work that will result.

What Is Control?

Farmers feel they are in control because they are within certain economic boundaries. They choose how to organize their work (based on skill and experience), what hours they will be on the job, when they will take breaks, and nobody orders them about (except for Mother Nature and sometimes the bank).

What sort of healthcare work environment would inspire feelings of control among nurses, technicians, orderlies, and other personnel? Perhaps the simplest answer to that would be one in which they *were* in control. But, one could argue, hospitals and other healthcare facilities require large groups of skilled workers to perform a wide variety of complex activities that must be coordinated and regulated. This requires "management" and therefore ward nurses or orderlies or lab technicians could never be "in control" in such a place. Therefore, they will inevitably feel a lack of control.

"Self-management" and "workers' control" has existed in some very large enterprises. The Mondragon cooperative with approximately 80,000 employees in the Basque region of Spain, where workers both own and manage the company, is one example,[86] but the business literature has other examples of "flattened" management structures to give control to the "shop floor." In the high technology industry for example, significant gains in productivity have often been the result. Interested readers can find a vast literature on the subject.

How might a hospital, or better, a hospital district transform itself into an organization in which its work force, nurses, aids, technicians, and orderlies performed in a work environment likely to reduce MSDs and provide productivity benefits and improved patient care? Would it require a radical makeover? In certain key areas, mainly those that are described as "management" significant change would be required, but in many respects, this new organization would probably not be much different than the "average" hospital that exists today. That is, most hospital personnel would do jobs almost identical to their current work, although their decision-making capacity would be fundamentally altered.

Let us return to the three characteristics of work that have been identified most often with MSDs and CVDs: low decision latitude, high demand, and low social support. No fundamental change is possible if "society" (and its management) is not convinced that a radical alteration of the social hierarchy is required; therefore, I would argue that societal support is the first step and I would argue there is evidence that it exists. In Canada, nurses top the list of the most trusted professionals in nearly every public survey.[87] With social support at the macro level, a regulating of job demands and support at the micro level is much more likely. Does this mean that all front line nurses are in charge all of the time? Of course not.

As mentioned above, perhaps the best way to think about this is the difference between "organic hierarchies" and ones imposed from without. This means that supervisors and managers must come from hospital wards. Perhaps it means they are chosen (or at least nominated) by their coworkers.

To illustrate how this might work, let us take a look at a health district in the province of Saskatchewan — where the provincial government first introduced Canada's universal system of healthcare after a bitter doctor's strike — although most of the remarks could also apply to a group of U.S. hospitals in the not-for-profit or public sector. To transform itself, the board of the health district would likely begin by inviting representatives of its workers (unions and professional associations) to discuss how the board could be restructured to give significant representation to workers. Let us say that the restructuring gave half the seats on the board to various representatives of the organization's workers. That board would then set up a committee to research various models of "self-management" and other ways of giving "control" to the people on the wards or laboratories or operating rooms. The next step would be to report the findings of that committee to the board and to the workers in order to educate those would be asked to participate in the design of the new system.

Perhaps various experiments within the facilities of the health district could be conducted to test what works. One intervention project could involve hospital nurses electing their head of nursing or an equivalent position. Another might see a hospital try a board of management that included representatives from each area of work as well as other stakeholders. What would be critical is a real sense of input and power by all those who work in every facility. Management structures would be changed to administrative structures. The goal would be to give "management" to everyone while maintaining a system of efficient performance of administrative tasks. This may sound vague but there are, in fact, many actually existing models to experiment with and choose from.

An important administrative task to strengthen would be the system of measuring patient outcomes and especially their link to on-the-job satisfaction, social support, workers' perceptions of job control,[88,89] and other work-related stressors such as violence.[90] Ultimately, better patient outcomes consisting of both objective measures such as morbidity and mortality and subjective measures such satisfaction with care or reported compliance with treatment, must be the litmus test of how healthcare work is organized.[91,92] The concluding statement of this section must be that by reducing musculoskeletal disease through giving real control of the workplace to people who work there patients will also benefit. Happier, healthier healthcare workers mean better care for our patients.

Practicalities

In 1990, Karasek et al.[21] estimated that U.S. companies were assuming occupational stress related costs from absenteeism, reduced productivity, compensation claims, health insurance, and direct medical expenses of about 150 billion dollars/year. In 1998 Goetzel and colleagues reported that depressed and stressed employees of six U.S. companies had 70% and 46% higher health expenditures.[93] In 1999, based on data from two national surveys, reported that workers with 30-day major depression had 1.5 and 3.2 more short-term disability days than other workers and that these days translated into a salary equivalent productivity loss of $182 and $395, respectively.[94] In 2002, Aiken et al.[56] reported that the training costs of every newly hired RN for medical/surgical wards were $42,000 (check) and $64,000 for RNs in critical care. The U.S. Advisory Board estimates the savings to a 500 bed hospital of reducing nurse turnover by 3% to be about $800,000 per year. But costs to the healthcare system are not just monetary since our product is patient care. A Canadian government report estimated that turnover and the number of part time employees (RNs are unable to find permanent jobs, even in 2002!) resulted in the average patient hospitalized for three days seeing over 80 different people.

So experiments in giving healthcare workers greater control might be highly practical in that potential savings in both money and quality of care are great.

INTERVENTION RESEARCH

Experts agree that we now have sufficient information to carry out intervention studies[95–97] implementing studies using multifaceted approaches.[98] While studies of interventions aimed at reducing the physical load through mechanical equipment, for example, have not been shown to be as effective as hoped,[99] combined interventions consisting of both physical/biomechanical and psychosocial components, are likely to be more effective.[100,101] Many of the studies investigating the effects of work redesign have been poorly designed.[102,–104] This is in part because identifying, measuring and analyzing individual (workers and patients) and organizational level factors, increases the complexity of this research.[105] It is therefore important that future investigations include appropriate evaluation programs so they can measure effects, if they do or do not, exist.[106,107] Just as nurses, doctors, and others in healthcare support the need to provide "evidence based" patient care,[108,109] the study of altered working conditions should contribute to the body of evidence needed to improve healthcare working environments internationally.[110] In addition to requiring rigorous methods, the research relationship between healthcare workers and researchers should incorporate the principles of social medicine, which emphasizes the importance of the whole process of transformation of working conditions.[111]

CONCLUSION

It would take a daring administration to attempt a workplace reorganization of the sort that I am proposing, but the benefits are potentially immense. While the courage required to undertake such a rethinking of management would be immense, it should be pointed out once more that, in fact, the actual ward level changes in work organization may not be revolutionary at all. The central premise of this chapter is, after all, modest.

Healthcare workers, their organizations and patients would benefit if management became more "democratic," giving more control to the actual people doing the actual work.

REFERENCES

1. Crook, J., Milner, R., Schultz, I.Z., and Stringer, B., Determinants of occupational disability following a low back injury: a critical review of the literature, *J. Occup. Rehab.*, 12, 277–295, 2002.
2. Workplace wellness must go further (Health Leadership C11), *Globe and Mail*, November 16, 2002.
3. Gardell, B., Worker participation and autonomy: a multilevel approach to democracy at the workplace, *Int. J. Health Serv.*, 12(4), 527–558, 1982.
4. Bernard, B.P., Ed., Musculoskeletal disorders and workplace factors: A critical review of epidemiologic evidence for work-related musculoskeletal disorders of the neck, upper extremity, and low back. National Institute of Occupational Health DHHS (NIOSH) (Publication No. 97–141), U.S. Department of Health and Human Services: Cincinnati, Ohio, 1997. Available at: http://www.cdc.gov/niosh/97–141pd.html
5. Berkman, L.F., Psychological distress: a matter of hearts and minds (Commentary), *Int. J. Epidemiol.*, 31, 256, 2002.
6. Viikari-Juntara, E. and Riihimaki, H., New avenues of research on musculoskeletal disorders, *Scand. J. Work Environ Health*, 25, 564–568, 1999.
7. Organization of Work. The National Occupational Research Agenda (NORA), Available at: www.cdc.gov/niosh/nrworg.html
8. American Nurses Association, *Quality Indicators, Definitions and Implications,* American Nurses Publishing, Washington, D.C., 1996.
9. Williams, S., Cooper, G.L., Measuring occupational stress: development of the pressure management indicator, *J. Occup. Health Psych.,* 3, 306–321, 1998.
10. Bongers, P., Kremer, A.M., ter Laak, J., Are psychosocial factors risk factors for symptoms and signs of the shoulder, elbow or hand/wrist?: a review of the epidemiological literature, *Am. J. Ind. Med.*, 41, 315–342, 2002.
11. Devereux, J.J., Vlachonikolis, I.G., and Buckle, P.W., Epidemiological study to investigate potential interaction between physical and psychological factors at work that may increase the risk of symptoms of musculoskeletal disorder of the neck and upper limb, *Occup. Environ. Med.,* 59, 269–277, 2002.

12. Bongers, P.M., de Winter, C.R., Kompier, M.A.J., and Hilderbrandt, V.H., Psychosocial factors at work and musculoskeletal disease: A review of the literature, *Scand. J. Work Environ. Health*, 19, 297–312, 1993.
13. De Beeck, R.O. and Hermans, V., Work-related low back disorders, European Agency for Safety and Health at Work, 2000, Available at: http://agency.osha.eu.int/publications/reports/204/en/lowback.pdf
14. Bernard, B.P., Ed., Musculoskeletal disorders and workplace factors: A critical review of epidemiologic evidence for work-related musculoskeletal disorders of the neck, upper extremity, and low back.
15. Koehoorn, M., Musculoskeletal disorders among health care workers: individual, biomechanical and work organizational factors, PhD thesis, Vancouver (BC) University of British Columbia, 1999.
16. Josephson, M., Lagerstrom, M., Hagberg, M., and Wigaeus-Helm, E., Musculoskeletal disorders and job strain among nursing personnel: a study over three year period, *Occup. Environ. Med.,* 54, 681–685, 1997.
17. Lartese, F. and Fiorito, A., Musculoskeletal disorders in hospital nurses: a comparison between two hospitals, *Ergonomics*, 37, 1205–1211, 1994.
18. Shannon, H.S. et al., Changes in general health and musculoskeletal outcomes in the workforce of a hospital undergoing rapid change: a longitudinal study, *J. Occup. Health Psych.*, 6, 3–14, 2001.
19. Marmot, M., Bosma, H., Hemingway, J., Brunner, E., and Stansfeld, S.. Contribution of job control and other risk factors to variations in coronory heart disease incidence, *Lancet,* 350, 235–239, 1997.
20. Theorell, T., Tsutsumi, A., and Hallquist, J., The Sheep Study Group. Decision latitude, job strain, and myocardial infarction: a study of working men in Stockholm, *Am. J. Public Health*, 88, 328–328, 1998.
21. Karasek, R., Theorell, T., Schwartz, J.E., Schnall, P.L., Pieper, C.F., and Michela, J.L., Job characterisitcs in relation to the prevalence of myocardial infarction in the US Health Examination Survey (HES) and the Health and Nutrition Examination Survey (HANES), *Am. J. Public Health*, 78, 910–918, 1988.
22. Karasek, R., Baker, D., Marxer, F., Ahlbom, A., and Theorell, T.. Job decision latitude, job demands, and cardiovascular disease: a prospective study of Swedish men, *Am. J. Public Health*, 71, 694–705, 1981.
23. Siegrist, J., Adverse health effects of high/low reward conditions, *J. Occup. Health Psych.,* 1, 27–41, 1996.
24. Peter, R., Siegrist, J., Hallqvist, J., Reuterwall, C., Theorell, T., and the SHEEP Study Group, Psychosocial work environment and myocardial infarction: improving risk estimatation by combining two complementary job stress models in the SHEEP study, *J. Epidemiol. Comm. Med.* 56, 294–300, 2002.
25. Ala-Mursula, L., Vahtera, J., Kivimaki, M., Kevin, M.V., and Pentti, J., Employee control over work times: associations with subjective health absences, *J. Epidemiol. Comm. Health*, 56, 272–278, 2002.
26. Kivimaki, M., Vahtera, J., Pentti, J., and Ferrie, J.E., Factors underlying the effect of organisational downsizing on the health of employees: longitudinal cohort study, *BMJ,* 320, 971–975, 2000.
27. Vahtera, J., Kivimaki, M., Pentti, J., and Theorell, T., Effect of change in psychosocial work environment on sickness absence: a seven year follow up of initially healthy employees, *J. Epidemiol. Comm. Health*, 54, 484–493, 2000.
28. Cheng, Y., Awachi, I., Coakley, E.H., Schwartz, J., and Colditz, G., Association between pshychosocial work characteristics and health functioning in American women, *BMJ*, 320, 1432–1436, 2000.
29. Baumann, A. et al., Commitment and care: the benefits of a healthy workplace for nurses, their patients and the system, Ottawa: Canadian Health Services Research Foundation. 2001. Available at: http//: www.changefoundation.ca
30. Akyeampong, E.B., Fact-sheet on absenteeism, *Perspectives on Labour and Income*, 13, 46–53, 2001.
31. Stansfeld, S.A., Fuhrer, R., Shipley, M.J., and Marmot, M.G., Work characteristics predict psychiatric disorder: prospective results from the Whitehall II Study, *Occup. Environ. Med.*, 56, 302–307, 1999.
32. Karasek, R, and Theorell, T., *Healthy Work: Stress, Productivity and the Reconstruction of Working Life*, Basic Books Publishers, New York, 1990.
33. Bourbonnais, R., Vinet, A., Vezina, M., and Gingras, S., Certified sick leave as a non-specific morbidity indicator: a case-referent study among nurses, *Br. J. Ind. Med.*, 49, 673–678, 1992.
34. Bourbonnais, R. and Mondor, M., Job strain and sickness absence in Québec, *Am. J. Ind. Med.*, 39, 194–202, 2001.

35. Bourbonnais, R., Comeau, M., and Vézina, M., Job strain and evolution of mental health among nurses, *J. Occup. Health Psych.*, 4, 95–107, 1999.
36. Hawton, K. and Vislisel, L.. Suicide in nurses (Review), *Suicide Life Threatening Behavior*, 29, 86–95, 1999.
37. Hawton, K. et al., Suicide in female nurses in England and Wales, *Psychom. Med.*, 32, 239–250, 2002.
38. Feskanich, D. et al., Stress and suicide in the Nurses' Health Study, *J. Epidemiol. Comm. Health*, 56, 95–98, 2002.
39. Theorell, T., Alhlberg-Hulten, G., and Jodko, M., Influence of job strain and emotion on blood pressure in female hospital personnel during work hours, *Scand. J. Work Environ. Health,* 19, 313–318, 1993.
40. Kivimaki, M., Leino-Arjas, P., Luukkonen, R., Riihimaki, H., Vahtera, J., and Kirjonen, J., Work stress and risk of cardiovascular mortality: prospective cohort study of industrial employees, *BMJ*, 325, 857–860, 2002.
41. Payne, N., Jones, F., and Harris, P., The impact of working life on health behavior: the effect of job strain on cognitive predictors of exercise, *J. Occup. Health Psych.*, 7, 342–353, 2002.
42. Brisson, C., Larocque, B., Moisan, J., Vezina, M., and Dagenais, G., Psychosocial factors at work, smoking, sedentary behavior, body mass index: A prevalence study among 6995 white collar workers, *J. Occup. Environ. Med.*, 42, 40–48, 2000.
43. Johansson, G., Johnson, J.V., and Hall, E.M., Smoking and sedentary behavior as related to work organization, *Soc. Sci. Med.*, 32, 837–846, 1991.
44. Hellerstedt, W.L. and Jeffery, R.W., The association of job strain and health behaviors in men and women, *Int. J. Epidemiol.*, 26, 575–583, 1997.
45. Landbergis, P.A., Schnall, P.L., Deitz, D.K., Warren, K., Pickering, T.G., and Schwartz, J.E., Job strain and health behaviors: results of a prospective study, *Am. J. Health Promotion*, 12, 237–245, 1998.
46. Schnall, P.L., Schwartz, J.E., Landbergis, P.A., Deitz, D.K., Warren, K., and Pickering, T.G., The relationship between job strain, alcohol and ambulatory blood pressure, *Hypertension*, 19, 488–494, 1992.
47. Muntaner, C., Anthony, J.C., Crum, R.M., and Eaton, W.W., Psychosocial dimensions of work and the risk of drug dependence among adults, *Am. J. Epidemiol.*, 142, 183–190, 1995.
48. Storr, C.L., Trinkoff, A.M., and Anthony, J.C., Job strain and non-medical drug use, *Drug Alcohol Depend.,* 55, 45–51, 1999.
49. Woodward, C.A. et al., The impact of re-engineering and other cost reduction strategies on the staff of a large teaching hospital, *Med. Care*, 37, 556–569, 1999.
50. Jamal, M. and Vishwanath, B.V., Job stress and burnout among Canadian managers and nurses: an empirical examination, *Can. J. Pub. Health*, 91, 454–458, 2000.
51. De Jonge, J., Bosma, H., Peter, R., and Siegrist, J., Job strain, effort-reward imbalance and employee well-being: a large-scale cross-sectional study, *Soc. Sci. Med.*, 50, 1317–1327, 2000.
52. Aiken, L.H. et al., Nurses' reports on hospital care in five countries, *Health Aff. (Millwood),* 20, 43–53, 2001.
53. Aiken, L.H., Sloan, D.M., and Lake, E., Satisfaction with inpatient AIDS care: a national comparison of dedicated units and scattered beds, *Med. Care,* 35, 948–962, 1997.
54. Arnetz, J.E. and Arnetz, B.B., Violence towards health care staff and possible effects on the quality of patient care, *Soc. Sci. Med.*, 52, 417–427, 2001.
55. Aiken, L.H., Smith, H.L., and Lake, E.T., Lower medicare mortality among a set of hospitals known for good nursing care, *Med. Care,* 32, 771–787, 1994.
56. Aiken, L.H., Clarke, S.P., Sloan, D.M., Sochalski, J., and Silber, J.H., Hospital nurse staffing and patient mortality, nurse burnout, and job dissatisfaction, *JAMA*, 288, 1987–1993, 2002.
57. Irvine, D. and Evans, M., Job satisfaction and nursing turnover among nurses: integrating research findings across studies, *Nursing Res.*, 44, 246–253, 1995.
58. Aiken, L.H. et al., Nurses' reports on hospital care in five countries, *Health Aff. (Millwood)*, 20, 43–53, 2001.
59. Aiken, L.H., Clarke, S.P., Sloan, D.M., Sochalski, J., and Silber, J.H., Hospital nurse staffing and patient mortality, nurse burnout, and job dissatisfaction,
60. Pinker, S., *How the Mind Works*, Norton Publishers, New York, 1997.
61. Karasek, R. and Theorell, T., Healthy work: stress, productivity and the reconstruction of working life.

62. Siegrist, J., Adverse health effects of high/low reward conditions, *J. Occup. Health Psych.,* 1, 27–41, 1996.
63. Selye, H., *Stress without Distress*, JB Lipincott, New York, 1974.
64. Brunner, E., Socioeconomic determinants of health. Stress and the biology of inequality, *BMJ*, 314, 1472–1476, 1997.
65. Theorell, T., How will future worklife influence health? *Scand. J. Work Environ. Health*, 23(suppl 4), 16–22, 1997.
66. Peter, R. and Siegrist, J., Chronic psychosocial stress at work and cardiovascular disease: the role of effort-reward imbalance, *Int. J. Law Psych.*, 22, 441–449, 1999.
67. Buerhaust, P., Staiger, D.O., and Auerbach, D.I., Implications of an aging registered nurse workforce, *JAMA*, 283, 2948–2954, 2000.
68. Baker, M. and Fortin, N., The gender composition and wages: why is Canada different from the United States? Statistics Canada, Analytical Studies Branch, Ottawa, 2000. Available at: http://www.statcan.ca/english/research/11F0019MIE/11F0019MIE2000140.pdf.
69. Spratley, E., Johnson, A., Sochalski, J., Fritz, M., and Spencer, W., The registered nurse population, 1980–2000. Findings from the National Sample Survey of Registered Nurses, U.S. Department of Health and Human Services, Division of Nursing, Washington D.C., 2000. Available at: ftp://ftp.hrsa.gov/bhpr/rnsurvey2000/rnsurvey00–1.pdf
70. Statistics Canada, CANSIMII, TABLE 051–0001. Accessed at: http://www.can-nurses.ca/_frames/resources/statsframe.htm.
71. Canadian Institute for Health Information. Canada's Health Care Providers. Ottawa: CIHI. 2001. Available at: http://www.cihi.ca
72. U.S. Bureau of Labor Statistics, 1999. Available at: http://www.bls.gov/oes/1999/oes_29he.htm#(3)
73. Spratley, E., Johnson, A., Sochalski, J., Fritz, M., and Spencer, W., The registered nurse population, 1980–2000. Findings from the National Sample Survey of Registered Nurses.
74. Statistics Canada, CANSIMII, TABLE 051–0001.
75. Pinker, S., *How the Mind Works*, 1997.
76. Wortsman, A. and Lockhead, C., Full-time equivalents and financial costs associated with absenteeism, overtime, and involuntary part-time employment in the nursing profession. Report commissioned for the Canadian Nursing Advisory Committee, Ottawa, 2002.
77. Karasek, R. and Theorell, T., Healthy work: stress, productivity and the reconstruction of working life.
78. Karasek, R. and Theorell, T., Healthy work: Stress, productivity and the reconstruction of working life.
79. Siegrist, J., Adverse health effects of high/low reward conditions, *J. Occup. Health Psych.,* 1, 27–41, 1996.
80. Siegrist, J., Adverse health effects of high/low reward conditions, *J. Occup. Health Psych.,* 1, 27–41, 1996.
81. Taylor, S.E., Cousino Klein, L., Lewis, B.P., Gruenewald, T.L., Regan, G.A.R., and Updegraff, J.A., Biobehavioral responses to stress in females: tend-and-befriend, not fight-or-flight, *Psychol. Rev.*, 107, 411–429, 2000.
82. Taylor, S.E., Dickerson, S.S., and Cousino Klein, L., Toward a biology of social support, In Snyder, C.R., Lopez, S.J. Eds., *Handbook of Positive Psychology.* Oxford University Press, London, 2002, Chap. 40.
83. Canadian Nursing Advisory Committee, Our health, our future: creating quality workplaces for Canadian nurses. Final report submitted to Health Canada, Ottawa, 2002. Available at: http://www.hc-sc.gc.ca/english/for you/nursing/cnac report/index.html.
84. Shannon, H.S., Mayr, J., and Haines, T., Overview of the relationship between organizational and workplace foctors and injury rates, *Saf. Sci..*, 26, 201–217, 1997.
85. Koehoorn, M., Lowe, G.S., Rondeau, K.V., Shellenberg, G., and Wagar, T.H., Creating high-quality health care workplaces, Canadian Policy Research Networks (CPRN) Discussion paper No. W/14, Ottawa, January 2002. Available at: http://www.cprn.com/docs/work/hcw_e.pdf.
86. Mondragon Cooperative Information (English). Available at: http://www.mondragon.mcc.es/ingles/menu_ing.html
87. Canadian Nursing Advisory Committee, Our health, our future: Creating quality workplaces for Canadian nurses. Available at: http://www.hc-sc.gc.ca.

88. Tomsen, S., Arnetz, J., Arnetz, B., Patient and personnel perspectives in intervention studies of the health care work environment, *Work Health*, 10, 36–48, 2000.
89. Wong, S.T., Outcomes of nursing care: how do we know? *Clin. Nurse Spec.*, 12, 147–151, 1998.
90. Arnetz, J.E. and Arnetz, B.B., Violence towards health care staff and possible effects on the quality of patient care, *Soc. Sci. Med.*, 52, 417–427, 2001.
91. Thomsen, S., Stress, satisfaction and quality: studies of oganization and individual well-being in health care, Dissertation, Karolinska Insitutet, Stockholm, 2000.
92. Kangas, S., Kee, C.C., and McKee-Waddle, R., Organizational factors, nurses' job satisfaction, and patient satisfaction with nursing care, *J. Nursing Adm.*, 29, 32–42, 1999.
93. Goetzel, R.Z., Anderson, D.R., Whitmer, R.W., Ozminkowski, R.J., Dunn, R.L., and Wasserman, J., The relationship between modifiable health risks and health care expenses, *J. Occup. Environ. Med.*, 40, 843–854, 1998.
94. Kessler, R.C., Barber, C., Birnbaum, H.G., and Frank, R.G., Depression in the workplace: effects on short-term disability, *Health Aff. (Millwood),* 18, 163–171, 1999.
95. Huang, G.D., Feuerstein, M., and Sauter, S.L., Occupational stress and work-related upper extremity disorders: concepts and models, *Am. J. Ind. Med.*, 41, 298–314, 2002.
96. Shannon, H.S., Robson, L.S., and Sale, J.E.M., Creating safer and healthier workplaces: role of organizational factors and job characteristics, *Am. J. Ind. Med.*, 40, 319–334, 2001.
97. Kristensen, T.S., Socioeconomic status and psychosocial work environment: results from a Danish national study, *Scand. J. Public Health*, 30, 41–48, 2002.
98. Garcia, A.M., Working conditions and sickness absence: the need for action oriented research, *J. Epidemiol. Comm. Health*, 54, 482–483, 2000.
99. Westgaard, R.H. and Winkel, J., Ergonomic intervention research for improved musculoskeletal health: a critical review, *Int. J. Ind. Ergon.,* 20, 463–500, 1996.
100. Westgaard, R.H. and Winkel, J., Ergonomic intervention research for improved musculoskeletal health: a critical review.
101. Theorell, T., How to deal with stress in organizations — a health prespective on theory and practice,
102. Jones, F., Bright, J.E.H., Searle, B., Cooper, L., Modeling occupational stress and health: the impact of the demand-control model on academic research and on workplace practice, *Stress Med.*, 14, 231–236, 1998.
103. Zwerling, C., Daltroy, L.H., Fine, L.J., Johnston, J.J., Melius, J., and Silverstein, B.A., Design and conduct of occupational injury intervention studies: a review of evaluation strategies, *Am. J. Ind. Med.,* 32, 164–179, 1997.
104. Briner, R.B. and Reynolds, S., The costs, benefits, and limitations of organizational level stress interventions, *J. Org. Behavior*, 20, 647–664, 1997.
105. Shannon, H.S., Robson, L.S., and Sale, J.E.M., Creating safer and healthier workplaces: role of organizational factors and job characteristics.
106. Robson, L.S., Shannon, H.S., Goldenhar, L.M., Hale, A.R., Guide to evaluating the effectiveness of strategies for preventing work injuries — how to show whether a safety intervention really works, Institute for Work and Health, NIOSH, CDC 2001. Available at: (http://www.iwh.on.ca/Pages/Publications/safetybook.htm)
107. Shannon, H.S., Robson, L.S., and Guastello, S.J., Methodological criteria for evaluating occupational safety intervention research, *Saf. Sci.*, 31, 161–179, 1999.
108. Parker, J.M., Evidence-based nursing: a defence, *Nursing Inq.*, 9, 139–140, 2002.
109. Guyatt, G.H., Haynes, R.B., McKibbon, K.A., and Cook, D.S., Evidence-based health care, *Mol. Diagn.,* 2, 209–215, 1997.
110. Briner, R.B., Improving stress assessment: toward an evidence-based approach to organizational stress interventions, *J. Psychosom. Res.*, 43, 61–71, 1997.
111. Laurell, C.A., Noriega, M., Martinez, S., and Villegas, J., Participatory research on workers' health, *Soc. Sci. Med.*, 34, 603–613, 1992.

Injured Nurse Story #16: The Writing on the Wall

by Quick Learner

I worked as a CNA in a local nursing home for a brief 6 months in the early 1990s. Unable to find work in my field in this rural and depressed economy, I'd seen weekly advertisements in the local newspaper for CNAs and knew that's where the jobs were. I enrolled for the course at the local community college, became certified, and was immediately hired.

At the nursing home, I was assigned to work the East Wing where the non-Alzheimer patients lived. Patient load was usually 11 to 16 patients per CNA and, when we were fully staffed, there would be four CNAs assigned to the wing. We were seldom fully staffed, which meant that someone would have to help on both ends of the floor.

It wasn't long before my back started hurting. Muscle aches, I thought, but I lived with pain for most of that 6 months. A friend gave me a 45-minute deep massage once, which relieved it for a while, until the first hour back at work later that day. I thought I was just weak and wimpy and did work in tandem with another aide. We assisted each other with heavy lifts and using the Hoyer to get patients up and back into bed.

But, sometimes there simply wasn't enough room to maneuver a wheelchair in the bathroom, lift the patient from the chair, and situate him on the toilet without doing the lift and turn from the side instead of from the front because you couldn't leave a patient who couldn't stand without assistance to move the wheelchair in order to better position yourself to do the lift. It was a catch-22 situation.

There was a more experienced CNA from the Alzheimer's unit that I would occasionally meet in the lunch room for coffee. She'd worked there for 10 years or more and was always upbeat. You could tell she loved the patients and the work. Then, she injured her back lifting patients, was out of work for a few weeks and came back just long enough to be let go for having injured her back.

That did it for me. I could read the writing on the wall and decided right then and there that I was leaving that kind of work before I, too, had a serious injury that might leave me crippled or in pain for the rest of my life.

1-56670-631-9/03/$0.00+$1.50

14 The Relationship between the Nursing Shortage and Nursing Injury

Elizabeth Shogren

CONTENTS

INTRODUCTION

There seems to be little debate that the United States is experiencing a growing nursing shortage or that a nursing shortage creates increased risk to patients. In the face of this overwhelming crisis, we continue to lose desperately needed registered nurses to workplace injury. Although the experts have examined this crisis, they have failed to recognize the need to reduce injury as the first goal in retention and recruitment. An unsafe workplace is not an incentive to become or remain a registered nurse.

An early study in 1995 by Shogren and Calkins involving 12 Minnesota hospitals demonstrated a probable connection between dramatic increases in injury and decreased length of stay (LOS), increased acuity, and reductions in nursing staff. In Minnesota, RN injuries rose a staggering 65% from 1990 to 1994. Injuries to other healthcare workers also increased at an alarming rate.

	1990	1992	% Increase	1990	1994	% Increase
Registered nurses	569	921	61.8%	569	940	65.2%
LPNs	106	142	33.9%	106	159	50.0%
Nursing assistants	196	312	59.7%	196	295	51.0%
Other UAP	205	359	75.1%	205	444	116.5%
Other professionals	62	133	114.5%	62	105	85.4%

The overwhelming majority of these injuries were musculoskeletal disorders (MSDs).

1990–1994 Nature of Injury/Illness	Number of Injuries/Illnesses
Traumatic injuries to muscles, tendons, ligaments, joints, etc.	1866
Blood and body fluid exposures (primarily needlestick/sharps)	1449
Other traumatic injuries and disorders	774
Surface wounds and abrasions	338
Disorders of skin and subcutaneous tissue	238
Open wounds	169

Source: OSHA 200 Logs compiled by identified hospitals.

1-56670-631-9/03/$0.00+$1.50

The U.S. Department of Labor defines MSDs as injuries or disorders of the muscles, nerves, tendons, joints, cartilage, and spine that occur as a result of awkward postures, repetitive motion, repeated impacts, or heavy, frequent, or awkward lifting. They do not include slips, trips, falls, or motor vehicle accidents.

Nationally, work-related musculoskeletal disorders (WRMSD) account for more than one-third of all lost work injuries. Back injuries account for more than half of the work-related MSDs. Hospitals and Skilled Nursing Care Facilities are among the top five industries that report WRMSDs. For example, in 1999 and 2000, Minnesota hospitals accounted for 6.1% of all WRMSD reported and Skilled Nursing Care Facilities accounted for an additional 5.5%.

In April 2002, the Bureau of Labor Statistics (BLS) continued to rank occupations of nurses aides/orderlies and registered nurses third and tenth respectively for sheer numbers of occupational injuries and illnesses involving time away from work. (Other occupations with high rates of injury include construction workers, assemblers, and truck drivers.)

The American Nursing Association (ANA) reports that more than any other work-related injury or illness, it is musculoskeletal disorders (MSDs) that are responsible for lost work time, protracted medical care, and permanent disability among healthcare workers. This is supported by the latest government statistics (U.S. Department of Labor, Bureau of Labor Statistics, 2002) which rank nurses aides/orderlies second and registered nurses sixth in occupations in the number of MSDs involving lost work time in 2000.

For nurses, this data translates to a reasonable fear of disabling injury. A 2001 ANAs Health and Safety Survey,[5] reports that over 70% are concerned about the chronic effects of stress and overwork and 59.4% of survey respondents list "disabling back injury" as one of the top three health and safety concerns.

Response Option	Total Selected	Percentage
Acute/chronic effects of stress and overwork	3329	70.5%
A disabling back injury	2803	59.4%
Getting HIV or hepatitis from a needlestick	2139	45.3%
Infection with tuberculosis or other infectious diseases	1762	37.4%
An on-the-job assault	1167	24.7%
Developing a latex allergy	990	20.9%
Fatigue-related car accident after a shift	891	18.8%
Toxic effects from exposure to chemicals, including adverse reproductive effects	314	6.6%
Exposure to hazardous drugs like chemotherapy, pentamidine, ribavirin, etc.	236	5%
Exposure to smoke from lasers or electro cautery devices	128	2.7%

In addition, 87.9% of survey respondents indicated that health and safety concerns have an impact on the type of nursing work performed and their continued practice in the field of nursing. Eighty-three percent reported working with back pain. Seventy-six percent indicated unsafe working conditions interfere with their ability to deliver quality care.

The healthcare industry has relied on people to do the work of machines. Nelson et al.[6] reports that the healthcare industry is gradually accepting the reality that manual lifting is a major contributing factor to the numbers and severity of injuries in healthcare organizations.

The cost of these injuries is not confined to the workers' compensation costs of medical care and wage replacement. There are costs associated with temporary replacement if the nurse is eventually able to return to work and permanent replacement if the nurse cannot return.

In August 2002, the Joint Commission on Accreditation of Healthcare Organizations (JCAHO) reported that, "according to a recent report from the Voluntary Hospitals of America (VHA), it costs approximately 100% of a nurse's salary to fill a vacated nursing position. For a medical/surgical nurse, that averages about $46,000; the cost is $64,000 for a critical care nurse.

Assuming a turnover rate of 20% — the current average turnover rate among healthcare workers — a hospital employing 600 nurses at $46,000 per nurse per year will spend $5,520,000 a year in replacement costs."

These costs to the employer should not overshadow the cost to the injured nurse. Workers' compensation replaces a fraction of his/her income. Eligibility for insurance coverage is frequently lost and the nurse must pay for insurance through COBRA. Accrued benefit hours are consumed and the ability to earn benefits is lost, not to mention the physical pain and emotional stress.

Uninjured nurses see this and understand all too clearly that it could easily be them. Caught between the pressures of needing to care for the patient and needing to protect themselves, the needs of the patient usually win out and the nurse frequently is injured in the process. It's easy to see why, when not meeting the patient needs has life threatening consequences.

Experts predict that this situation could become a major public health problem over the coming decades. In April 2001, the Hospital and Healthcare Association of Pennsylvania concluded that "the impending shortage that will occur as the working population ages has the potential to threaten the health of the public by decreasing access to services and quality of care in just over ten years if the current trends in shortages continue."

Currently (2003), there is a shortfall of nearly 200,000 full time employees (FTEs). This is less than a quarter of the projected shortfall for the year 2020, but we already can see the effect the shortage has, and will have, on patient care.

In relation to patient care, a nursing shortage means less nursing time for each patient and conditions that jeopardize safe patient care. This effect has been documented in recent studies.

Needleman et al.[1] reports an association between the proportion of total hours of nursing care provided by registered nurses, or the number of registered nurse hours per day, and six adverse outcomes among medical patients, including "failure to rescue." A higher level of staffing by registered nurses for surgical patients corresponds to decreased rates of failure to rescue.

Failure to rescue is defined as the death of a patient with one of the following five life-threatening complications: pneumonia, shock or cardiac arrest, upper gastrointestinal bleeding, sepsis, or deep venous thrombosis.

The Joint Commission on Accreditation of Healthcare Organizations (JCAHO) issued a 40-page document in May 2002, *Health Care at the Crossroads*, that reported that nurse staffing shortcomings were at the root of one in four reported incidents of unexpected deaths or serious injuries caused by errors in hospitals. JCAHO is the principal accrediting body for quality in healthcare and evaluates healthcare organizations in the United States on a continuing basis

Other studies from the American Hospital Association, the Voluntary Hospitals of America, the American Nurses Association, and the Robert Wood Johnson Foundation were cited by JCAHO in reaching its conclusions.

In reviewing the available literature, JCAHO concluded that hospitals with turnover rates of less than 12% had lower death rates; patients in hospitals with fewer intensive care nurses were more likely to have longer stays and postoperative complications; and that higher nurse staffing ratios correlated with a 12 to 39% reduction in certain adverse outcomes.

Even *Consumer Reports*[2] conducted a consumer satisfaction survey, published in the January 2003 edition, that concludes that of all the factors measured, satisfaction with care and attention from nurses, doctors, and other staff made the most difference in overall satisfaction with hospital care. In addition, only 2% of the survey respondents who reported attentive nursing care ended up with a serious health complication compared to 8% who found it more difficult to get nursing attention.

JCAHO cited additional contributing factors to the nursing shortage. They are: consequences of reimbursement incentives, decreased length of stay, and higher patient acuity. Obviously alarmed by the impact of the shortage, JCAHO issued a call to action.

JCAHO's nurse shortage solutions target three areas as shown in the following table.

Creating a culture of retention
Minimizing paperwork burdens
Zero tolerance for abusive behavior toward nurses
Delegation of authority to nurses for care decisions
Limiting mandatory overtime to emergencies only
Bolstering nursing education
Emphasizing team-training approaches
Establishing residency programs for nurse graduates
Creating career ladders according to education, experience
Incentives to invest in nursing
New federal funding for investment in nursing
Basing continued funding on achievement of goals
Aligning reimbursement incentives to reward effective staffing

Source: JCAHO report, *Health Care at the Crossroads.*

At the current level of shortage, the average national vacancy rate for RNs is 10.2% as reported by the American Organization of Nurse Executives (AONE) in the 2001 Acute Care Hospital Survey of RN Vacancy and Turnover Rates. A vacancy rate is established by dividing the average number of vacant FTE positions by the average number of budgeted FTE positions.[3]

10 vacant FTE positions ÷ 100 budgeted FTE positions = 10% vacancy rate.

AONE goes on to state that the West (12.2%) and the South (11.0%) exceeded the national average. In all areas of the country, RN vacancy rates were the highest in critical care (14.6%), medical/surgical units (14.1%), and emergency rooms (11.7%).*

A vacancy rate is a key indicator of the minimum nursing shortage because the employer determines the number of RNs needed as part of a budgeting process. Budget constraints can have the effect of driving the needed FTEs down. Frontline workers often believe that budgeted FTEs understate the actual needed FTEs.

AONE also reported high turnover rates for RN positions. A turnover rate is determined by dividing the number of resignations or terminations in a budget year by the average number of direct and indirect RN FTE positions for the same year.

10 resignations/terminations ÷ 100 direct/indirect RN FTEs = 10% turnover rate.

The national turnover rate for 2000 was 21.3%. The South was the highest at 24.0%, the West 22.2%, the Midwest 20.2%, and the Northeast 17.4%.[3]

AONE also asked its members to report on the most effective recruitment and retention strategies.**

* AONE *Perspectives on the Nursing Shortage: A Blueprint for Action* (2000).

** The HSM Group. Ltd. (HSM) created 2001 for the American Organization of Nurse Executives Acute Care Hospital Survey of RN Vacancy and Turnover Rates in 2000.

AONE Summary of Recruitment and Retention Methods

Most Effective Methods Used to Improve Recruitment and Retention (open-ended question, multiple responses accepted)	Percentage of Respondents Citing Method
Salary (increased/competitive salaries, agency/higher wages paid for in-house staff to work extra shifts, weekend incentive, higher shift differential)	29%
Nursing schools/colleges (recruitment at schools, preceptorships, externs, clinical rotations, our own school of nursing, recruit/hire new graduates, student nurse positions)	20%
Staff involvement/outreach (efforts to improve employee satisfaction, positive culture/professional environment, staff input/meetings, team/committee involvement, feedback mechanisms, staff surveys, appreciation/respect, relationship building, visible/available managers, shared governance, giving nurses a say in care/operations decisions)	19%
Bonuses paid (sign-on, referral, retention)	18%
Scheduling flexibility (varied shift lengths, flexible hours, self-scheduling, more part-time availability, seasonal employment, low/no mandatory overtime and/or floating, increased number of 12-hour shifts)	18%
Scholarships and education funding (includes employee and non-employee funding, work-while-in-school programs, tuition assistance)	15%
Advertising (written ads, newspaper/journals/trade publications, web site/ads, direct mailing, radio ads)	14%
Word of mouth (staff referrals, networking, etc.)	10%
Job fairs/open houses/career days	9%
Benefits (improved benefits, more benefits to part-time staff, retirement benefits)	9%
Staffing (better nurse/patient ratios, support staff)	5%
Nurse recruiters/formal recruiting process (dedicated recruiters, on-the-spot interviews, telephone screening, personalized recruiting)	5%
Job rotation/transfers/training (includes ability to transfer, in-house training for specialty areas)	5%
Career/clinical ladders (credentialing programs, professional growth opportunities)	3%
Good location	2%
Pay relocation costs	2%
Reputation (in the community, patient satisfaction)	2%
Student or community outreach/education (job shadowing, high school student outreach)	2%
Foreign recruitment (international, wide-area recruitment)	1%
Orientation process	1%
Frontline management leadership	1%
Time off (increased paid time off)	1%
Full-time positions offered/guaranteed hours	<1%

Even though AONE and JCAHO concluded the recruitment and retention of registered nurses is critical, they do not include strategies to improve workplace safety as a way to retain current staff.

This is all the more curious when a common sense approach to combating a resource shortage of any type is to conserve and protect available assets.

Even though there is a steady, but slow, growth in the registered nurse (RN) supply, the rate of growth is outstripped by a booming demand. The projected shortage of RNs results from a projected 40% increase in demand between 2000 and 2020, compared to a projected 6% growth in supply. In real numbers, this is a shortage of 808,416 RN FTEs. Since a significant number of nurses work part-time, the actual number of nurses will be higher than the FTEs (see Figure 14.1).

This projected shortfall of 29% is based on a number of contributing factors. Certainly, the most significant is the fact that America's nurses are aging. Over 15% of RNs are currently in the 50 to 59 age group and more than 35% fall in the 40 to 49 age group.[4] The majority of these nurses will reach retirement age by the year 2020.

Other nurses will simply leave the profession because of increasing job dissatisfaction, emotional exhaustion, and injury.

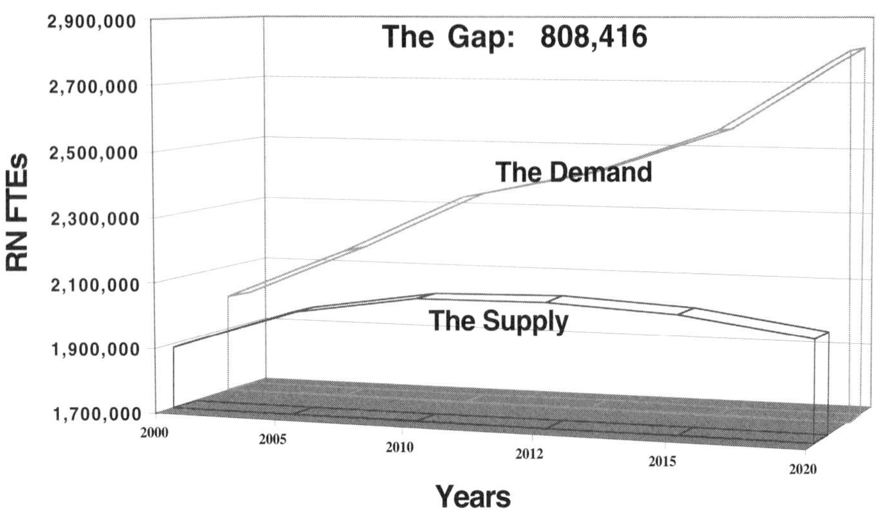

FIGURE 14.1 Nurse shortage gap.

Fueling the demand is the overall aging of America, especially the impact of the baby boomer generation and its need for healthcare services, increased managed care/HMO, and breakthroughs in diagnostics, treatments, and pharmaceuticals.

The increase in demand exists across all services.

The healthcare industry cannot afford to let this situation continue even without a nursing shortage. On both an ethical and financial plane, there is an obligation to take action. In the face of a nursing shortage, it becomes imperative that employers recognize that injury and risk of injury translates into fewer nurses.

Along with all the other retention strategies that are considered, it is time to move health and safety concerns, and especially reduction of back injures, to the list — to the top of the list.

REFERENCES

1. Needleman, J. et al., *Staffing and Patient Outcomes in Hospitals, Final Report*, U.S. Department of Health and Human Services, Washington, D.C., February 28, 2001.
2. *Consumer Reports*, January 2003.
3. AONE *Perspectives on the Nursing Shortage: A Blueprint for Action* (2000).
4. The HSM Group. Ltd. (HSM) created 2001 for the American Organization of Nurse Executives Acute Care Hospital Survey of RN Vacancy and Turnover Rates in 2000.
5. American Nurses Association. Nursing World Health Safety Survey, September 2001, pp. 2–3.
6. Nelson, A., Patient Care Ergonomic Resource Guide Safe Patient Handling and Movement, Patient Safety Center of Inquiry, Tampa, FL, Oct. 2001, p. 29.

15 Preventing Back Injuries to Healthcare Workers in British Columbia, Canada and the Ceiling Lift Experience

Chris Engst, Rahul Chhokar, Aaron Miller, and Annalee Yassi

CONTENTS

INTRODUCTION

In British Columbia, Canada, the healthcare sector injury rate from 1996 to 2000 was higher than the average for all industries combined. In 2000, the injury rates for the acute care and long-term care sectors were 7.6 and 9.2, respectively, and the injury rate for all industries in British Columbia was 3.9 (Workers' Compensation Board of British Columbia, 2001). From 1994 to 1998, overexertion from patient handling activities was the most common cause of injury, representing 38% of all claims (Workers' Compensation Board of B.C., 2000a) (see Figure 15.1).

The purpose of this chapter is to describe the collaborative efforts that have taken place in British Columbia to prevent injuries to healthcare workers. Specifically, the formation of the Occupational Health and Safety Agency for Healthcare (OHSAH) in British Columbia to develop

1-56670-631-9/03/$0.00+$1.50

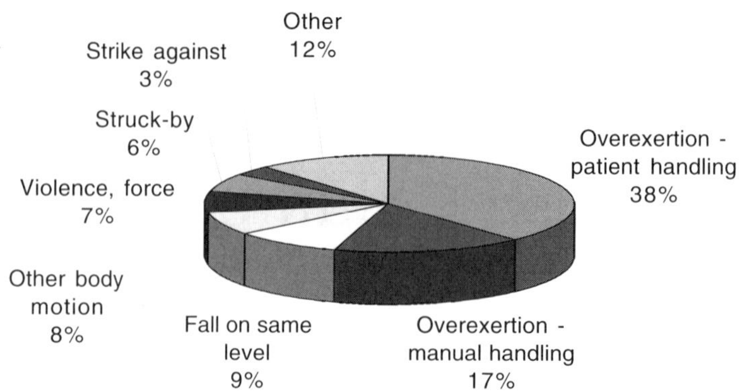

FIGURE 15.1 Accident types among workers in the British Columbia healthcare industry from 1994 to 1998. (From the Workers' Compensation Board of British Columbia. *Healthcare Industry: Focus Report on Occupational Injury and Disease*, 2000.)

best practices in health and safety, implement and evaluate pilot programs, and facilitate the sharing of best practices (Yassi et al., 2002). Existing knowledge was synthesized and primary research conducted to develop reference guidelines for patient handling, an integrated prevention and early return to work program, an on-line health and safety resource tool, and extensive evaluations on the effectiveness of overhead ceiling lift systems.

A memorandum of understanding (MOU) between the Association of Unions and the Health Employers Association of British Columbia (HEABC) was signed to prevent unsafe manual lifting of patients, and to facilitate the implementation of evidence-based practices. This province-wide agreement paved the way for the formation of a new initiative to secure funds from the Ministry of Health to ensure sufficient numbers of ceiling lifts and electric beds in all facilities. A steering committee was established to develop a funding framework for this program and materials were developed to assist in the implementation process. The provincial Ministry of Health Services in turn provided $15 million for mechanical lifting devices and electric beds, and the Workers' Compensation Board (WCB) of British Columbia provided an additional $6 million.

These collaborative efforts between unions and employers have made health and safety a priority in the British Columbia healthcare industry. This success, in forging new policies and programs as well as the bringing together of key players, has turned heads in this province as well as in other regions across the country.

PREVALENCE OF MUSCULOSKELETAL INJURIES IN HEALTHCARE WORKERS

A high rate of musculoskeletal injuries (MSIs) among healthcare workers has been well documented in the literature (Cato et al., 1989; French et al., 1997; Fujimura et al., 1995; Garg et al., 1992; Harber et al., 1985; Leighton and Reilly, 1995; Smedley et al., 1995). Lifetime prevalence rates of back pain greater than 70% have been reported in healthcare workers (French et al., 1997; Fujimura et al., 1995) and higher incidence rates of MSIs have been observed in healthcare workers compared to the general population (Pheasant and Stubbs, 1992) and to other occupational groups (Jensen, 1987; Ono et al., 1995). In British Columbia, the evidence is similar (Yassi et al, 2002).

Nurses who "frequently" work with patients have been shown to have a higher prevalence of back injuries than those who work with patients "infrequently" (Jensen, 1990), and nurses who have been injured commonly report patient handling as a major cause of their injury (Cato et al., 1989; Ferguson, 1970; Harber et al., 1985; Hollingdale and Warin, 1997; Leighton and Reilly, 1995; Ono et al., 1995; Yassi et al., 1995). Patient handling tasks have been rated more stressful to the back than nonpatient handling tasks, both for perceived stress by nursing personnel and through biomechanical studies (Owen et al., 1992). Among the types of tasks commonly associated with patient handling, there is extensive evidence to suggest that manual lifting is a major risk factor for MSI. Nurses who lift more frequently are at increased risk for MSIs (Larese and Fiorito, 1994; Owen and Damron, 1984; Venning et al., 1987) just as nurses who frequently lift heavy objects (Engels et al., 1998; Ferguson, 1970; Josephson et al., 1998). The repeated mechanical stress that is associated with heavy lifting has been related to increased risk for herniated disc, vertebral endplate fracture, and degenerative spinal conditions (Jorgensen et al., 1994; Morrison et al., 1996).

THE OCCUPATIONAL HEALTH AND SAFETY AGENCY FOR HEALTHCARE IN BRITISH COLUMBIA

A Collaborative Evidence-Based Approach

In British Columbia, shared concern over spiraling injury rates and their financial impact brought the unions representing healthcare workers together with the employers to devise an alternative approach to preventing injuries. Thus, in 1999, the Occupational Health and Safety Agency for Healthcare (OHSAH) in British Columbia was formed. The agency's board of directors included healthcare leaders representing both the workforce (including representatives of nursing staff, therapists, technicians, support staff, and others) and the employers, in close collaboration with the Ministry of Health and the WCB. Active board involvement was integral from the outset: subcommittees of the board identified training needs for the sector, criteria for the disbursement of funding, and principles by which to operate prevention and return-to-work initiatives.

The newly created agency was seen as a combination of research and decision-making with the goal being collaborative identification and implementation of evidence-based best practices. OHSAH recognized the need for concerted efforts to reduce back injuries to healthcare workers as its top priority, particularly those injuries as a result of patient handling.

Best Practices for Safe Patient Handling

In conjunction with employers, front-line workers and union representatives, occupational health and safety practitioners, and internationally renowned researchers, OHSAH developed the Reference Guidelines for Safe Patient Handling (OHSAH, 2000a). These guidelines were developed through an extensive national peer review process funded from a federal research grant (Lee and Yassi, 2002; Yassi et al., 1998), while utilizing information from local best practices. The purpose was to identify patient handling risk factors and best practices for patient handling situations from the literature. This document provides a review of the legal and regulatory requirements within the context of MSI prevention and safe patient handling practices in British Columbia, and is a practical guide to identifying, assessing, and controlling risk factors associated with patient handling (see Figure 15.2 and Figure 15.3). A handbook on safe patient lifts and transfers (OHSAH, 2000b) accompanies the reference guidelines, illustrating over twenty patient handling tasks. The handbook serves as a quick and easy reference guide for staff or trainers. Within its first year of circulation, over 12,000 copies were disseminated to healthcare workers throughout the province.

Integrated Prevention and Early Active Return-to-Work

The healthcare industry is the number one source of days lost and claims in British Columbia (WCB of British Columbia, 2002). The literature has shown that "usual activities" is the best advice following back injuries (Malmivaara et al., 1995). For attending physicians and other practitioners to encourage usual activities and/or early return to work, however, the workplace must be ergonomically sound,

Step 1: Identifying and assessing potential risk factors:
Environment: flooring, obstacles, space, equipment, potentially confusing wall/floor patterns, distance to be moved, lighting, noise, temperature
Organization: education and training, availability of assistance, workload, work flow
Equipment: availability, cleanliness and condition, appropriateness to task, compatibility with environment, adequate caregiver training in equipment use, patient comfort and safety levels
Caregiver: skills, education and training, fitness and physical capabilities, medical and emotional status, clothing and accessories, physical force, posture, repetition, duration, contact stress, psychosocial stresses
Patient: care plan (checklist, pictogram) available with current handling procedures, communication level, cognitive status, behavioural and emotional status (history of violence or other current risk factors), medical status, physical and sensory status, clothing, assistive devices, ability to assist
Step 2: Deciding on the appropriate patient handling technique:
Check care plan (refer to care plan section) and history of previous incidents.
Consider risk assessment results and match with care plan.
Consider facility patient handling policies and procedures.
Determine appropriate patient handling technique:
Safe and comfortable for patient
Provides maximum patient independence
Causes minimal biomechanical load on caregiver(s) with maximal safety (good body mechanics, appropriate equipment, and appropriate number of staff)
Step 3: Preparing for the patient handling task:
Preparing the environment
Preparing the assistant(s)
Preparing the equipment
Preparing the care giver
Preparing the patient
Step 4: Performing the patient handling task.
Step 5: Evaluating the completed patient handling task.

FIGURE 15.2 Risk identification, assessment, and control in patient handling. (From The Occupational Health and Safety Agency for Healthcare in British Columbia, *Reference Guidelines for Safe Patient Handling*, 2000.)

and not likely to cause reinjury, i.e., workplace modifications will be needed if usual activities were the cause of injury.

To facilitate a timely return to work, the PEARS Program was created to reduce the incidence, morbidity, time loss and related costs due to workplace MSI, specifically by focusing on what could be done at the worksite to make it easier for the injured worker to stay at work or return promptly. The PEARS Program focuses on *p*reventing injuries through hazard assessment and workplace modifications, *e*arly intervention including encouraging early reporting of signs and symptoms, *a*ctive involvement of all key parties to attempt to modify the workplace to make it safer to return, and hence *r*eturn to work of the injured worker *s*afely.

Through a collaborative approach between board members, stakeholders, and researchers, the PEARS program focuses on integrating primary and secondary prevention of injuries, specifically, preventing injuries before they occur, and preventing disability by intervening early. It requires a high level of support from healthcare providers, managers, coworkers, unions, and insurers. Key principles were formed to protect the rights of workers while emphasizing best practices for the workforce (Yassi et al., 2002).

The PEARS Program is being piloted in two large hospitals, using two comparable hospitals as control groups. Outcomes that will evaluate the program's success include injury rates, time loss and costs, and staff and patient satisfaction.

An On-Line Tool to Promote Injury Prevention

Stakeholders in all jurisdictions in Canada recognize the need to reduce the high rates of workplace injuries and illness among healthcare workers, and to manage more effectively the

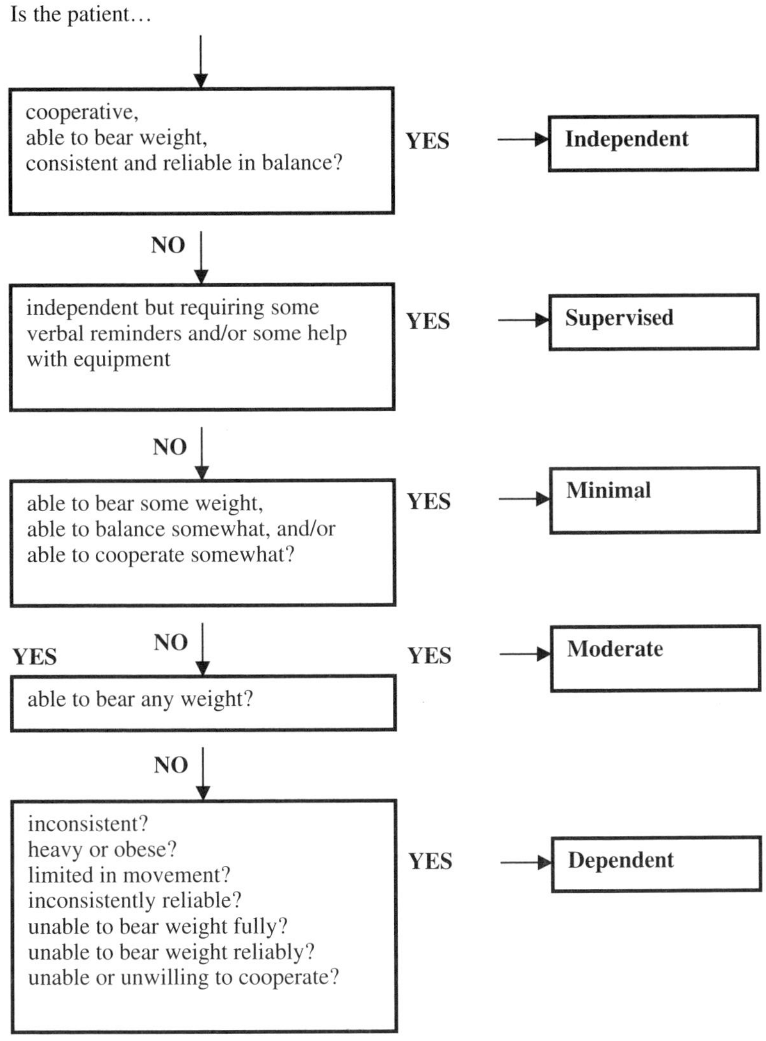

FIGURE 15.3 Patient transfer assessment flow-sheet. (From The Occupational Health and Safety Agency for Healthcare in British Columbia, *Reference Guidelines for Safe Patient Handling*, 2000.)

safe and rapid return of injured healthcare workers to their jobs. While the magnitude of the problem is enormous, considerable scientific evidence has accumulated over the last several years as to how to best manage and prevent these injuries. However, this information is not readily accessible in an easy-to-use fashion for those who need it. Through funds from the Health Evidence Application and Linkage Network (HEALNet), a National Centre of Excellence funded by CIHR and SSHRC, a website was designed to promote evidence-based approaches to the prevention and management of injury disability among healthcare workers (Lee and Yassi, 2002). This website was built using an interactive approach with comprehensive information to address occupational health and safety injuries in healthcare workers.

OSHTips, short for Occupational Safety and Health Tips, contains information on policy and procedure development, MSI program implementation, patient handling, injury tracking, work organization, and early return to work programs. Users can search by department to find information relevant to their work environment.

TABLE 15.1
Principles: Prevention and Early Active Return to Work Safely (PEARS)

1. Preventing disability must be seen as an extension of preventing the injury.
2. The focus of post-injury intervention must be on workplace accommodation.
3. All alternate or modified work assignments must be meaningful.
4. The program should build on previous experience within the workplace.
5. There must be an evidence-based education component and communication plan delivered for each of the stakeholder groups.
6. There must be recognition of and respect for existing patient-doctor relationships.
7. The program must be entirely voluntary.
8. The program must be designed for rapid and easy implementation.
9. The program should be independent of WCB claims processing.
10. Income continuity as part of this program should begin upon the injured worker's entrance into the program, and continue as long as the worker is participating in the program.
11. Provisions should be made for in-house rehabilitation wherever possible, either on-site or organized away from the workplace.
12. Union representatives must be involved in all stages of the design and implementation of the project, including decisions regarding accommodation of the injured worker.
13. The types of injuries to be the focus of intervention should, initially, be acute musculoskeletal injuries.
14. The scope and parameters of the program should be as broad as possible, within the confines of the resources available.
15. All injuries must be carefully tracked, and outcomes clearly identified.

The usefulness of the website is currently being evaluated to determine whether the synthesis of scientific evidence into usable web-based formats can have a significant impact on the rates of workplace injuries and illness among healthcare workers. It is hoped that this website will provide decision-makers, front-line workers, and health and safety professionals with a new means of accessing the most up to date health and safety literature available.

Ceiling Lifts: Preventing Back Injuries with better Lifting Equipment

Many researchers and health and safety practitioners have recommended replacing manual patient handling with mechanical options through the introduction of mechanical floor lifts (Blue, 1996; Laflin and Aja, 1995; Ljungberg et al., 1989; Marras et al., 1999; Varcin-Coad and Barrett, 1998). Studies examining the effectiveness of mechanical lifting equipment have found decreased injury rates, as well as decreased manual lifting and twisted trunk positions (Gingher et al., 1996; Holliday et al., 1994; Takala and Kukkonen, 1987), however, some potential for increased risk of injury due to cumulative loading (which may result in chronic MSI conditions) has been noted if lifts are used inappropriately or when not indicated (Daynard et al., 2001). Thus, the introduction of mechanical lifting equipment, such as floor lifts and assistive devices, may not be enough given that back injuries from patient handling are still highly evident in the workplace (Daynard et al., 2001; Garg et al., 1991; OHSAH, 2000a; WCB of British Columbia, 2000b). Ceiling-mounted lifts provide an alternative means of performing these tasks. With the advent of the ceiling lift, most physical barriers in a standard room are eliminated because ceiling lifts utilize space above the clearance requirements of most objects (see Figure 15.4a and Figure 15.4b).

Ceiling lifts have been used in eastern Canada for many years with great success (Holliday et al., 1994; Villeneuve, 1998). In Quebec, specified standards that accommodate ceiling lifts are included as part of the building code for long-term beds in all new facilities. However, even with the installation of ceiling lifts in eastern Canada, and now in British Columbia, there have been very few scientific studies evaluating their full effectiveness, which led OHSAH to conduct several rigorous evaluations of ceiling lift systems to better understand their effectiveness in various settings.

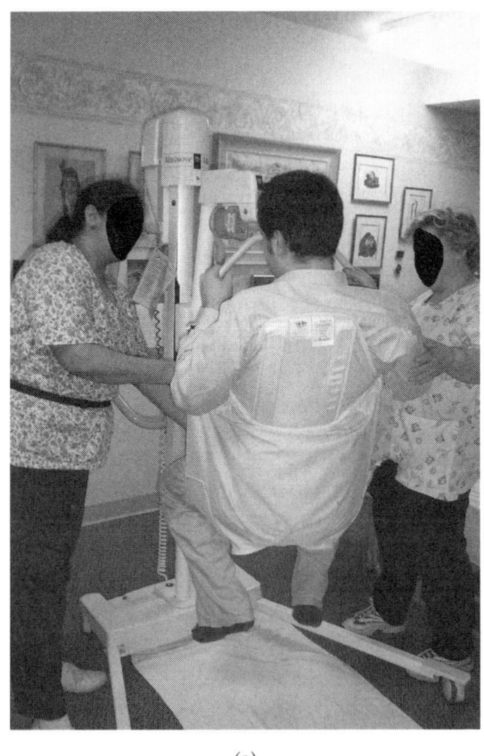

(a)

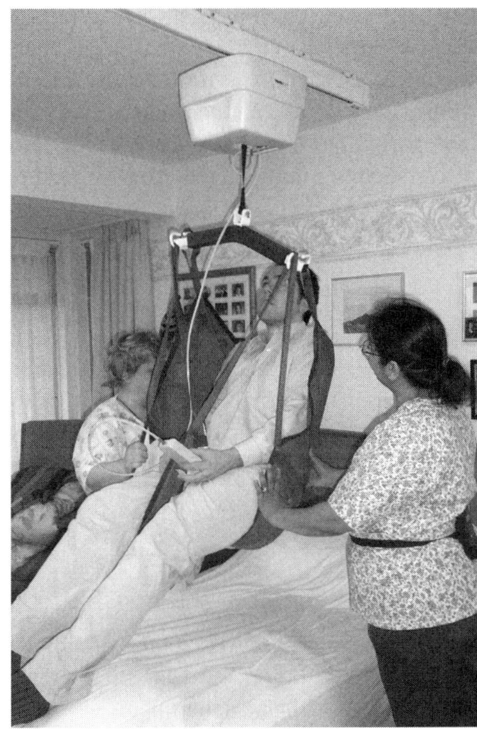

(b)

FIGURE 15.4 Conventional (a) floor and (b) ceiling lift models.

EVALUATION OF THE COST-BENEFIT AND EFFECTIVENESS OF CEILING LIFTS

CASE STUDY 1

A retrospective evaluation of ceiling lifts was conducted in an extended care unit of a hospital to evaluate the effectiveness of ceiling lifts in decreasing staff MSI and to determine the economic viability of this new program. The WCB of British Columbia provided funding for the installation of 65 units over 124 beds in 62 rooms.

Staff completed a questionnaire assessing the prevalence of MSI symptoms and satisfaction with the system, and a resident questionnaire was administered to residents and/or family members to assess comfort and satisfaction (Ronald et al., 2002). The rate of MSI due to lifting/transferring patients was significantly reduced (58% reduction, $p < .05$) postintervention. However, rates of total MSI and MSIs due to repositioning did not significantly decline because ceiling lift repositioning slings were found not to be suitable. Costs and benefits attributable to the project were identified and measured for a one-year period preceding and following the intervention. As a result of the intervention, there was a considerable reduction in the cost of compensation claims, by 69% for lift and transfer injuries, and by 42% for total MSIs. It was concluded that these direct savings predicted a payback within 4 years, and more quickly if the effect of indirect savings or rising compensation costs is considered. Over the estimated 12-year life span of the equipment, the present value of the accumulated claim cost reductions exceeded the investment cost by a factor of 2.5 to 1, representing an internal rate of return of 8.1% per annum (Spiegel et al., 2002).

A follow-up evaluation was conducted by OHSAH 3 years after the original study to determine the long-term effects of ceiling lifts and help validate the original findings (OHSAH, 2003b). Few studies have investigated the long-term effects of ceiling lifts on reducing patient handling injuries. The follow-up evaluation included 3 years of injury data that was not previously available, allowing for a comprehensive 7-year analysis from 1995 to 2001.

Injury data were separated into a 3-year preintervention period (1995 to 1997), 1-year intervention period (1998), and 3-year postintervention period (1999 to 2001). Data for each period were averaged to minimize the effect of year-to-year variation on the data comparisons.

Results of the follow-up study indicated a 40% reduction in total claims costs, 82% reduction in lift/transfer claims costs, and 83% reduction in lost hours for lift/transfer injuries postintervention. However, over this same time period, there was a 34.5% increase in lost time due to repositioning.

The hospital realized that repositioning injuries were not positively impacted by the use of ceiling lifts, and therefore initiated the use of a new repositioning draw sheet. Since implementing the use of a draw sheet, there has only been one repositioning injury, which resulted in minimal claims costs.

These results provide strong evidence of the long-term effectiveness of ceiling lifts for lifting and transferring residents. In addition to reduced lift/transfer injuries and cost-savings, staff commented that the use of ceiling lifts improved resident quality of life. Ceiling lifts allowed residents to be out of bed more often and also helped to improve resident comfort during transfers.

Case Study 2

A second study was conducted to evaluate the effectiveness of portable ceiling lift devices in diagnostic imaging (OHSAH, 2003a). In this study, risk of staff MSI, frequency of patient handling methods, staff discomfort levels and fatigue, and job satisfaction were evaluated.

Two nuclear medicine departments were retrofitted with ceiling lift tracks in each of four camera rooms and each department was provided with one portable ceiling lift. A portable ceiling lift, as opposed to a fixed ceiling lift, was deemed more cost-beneficial in nuclear medicine because only some of the patients required staff assistance and therefore the portable motor could be moved from room to room as necessary to lift and transfer patients onto the imaging table.

Results included reduced perceived discomfort while repositioning and lifting patients with ceiling lifts. In addition, staff perception of safety from injury during lifts/transfers improved significantly ($p < .05$). When asked to report on areas of discomfort while moving a patient from stretcher/bed to pallet and vice versa, there were significant decreases in discomfort for the head, neck, shoulders, and arms when using ceiling lifts. Significant decreases in staff discomfort in the shoulders and arms, and lower back were found when repositioning patients, contrary to the results found in Case Study 1.

All of the staff agreed that using the ceiling lifts made lifting patients easier and 88.9% of staff agreed that ceiling lifts made their job easier to perform overall. However, only 22% of staff members thought that using the ceiling lifts required less time than other devices.

Use of ceiling lifts did not show a significant reduction in patient handling MSIs because of a low prevalence of injury prior to the installation of lifts. Hence, levels of pain or discomfort and perceived risk of injury were seen as the most appropriate indicators for this study.

Case Study 3

A third intervention study of overhead lifts was conducted in an extended care unit with similar objectives to Case Study 1, however, this study did not occur retrospectively. Ceiling lifts were installed in a 75-bed extended care unit (ECU 1) and an identical unit (ECU 2) served as the control. Staff on ECU 1 were given a one-hour training session that provided hands on training using the lifts for a variety of patient-handling techniques. This intervention was evaluated pre- and post-intervention using repeated measures ANOVA, and included interviews, questionnaires and injury data (Engst et al., 2003).

Interviews revealed that staff risk of injury decreased and job satisfaction increased 1 year after the introduction of the ceiling lifts. Results of the questionnaire indicated that a greater proportion of staff in ECU 1 used the ceiling lifts to lift and transfer residents from bed to chair than manually or with floor lifts. When examining the perception of pain and discomfort when lifting and transferring residents, there was a significant difference between the two units ($p < .05$); ECU 1 staff exhibited a decreased perception of discomfort when using the ceiling lifts to lift and transfer, while the perception of pain or discomfort in ECU 2 increased.

When staff from ECU 1 repositioned residents in bed, 54.2% did not use the ceiling lift, as they preferred to do it manually alone or with coworkers. These trends were similar to those observed by OHSAH (2003b) and Villeneuve (1998).

Examining preferred methods of transfer, 71.4% of staff in ECU 1 ranked the ceiling lifts number one, while staff in ECU 2 preferred lifting or transferring residents with coworkers (51.7%).

Total compensation due to patient handling injuries was reduced by 23% in ECU 1 and increased by 12% in ECU 2. Compensation costs involving lifting and transferring claims were reduced by 68% in ECU 1, and increased by 68% on ECU 2. Compensation costs due to repositioning claims increased by 53% in ECU 1 and decreased by 34% in ECU 2.

From this study, it was concluded that MSI rates associated with patient lifting and transferring declined with ceiling lifts. Ceiling lifts also reduced the need for storage space, reduced the risk of staff injury, and were found to be safer to use than mechanical floor lifts. As well, job satisfaction and control increased and the ceiling lift was the preferred piece of equipment to use when lifting and transferring patients.

The results of these evaluations provide substantial evidence that a ceiling lift program can produce strong economic benefits, reduce risk of staff MSI, and improve staff satisfaction.

A POLICY FOR "NO UNSAFE MANUAL LIFTING" IN BRITISH COLUMBIA

In an effort to introduce engineering controls to patient handling procedures, numerous healthcare organizations have adopted "no unsafe manual lifting" policies (Willis, 1996). "The aim is to eliminate hazardous manual lifting in all but exceptional or life-threatening situations" (Royal College of Nursing in the United Kingdom). Patients should be encouraged to assist in their own transfers, and handling aids must be used whenever they can reduce the risk of injury. Handling patients manually may continue only if they do not involve lifting most or all of a patient's weight. Care must also be taken when supporting a patient, and pushing and pulling should be kept at a minimum. Staff should assess the capabilities of a patient to decide on which, if any, handling aids are suitable (Willis, 1996).

In March of 2001, HEABC and the Associations of Unions signed a memorandum of understanding (MOU) that outlined the actions that would be taken to eliminate manual handling of patients/residents in British Columbia. This was an exciting initiative aimed at preventing patient handling injuries.

The MOU stated that all parties involved agree to establish a goal of eliminating all unsafe manual lifts of patients/residents through the use of mechanical equipment, except where the use of mechanical lifting equipment would be a risk to the well-being of patients/residents (OHSAH, 2001). As well, the employers agreed to make every reasonable effort to ensure the provision of sufficient trained staff and appropriate equipment to handle patients/residents safely at all times, and specifically to avoid the need to manually lift patients/residents when unsafe to do so. It was stated that the use of mechanical equipment must not introduce a risk to the well-being of the patient/residents and that sufficient staff must be made available to lift patients/residents safely.

To achieve this goal of implementing a "no unsafe manual lifting" policy, OHSAH worked in partnership with the WCB, the Ministry of Health and others to establish a financing framework to make funds available to purchase necessary mechanical equipment. In accordance with this memorandum, OHSAH was also responsible for the development and distribution of clear industry guidelines for safe patients/residents handling. The signing of a "no unsafe manual lifting" policy was a bold step for the province, and a great example of the commitment of both unions and employers to prevent injuries to healthcare workers.

LIFTING EQUIPMENT PROCUREMENT PROJECT

CEILING LIFT PROGRAM

The "no manual lift" policy was important in creating a new initiative for securing funds from the Ministry of Health to ensure adequate lifts in facilities. The provincial Ministry of Health Services provided $15 million in funding in 2001 to 2002 for the purchase of mechanical lifting devices and electric beds. The Ministry of Health Services agreed to allow the carry-over of funds from this fiscal year to the next, to grant time for the purchase of equipment that the program requires. In addition to this $15 million, the WCB provided $6 million as a "targeted rebate." This rebate was taken from a surplus of $20 million in the WCB acute care rate group. Since 98% of the rate group is comprised of the six British Columbia Health Authorities, the WBC agreed that this "targeted rebate" be reinvested by the Health Authorities over a 2-year period (January 2003 to January 2005) into any element of a "no manual lift" patient handling program.

A multi-agency Overhead Lift Program Steering Committee gave general direction as OHSAH developed a funding framework and program materials to support the installation and use of patient/resident ceiling lifts in healthcare facilities across British Columbia.

Program materials were developed to aid facilities in budget estimation, the selection of equipment suppliers and distributors, and on the configuration, use, and maintenance of ceiling lifts in a variety of settings. The program also contained material produced by the Okanagan-Similkameen Health Region (OSHR), given that it had implemented ceiling lift programs. OSHR's material was an excellent application of the management model being developed for this program.

OHSAH is producing a "report card" on the program's progress to date, including specific recommendations for the coming year. Through full participation of the local joint Occupational Health and Safety Committees, OHSAH evaluated the progress of facilities working towards the goals on the MOU and made recommendations to the Ministry of Health that all new facilities be equipped with appropriate lifting equipment. The report card includes: interviews with Directors of Health and Safety on the status of ceiling lift and electric bed installations; patient-handling injuries, bed counts, equipment inventories, and training programs; and in-depth target facility interviews to better understand the processes used to implement ceiling lift programs.

CONCLUSION

With injury rates continuing to increase at a rate higher than the provincial average, the healthcare industry of British Columbia has made innovative steps forward in the last few years to prevent injuries — particularly back injuries. OHSAH's establishment in 1999 was a step toward achieving this goal. Understanding the need for preventative measures, OHSAH was instrumental in the evaluation and promotion of ceiling lift devices as a method for patient handling, and worked with stakeholders to successfully secure $21 million in funding for the Health Authorities for ceiling lifts and electric beds. Achieving province-wide support, in the form of a "no unsafe manual lift" policy, was important for standardizing practices for safe patient handling and impacting injury rates at a broad level. The PEARS Program, Reference Guidelines for Safe Patient Handling, and OSHTips have also been important resources. As discussed by Yassi et al., (2002), the British Columbia experience illustrates the importance of a collaborative evidence-based approach to this complex problem.

REFERENCES

Blue, C.L. Preventing back injury among nurses, *Orthopedic Nursing,* 15(6), 9–20; quiz 21–22, 1996.

Cato, C., Olson, D.K., and Studer, M., Incidence, prevalence, and variables associated with low back pain in staff nurses, *AAOHN J.,* 37(8), 321–327, 1989.

Daynard, D., Yassi, A., Cooper, J.E., Tate, R., Norman, R., and Wells, R., Biomechanical analysis of peak and cumulative spinal loads during simulated patient-handling activities: a substudy of a randomized controlled trial to prevent left and transfer injury of health care workers, *Applied Ergonomics,* 32, 199–214, 2001.

Engels, J.A., Van der Beek, A.J., and Van der Gulden, J.W.J., A LISREL analysis of work related risk factors and health complaints in the nursing profession, *Int. Arch. Env. Health*, 71, 537–542, 1998.

Engst, C., Chhokar, R., Miller, A., Tate, R., and Yassi, A., Effectiveness and cost-benefit of overhead lifting devices in reducing the risk of injury to care staff in extended care facilities, (in preparation for submission to *Ergonomics*), 2003.

Ferguson, D., Strain injuries in hospital employees. *Med. J. Aust.,* 1(8), 376–379, 1970.

French, P., Flora, L.F., Ping, L.S., Bo, L.K., and Rita, W.H., The prevalence and cause of occupational back pain in Hong Kong registered nurses, *J. Adv. Nursing,* 26(2), 380–388, 1997.

Fujimura, T., Yasuda, N., and Ohara, H., Work-related factors of low back pain among nursing aides in nursing homes for the elderly, *Sangyo Eiseigaku Zasshi,* 37(2), 89–98, 1995.

Garg, A., Owen, B., Beller, D., and Banaag, J., A biomechanical and ergonomic evaluation of patient transferring tasks: wheelchair to shower chair and shower chair to wheelchair, *Ergonomics*, 34(4), 407–419, 1991.

Garg, A., Owen, B.D., and Carlson, B., An ergonomic evaluation of nursing assistants' job in a nursing home, *Ergonomics,* 35(9), 979–995, 1992.

Gingher, M.C., Karuza, J., Skulski, M.D., and Katz, P., Effectiveness of lift systems for long term care residents, *Phys. Occup. Ther. Geriatrics,* 14(2), 1–11, 1996.

Harber, P., Billet, E., Gutowski, M., Soohoo, K., Lew, M., and Roman, A., Occupational low-back pain in hospital nurses, *J. Occup. Med.,* 27(7), 518–524, 1985.

Holliday, P.J., Fernie, G.R., and Plowman, S., The impact of new lifting technology in long term care: a pilot study, *AAOHN J.,* 42(12), 582–589, 1994.

Hollingdale, R. and Warin, J., Back pain in nursing and associated factors: a study, *Nursing Standards,* 11(39), 35–38, 1997.

Jensen, R.C., Disabling back injuries among nursing personnel: research needs and justification, *Res. Nursing Health*, 10(1), 29–38, 1987.

Jensen, R.C., Back injuries among nursing personnel related to exposure, *Appl. Occup. Env. Hygiene,* 5(1), 38–45, 1990.

Jorgensen, S., Hein, H.O., and Gyntelberg, F., Heavy lifting at work and risk of genital prolapse and herniated lumbar disk in assistant nurses, *Occup. Med.,* 44, 47–49, 1994.

Josephson, M., Vingard, E., and Music-Norrtalje Study Group, Workplace factors and care seeking for low back pain among female nursing personnel, *Scand. J. Work Env. Health,* 24(6), 465–472, 1998.

Laflin, K. and Aja, D., Health care concerns related to lifting: an inside look at intervention strategies, *Am. J. Occup. Ther.,* 49(1), 63–72, 1995.

Larese, F. and Fiorito, A., Musculoskeletal disorders in hospital nurses: a comparison between two hospitals, *Ergonomics*, 37(7), 1205–1211, 1994.

Lee, E.J. and Yassi, A., Promoting evidence-based decision making in injury prevention and return to work programs in the health sector, *E-Health,* Vancouver, British Columbia, 2002.

Leighton, D.J. and Reilly, T., Epidemiological aspects of back pain: the incidence and prevalence of back pain in nurses compared to the general population, *Occup. Med. (London),* 45(5), 263–267, 1995.

Ljungberg, A.S., Kilbom, A., and Hagg, G.M., Occupational lifting by nursing aides and warehouse workers, *Ergonomics*, 32(1), 59–78, 1989.

Marras, W.S., Davis, K.G., Kirking, B.C., and Bertsche, P.K., A comprehensive analysis of low-back disorder risk and spinal loading during the transferring and repositioning of patients using different techniques, *Ergonomics*, 42(7), 904–926,1999.

Malmivaara, A., Hakkinen, U., Aro, T., Heinrichs, M.L., Koskenniemi, L., Kuosma, E., Lappi, S., Paloheimo, R., Servo, C., Vaaranen, V., and Hernberg, S., The treatment of acute low back pain- bed rest, exercises, or ordinary activity? *New Engl. J. Med.*, 332(6), 351–355, 1995.

Morrison, J.B., Martin, S.H., Robinson, D.G., Roddan, G., Nicol, J.J., Springer, M.J-N., Cameron, B.J., and Albano, J.P., Development of a comprehensive method of health hazard assessment for exposure to repeated mechanical shocks. Proc.U.K. Group Meeting on the Human Response to Vibration, Motor Industry Research Association (MIRA), Nuneaton, U.K., 1996.

OHSAH, Reference guidelines for safe patient handling, OHSAH, Vancouver, British Columbia, 2000a.

OHSAH, Safe patient & resident handling: acute & long term care sectors handbook, OHSAH, Vancouver, British Columbia, 2000b.

OHSAH, Memorandum of understanding with regards to manual patient handling between the Association of Unions and Health Employers Association of British Columbia, OHSAH, Vancouver, British Columbia, 2001.

OHSAH, Evaluation of portable ceiling lifts for patient handling in diagnostic imaging, project update, OHSAH, Vancouver, British Columbia, 2003a.

OHSAH, The ceiling lift project at St. Joseph's General Hospital: follow-up evaluation August 2002, project update, OHSAH, Vancouver, British Columbia, 2003b.

Ono Y., Lagerstrom, M., Hagsberg, M., Linden, A., and Malker, B., Reports of work related musculoskeletal injury among homecare service workers compared with nursery school workers and the general population of employed women in Sweden, *Occup. Env. Med.,* 52**,** 686–693, 1995.

Owen, B.D. and Damron, C.F., Personal characteristics and back injury among hospital nursing personnel, *Res. Nursing Health,* 7, 305–313, 1984.

Owen, B.D., Garg, A., and Jensen, R.C., Four methods for identification of most back-stressing tasks performed by nursing assistants in nursing homes, *Int. J. Ind. Ergonomics,* 9, 213–220, 1992.

Pheasant, S. and Stubbs, D., Back pain in nurses: epidemiology and risk assessment, *Appl. Ergonomics,* 23(4), 226–232, 1992.

Ronald, L.A., Spiegel, J., Tate, R., Yassi, A., and Mozel, M.R., Effectiveness and cost-benefit of installing mechanical ceiling lifts to reduce musculoskeletal injuries in an extended care hospital unit, *AAOHN J.,* 50(3), 120–127, 2002.

Smedley, J., Egger, P., Cooper, C., and Coggon, D., Manual handling activities and risk of low back pain in nurses, *Occup, Env. Med.*, 52(3), 160–163, 1995.

Spiegel J., Yassi A., Ronald L.A., Tait D., Hacking P., and Colby T., Implementing a resident lifting system in an extended care hospital: demonstrating cost benefit, *AAOHN J.,* 50(3), 128–134, 2001.

Takala, E.P. and Kukkonen, R., The handling of patients on geriatric wards, *Appl. Ergonomics*, 18(1), 17–22, 1987.

Varcin-Coad, L. and Barrett, R., Repositioning a slumped person in a wheelchair. A biomechanical analysis of three transfer techniques, *AAOHN J.,* 46(11), 530–536, 1998.

Venning, P.J., Walter, S.D., and Stitt, L.W., Personal and job related factors as determinants of incidence of back injuries among nursing personnel, *J. Occup. Med.,* 29(10), 820–825, 1987.

Villeneuve, J., The ceiling lift: an efficient way to prevent injuries to nursing staff, *J. Healthcare Saf., Compliance Inf. Control,* January, 19–23, 1998.

WCB of British Columbia, Healthcare industry focus report on occupational injury and disease, WCB, Vancouver, British Columbia, 2000a.

WCB of British Columbia, Joint occupational health and safety committee: reference guide and workbook, WCB, Vancouver, British Columbia, 2000b.

WCB of British Columbia, CU 766001 Acute Care statistical profile: 1996–2000, WCB, Vancouver British Columbia, 2001.

WCB of British Columbia, WCB current and planned activities in healthcare, WCB, Vancouver British Columbia, 2002.

Willis, J., Back care: preventing injury, *Nursing Times,* 91(43), 44–46, 1996.

Yassi, A., Khokhar, J., Tate, R., Cooper, J., Snow, C., and Vallentyne, S., The epidemiology of back injuries in nurses at a large Canadian tertiary care hospital: implications for prevention, *Occup. Med. (London)*, 45(4), 215–220, 1995.

Yassi, A., Ostry, A.S., Spiegel, J., Walsh, G., and de Boer, H.M., A collaborative evidence-based approach to making healthcare a healthier place to work, *Hospital Quarterly,* 5(3), 70–7, 2002.

Yassi, A., Lau, F., Cooper, J., Moehr, J., and Lemieux-Charles, L., Promoting evidence-based decision making in injury prevention and return-to-work programs in the healthcare sector, Health Evidence Application and Linkage Network, 1998.

Injured Nurse Story #17: An Advocate for the Ill, Injured, or Disabled Nurse: It Started with One

by Cynthia Barnes

The advocacy role within my hospital corporation had roots in contract language that gave protection of seniority and priority hiring rights to nurses with work-related injuries. Support for nurses with nonwork-related illness and disability was added in the 1998 to 2001 contract. The advocate role for a designated bargaining unit representative was developed and written into the Local Unit Guidelines of the United Hospital Bargaining Unit of the Minnesota Nurses Association. The advocate for ill, injured, or disabled nurses is a bargaining unit specialized appointment, approved by the bargaining unit representatives (union stewards). (Role description follows.) Time is paid according to agreements for all official labor-management activities. Because I had been doing the advocacy role as a bargaining unit "rep" for more and more nurses, my appointment was quickly approved.

Like many nurses who have been around hospital nursing a number of years, I remembered nurses who just disappeared following a back injury. Back then, many believed that we all had a husband somewhere that would be able to support us just fine. There was no support or help in a job search and the workers' compensation system seemed punitive. Many of us who had injuries, especially back injuries, ran to our private physicians for fear of losing our jobs. I was one of those who had a back injury, was hospitalized and, then, any time thereafter, called in sick with "the flu" just to avoid the system. Over these last few years, there have been some dramatic changes, and yet, many issues remain, so many hurdles to climb over for the injured and now vulnerable nurse.

My path to the advocacy role really started with one. As the bargaining unit "rep" on my intensive care unit, I had been helping one of our nurses with some practice concerns. This nurse had very high standards and worked very hard to maintain these standards of nursing practice. Just as I began helping her work with management through these concerns, she was injured by a patient. She never made a full recovery. After she was injured, we worked through issue after issue, learning a lot about the workers' compensation system and the lack of support and opportunities within the corporation.

Along the way, the rehab center in another corporate hospital was developing an internal placement system for the corporation. Since this program was new, I accompanied the nurse to some of these meetings to support her in dealing with issues that arose. This internal placement system continued to improve and eventually became a very valuable asset. We also got a full time disability case manager at Employee Occupational Health and a change in the Workers' Compensation Administrator.

However, these important decisions were not in time for her. She left our hospital having exhausted all "light duty" positions. Even with the help of a qualified rehabilitation consultant (QRC), along with the internal placement service and the disability case manager, she found nothing within the corporation that was within her restrictions. Then, she had difficulty finding a position outside the corporation. She was devastated. She had gone from being a stellar employee to being

1-56670-631-9/03/$0.00+$1.50

cast out. She eventually found a position that paid well, but underutilized her intelligence, skills, and energy. Then I lost track of her. I assumed she needed some time to heal from the extraordinary stress that had been a constant in her life. Sometime later, seeming to have found a more suitable position, she had a cardiac arrest while at work and died, leaving a young son and husband.

Many months later, I was asked to accompany our Minnesota Nurses Association staff specialist, Bettye Shogren, an expert in workplace injuries, to present to the Massachusetts Nurses Association how far we had come in the return-to-work process. I was speaking about my role as an advocate and where we had started. As I started talking about that first nurse, I started to cry. I was unprepared and surprised. I told the audience that I had no idea I would react this way and we all would have to be OK with it and continue. What an epiphany. What a battle that had been and how right she had been on the issues. How I wished we could have done more to help her. What a motivation to continue.

By the end of 1999, all seven nurses, who would not be able to return to bedside nursing, had found positions within the corporation. Relationships with all parties had improved greatly. While nurses were still struggling with many issues, they were returning to work. Through rehabilitation programs and work hardening, by gradually increasing hours on their nursing units, many nurses eventually returned to a full schedule. Working with the disability case manager, the assigned QRC, and managers of the units and care centers, most injured nurses are returning to work. I had expected to see my work decreasing as improvements were made and issues resolved. Yet, the nurses continued being injured at rates nearly double the state and national norms. As nurses got back to work, more were leaving the bedside temporarily or forever.

Sadly, not everyone with a work-related injury has been able to maintain employment, as some injuries are especially severe. The grieving from leaving the bedside is tremendous, being such a part of a nurse's identity. These people are facing concerns about affordable health insurance, loss of income, the reduced social security benefits that follow, and an uncertain future. Their injuries are life-changing.

These nurses often have complex questions about their earned benefits. With a complex pay and benefit scale, and an automated time collection and benefits system at a centralized Human Resources Center, errors have been made and misunderstandings occur. Often there are questions of contract interpretation regarding these benefits. One of these nurses, with great emotional risk to her, allowed me to save and play the recorded messages of frustration she had left for me.

These messages were played at the contract negotiating sessions in 2001. We nurses believe that hearing the emotional toll that a nurse can endure while trying to get information and resolution to questions regarding salary and benefits played an important part in gaining strong health and safety language in the 2001 to 2004 contract.

The other huge impact was the presentation of the business case. The corporation was spending millions on workers' compensation and even more in indirect costs; no one could deny that we had a problem. Included in this report by the corporate safety officer was an unforgettable statement: When a nurse is injured, "there is one less nurse and one more patient." In the midst of a nursing shortage that is a very powerful statement. Clearly, prevention with a safer workplace was where we needed to go. We had a common goal.

Nearing the end of 2002, tangible progress is being made, and not only with work-related injuries. Employment for nurses with nonwork-related conditions has been very difficult; there is no "light duty" budget for them and work-related injured nurses have the first priority for light duty jobs. Some of the concerns and restrictions are very different as well, such as waiting for a white blood count to increase before the nurse can have patient contact. Long-term disability insurance is available after 90 days, and by this time, many have exhausted their sick and vacation benefits. With input from our bargaining unit nurses and Minnesota Nurses Association staff and others, we have a new long-term disability insurance carrier that has improved service greatly. The 1998 to 2001 contract determined the right of these nurses to have the same services of return to work and the advocate to support them through the process. Best of all, the employee occupational health leaders are working to bring work-related and nonwork-related injuries, illness, and disability under one umbrella.

We have come a long way. When I started this work, I thought we would work through many issues and then I would gradually work myself out of a job. That doesn't seem to be the case. But, with the new value on safety, corporate-wide support, and work happening before our eyes, I see a different future for those nurses who follow us. No more disappearing nurses. It will be a nurse's choice to leave the bedside.

An advocacy role can take an emotional toll and one must be very careful with personal and professional boundaries. Though much of the job of advocate is "on the job training," there are continuous learning opportunities. There are classes on workers' compensation, Family Medical Leave Act (FMLA), the Americans with Disabilities Act (ADA), and the Occupational Safety and Health Administration (OSHA). The expert staff at the state Department of Labor and Industry is a great resource, as well as all the related web sites, including the site for the Social Security Administration. There is always something new to learn, some new twist or turn to navigate through the process. I have a thick notebook to pass on, but the role of advocate will always have something new to ponder.

I am grateful to have been on this journey. Communication has been so important and I am thankful for the support of the bargaining unit leaders, members and staff, the Employee Occupational Health staff, especially the Disability Case Manager, and the hospital and corporate administrative leaders, all of whom make the role of advocate possible. Most of all, it has been an honor to share in the often difficult lives of these nurses. I hope I have helped in some way.

JOB DESCRIPTION

From:
United Hospital
Minnesota Nurses Association Registered Nurse
Local Unit Structure Guidelines

Article VI: Specialized Appointments

Advocate for ill, injured, or disabled nurses at United Hospital established in contract negotiations for 1998–2001 contract.

1. Will help to establish effective communication between Employee Occupational Health (EEOH) and/or the hospital and the individual nurse.
2. If necessary, be a part of conflict resolution between the nurse and EEOH and/or the hospital.
3. Insure that the contractual rights of the nurse are upheld.
4. Participate as needed or requested on any committee or task force concerning Health and Safety, work and non-work related injuries, disability, or retraining issues.
5. Is, or will become, familiar with the services at United, Allina Health system, local or state agencies that may be of help in working with ill, injured, or disabled bargaining unit nurses.
6. Maintains communication with leaders of bargaining unit, MNA business agent, unit representatives, and nurse as appropriate.
7. Provide, as needed or requested, education to bargaining unit reps on issues affecting ill, injured, or disabled nurses at United Hospital.
8. Work with affected bargaining unit member with processing long-term disability forms and ensure compliance with contract, pg. 58, B.7, that the human resources person advocates for the employee.

Injured Nurse Story #18: More Valuable than Machines

by Maggie Flanagan

An estimated 38% of nurses experience a work-related back injury during their career as a nurse. However, not all nursing injuries are due to lifting patients. Workstations also can be damaging for nurses, especially in areas that are fast-paced like intensive care units. Nurses who must work hard and fast, working long shifts without breaks (isn't that all of us, now days?) are at high risk for cumulative trauma injuries. (Please realize that even though I refer to "nurses" as I write, all healthcare workers can be affected by what I describe.) These injuries commonly start with subtle aches, pains, numbness and/or weakness that often go away after a day or two off. What is terrible about these types of injuries is that their associated symptoms can become chronic. It is as if every shift is "a down payment" for an injury that eventually won't go away. "It's yours for keeps!"

Cumulative trauma injuries affect the hardest workers, too. These are the nurses who rise to the occasion of poor staffing or high acuity patients and who do not stop for breaks when patients are in need. (What nurse could do anything less?) Dehydration is also suspected as a factor in these type of injuries, which is "a double whammy" for nurses who cannot drink fluids in patient care areas, much less take a break. In fact, these injuries target those who work without breaks, stay extra hours or work extra shifts. What a price to pay for being "a good nurse" and helping out!

Commonly, workstations in hospitals do not take into account the human who delivers care. The typical patient care area in an acute healthcare setting is small, just enough room for a patient's bed and barely much more. The machines are then crowded in around the patient. The nurse is rarely thought of when the set-up is established. Machines take priority over the humans in the typical ICU workstation.

The flooring in nursing areas is often such poor quality, such that many nurses complain of lower extremity aches and pains. Especially in areas that require a lot of walking and standing, such as emergency departments, surgical suites and "floor" units like medical-surgical and orthopedics. Unnecessary pain indeed, since flooring that is padded is acceptable and available for these areas. Not only are these types of flooring worker-friendly, they can be very easy to clean!

For me, it was the neonatal ICU workstation that set me up for neck, back, and shoulder injuries that I did not think would become chronic. I did not realize what the common ergonomic hazards were in my various workstations during the years that I worked as a neonatal nurse. I tell my story to get the warning out: what you don't know, can hurt you. And to let other nurses know that the neonatal ICU is not a place of refuge for back-injured nurses. It can be a place to injure the rest of your back, and your shoulders and neck. Just ask around NICUs, and see what those nurses tell you about the aches in their upper bodies, their shoulders especially, as well as their lower backs.

The NICU workstation hazards are very obvious to anyone with even a small amount of ergonomic education. However, the training I received over the years from my employers did not include any such ergonomic information. It was about how to lift properly: use your knees, keep the load close to you, keep your back straight. Rather ironic, being that neonatal patients average two to three pounds in weight. The limited amount that neonatal nurses lift these tiny patients would be highly unlikely to cause an injury!

1-56670-631-9/03/$0.00+$1.50

Now I know that many workstation ergonomic hazards are simple and inexpensive to fix. Even with new units, many employers still disregard the human factor in their physical redesigns. Ridiculous, because good ergonomics increases efficiency and improves patient and worker safety. Education about reducing and eliminating many common nursing ergonomic hazards is not complex or difficult to understand. Yet nurses are not warned about these subtle but eventually disabling injuries. Commonly, nurses are not considered when units are redesigned. It would be relatively easy to eliminate or minimize many of these hazards.

I don't want this to happen to anyone else. I don't want this to happen to someone, like it happened to me. Dedicating my working life to taking care of others' children, becoming unable to take care of my own. The reality of workplace injuries goes beyond the lost workdays and dollars spent. The reality follows you home. The pain is with you as you fall asleep, the last thing you feel as you drift away. And pain is the first thing that you feel as you awake. Because of pain, you become isolated from those you love. Pain is now your constant companion. It is with you "24/7." Before shift, after shift, days off, holidays.

No one comes home with you to help take up where you now leave off, when it comes to taking care of your family. Your family life is devastated. You become hopeless. You have lost your career, your health, your family life, in one swoop. You fear simple every day tasks, knowing that they can cause pain, or worse, set you back by re-injuring you. Your children now take care of you. They watch too much TV because you are in pain, on the couch. They are losing their childhood. They are young enough that they will never remember you as you once were: active, strong, confident, healthy, playful, enthusiastic about life.

Years of repetitive stress to my back, neck and shoulders caused me pain. But working weekend shifts allowed me 5 days to recover. I did not understand that this pain could become chronic. I did not understand that it could set me up for further injury. But it did. The day did finally come…

That day was a shift in the neonatal ICU that I will never forget, for many reasons. My patient had undergone emergency surgery. He was fighting for his life with all the reserves that a 2-pound infant could muster. Earlier that day, I had pushed him to an operating room on a large neonatal open warming bed that was especially modified for surgery. Now, after pushing him back into the NICU to recover, I vigilantly monitored him. I had made sure that a pain medication "IV drip" awaited him upon his return from surgery. He did not need to feel post surgical pain as he now fought for his very life.

"Frequent vital signs" are part of monitoring patients postoperatively, as every nurse knows. I was soon taking my young patient's vital signs every 15 minutes, which required twisting and bending my neck and back to see my patient's skin temperature reading on the back of the surgical warming bed. The temperature monitoring equipment on that bed had been reversed, so that the patient's temperature could be observed from the back of the bed while the operation was being performed. This allowed the neonatal nurse to stay out of the way of the surgical team. (Monitoring a neonate's temperature is very important because operating rooms are often cool for the functional comfort of the surgical team, thus placing neonates at risk for cold stress.)

The baby's vital signs postoperatively were stable. Everything was in order. He was resting without any indication of postoperative pain. The ventilator was breathing for him. His postoperative laboratory values were normal. His surgical wound assessment showed no concerns. I finally had a chance to catch up on charting for a very busy day that started at 7a.m., and now was close to 5 p.m.

As I began to chart, I noticed that day's charge nurse approaching the empty bedspace next to me. She was moving toward a huge cardio-pulmonary monitor, located on a five-foot high recessed shelf, like all the other monitors in the NICU. She looked like she wanted to move the nearly 75-pound monitor. This same nurse had told me only 2 weeks prior how much back pain she was having at work — enough pain to prompt her to request an ergonomic evaluation of our unit. She stood inches taller than me. I have heard of taller nurses leaving NICU's because of "the one size fits all" equipment/workstation begins to take its physical toll on them. Her ergonomic evaluation had recommendations that she was told would not be possible due to the expense to the unit. (I would later learn that many ergonomic adjustments in our unit would cost nothing! When machines are considered expensive and irreplaceable, does that mean nurses are considered cheap and disposable?)

"STOP! Don't move that monitor by yourself! You will hurt yourself!" I said, alarmed.

"I have no choice!" the charge nurse replied. "I have called all around this hospital and no one is available to help move this monitor."

"But why do you have to move it?" I asked.

"Because we have a critically ill baby that needs this monitor. If this monitor is not moved, we will have to move the baby to this bedspace, instead."

Now I knew that moving critically ill babies is to be avoided at all costs. I have seen babies not recover after being moved. Never wanting to put a baby in harm's way, I said: "Here, let me help, so you don't get hurt." Previously, I had seen male staff members, by themselves, move these monitors in the unit, without any obvious physical injury. If they could move these monitors alone, surely the two of us could manage to move one.

We moved the monitor slowly and carefully, using the best body mechanics possible. We then placed it on a rolling cart, and the charge nurse moved it to the bedspace of the critical baby. The charge nurse then asked me to help her put the monitor up on the shelf. This time the move did not go smoothly because the five-foot shelf was recessed back from a counter at my waist-level, I had to extend my arms to put the monitor on the upper shelf. Possibly more weight shifted my way because I was the shorter person doing the lift. My arms were shaking as I used all my strength to put the monitor in place. I could feel my back and neck and shoulders tighten up. Once the move was over, I immediately went back to my recovering patient, nearby.

He was remaining stable, and appeared to be recovering without complications. Still requiring frequent assessments and vital signs, I was once again twisting and bending to look at the skin temperature monitor on the back of the surgical bed. Soon after that, I began to have severe spasms, especially in my mid and upper back. The same charge nurse came by for an update on my patient, as we were nearing the end of our shift. I told her about my spasms and filled out an injury report. I was not scheduled to work again for five days. I assumed that going home and resting for the week would take care of my pain and spasms. It always seemed to before…

After all, I had prior back strain after taking care of infants that required prolonged bending. I also had neck pain from looking quickly at alarming equipment located in difficult to visually-access places. (Monitors commonly were located behind me while I attended to patients, instead of in front of the patients and myself — which would have made the most sense ergonomically, and would have also allowed me to monitor patients more effectively.)

I also had shoulder pain from long, frequent and high reaches. These reaches were continuous in our unit, since we had constant false alarms from monitors on the same five-foot tall recessed shelves. In effort to keep these alarms from damaging the hearing or stressing our most fragile patients, neonatal nurses are constantly reaching to silence these annoying false alarms. Many times one could not let go of a patient or stop what one was doing to answer these alarms, which would then require a long, high, impossible reach to be performed.

So rest I did, after that lifting injury. Resting as much as one can, being a mother of a 3-year-old and 5-year-old. My husband worked during the week, so it was "mommy time" at home during the week. The spasms had started to subside, but the pain did not completely resolve. Saturday, my first day back to work, the pain began to increase as the shift progressed. I thought about calling in sick on Sunday, thinking I was not ready to be back at work, now realizing that I was still injured. I mentioned this to a co-worker, who promptly told me how short-staffed the unit would be if I called in sick.

Assignments in the NICU were always tight anyhow. My typical 12-hour shift included an unpaid hurried 30-minute lunch break. Rarely would I get more than that, and never would I get the two 15-minute rest periods that we were entitled to, if our patient load would allow. It was often difficult to get a bathroom break. And no one would cover for nurses on their breaks, meaning that you would become responsible for more patients, while the nurses you worked with took their breaks. Nurses commonly worked past the point of hunger, thirst, and fatigue. All the while, being required to be in top mental and physical condition to respond to, and take care of potentially life-threatening patient events.

Hearing about the short-staffing on Sunday, I decided that I would "tough it out" for one more day. After that, it would be five more days to rest and heal. However, Sunday completely undid me. One of my patients, with whom I had a very busy but manageable day on Saturday, became septic and extremely unstable. I responded to her quickly and continuously in the ergonomically hazardous workstation she was in. Early in the shift, I even suctioned her in her parents' laps, trying to allow them time to hold her, before it was obvious how sick she was becoming. I ended up suctioning and replacing her oral-pharnyx indwelling suction tube countless times, as her thickening secretions began to occlude her airway. She finally ended up requiring a septic work-up, intubated and on a ventilator.

The charge nurse for that day took over my patient's care as I left the room in tears, begging for a much-needed break. She later asked me, angrily, "what was wrong" with me. I was so worried about my patient and on such "an adrenaline rush" at that point, I didn't even realize how much pain I was in. All I knew, while talking to the charge nurse in our break room, was that I was overwhelmed and unable to stop crying. As the adrenaline began to subside, the pain soon became evident. After that day, I did not return to bedside nursing for six months. The down payment was made. The damage was done.

As if the physical damage wasn't enough, this was only the beginning of the descent into the labyrinth of HELL of a work-related injury. Just when you are physically and emotionally the most vulnerable, then you need to navigate a system that discourages you, calls your integrity into question, blames you for all your troubles. Recounting this story is distressful, but it becomes even more so, when I remember how I was treated by my employer and their workers' compensation insurance company. In the short term, workers' compensation benefits do compensate injured workers. In the long term, the workers' compensation system works to protect the employer by limiting how much compensation injured workers receive.

My injury flared up constantly when I tried to do anything with my arms forward. It was the same movement that I used hundreds of times at work every day, to answer alarms on the monitors on those 5-foot high shelves. It was the exact same movement that I used to move the monitor on that fateful day. My employer offered me temporary light duty jobs such as filing and answering phones. I refused and explained that there was not much I could do with my arms. Even typing was painful.

At home I could not dress my children. I could not bath them, comb their hair or brush their teeth. I could barely prepare food for them. Lucky for me, they could feed themselves. I could not pour them milk from a gallon container. I could barely reach into the refrigerator to pull food out for them. The chores involving the vacuum cleaner, the dishwasher, and the washer/dryer were definitely out. The one time we went to the store, my 5-year-old had to push the shopping cart. I did not go shopping again for a long time. When we went to the nearby playground, I could only watch them. My husband became a single parent, essentially. He worked full-time and did everything else. He never lost faith in me, and lucky for me, my marriage survived. I know of relationships that did not endure a partner being disabled. It really bothered him to see me in pain.

In the beginning of my acute injury, I took narcotic pain medications, but only when my husband was home to watch the kids. I was able to stop using the narcotics after some time. My physician tried other different medications for pain treatment, but I did not get much relief from them. I eventually realized that I was having a drug reaction from all the nonsteroidal anti-inflammatory drugs (NSAIDs) I was on. It would seem that I had finally become allergic to them.

I was finally allowed to have an MRI of my upper back and neck, which showed several discs were damaged in my neck and thoracic area. It took me three months before I had that procedure, because with a work-related injury, your employer's insurance company can "call the shots." With a work-related injury, eventually the worker becomes evaluated to see if they are malingering. The independent medical evaluators that I was sent to by my employer were anything but "independent." (Their goal is to get injured workers back on the job, even if it takes denying that anything is wrong with them.) The two doctors I was evaluated by minimized my physical condition. They denied my pain. Ironic, because as a healthcare professional, you are taught that the patient is "the expert" when it comes to describing her own pain. Ironic, because healthcare professionals are taught that patients do not exaggerate their pain. It would seem that these independent medical evaluators were not current on modern pain research. When I explained to one of the doctors that my ergonomic evaluation recommendations were not being followed by my unit manager, and that she was placing more nurses in harm's way, the doctor wrote in his report that it was my poor attitude, and not my physical condition that was my major problem. (I don't know how those doctors can sleep at night.)

I had a skilled physical therapist who worked with me, and was very helpful as part of my recovery. I eventually went into a back rehabilitation program. I was afraid that I could get re-injured in a rehab program, and was very cautious in entering into one. I did have to work through the pain, but my back did get stronger. However, my back and neck still hurt. Using my hands above my shoulders continued to be painful. Taking my jacket off over my head, in my car, would send spasms of pain in my shoulders and upper back. I did have setbacks in my recovery, including flare-ups caused by the therapy programs.

The back rehab program required me to find daycare for my children, which was difficult. I had accepted part-time work several years prior when my employer asked, wanting to downsize nursing positions. Now that I was an injured worker, I was told that I would have to go to fulltime rehabilitation, no matter what my prior work schedule was or what my family needs were. I was told point-blank by an insurance adjuster that my family "was no longer my priority."

After 6 months, I was released to return to work by my primary physician. My part-time day shift job was gone by then. I was offered a full-time day shift position — which I did not believe I had the stamina to do, much less how it would impact my family. Or part-time night shift — which after working up through the ranks from night shift to finally day shift, after working for this employer for nearly 10 years, was hard to believe that was my only option. Hard to believe, in the midst of a national/global nursing shortage. Hard to believe, because at that time, my unit had unfilled nursing positions and the hospital had over 80 positions open for nurses on a regular basis. I went back into a per-diem slot without benefits. After a year, my union seniority was enough to get my part-time day shift position back, benefits included. Financially we lost thousands, due to this injury.

Some of the doctors I saw in the course of the treatment grew frustrated with me, seemingly tired of the complaints of pain and hopelessness. One doctor, whose only suggestion was cortisone injections to numb my pain trigger points, threw up his hands and walked out of the room when I refused. I had asked for written information about the injections first. He said that he didn't have any. I asked why would I want to numb the pain and allow myself to re-injure the very areas that my body was trying to protect? Isn't my body using pain to tell me to stop what I was doing? He had nothing else to offer me and nothing else to say.

Another doctor told me that no one gets injured discs in their thoracic area after I told her I thought that is what had happened. The MRI proved me right, unfortunately. To me it also proved that years of cumulative trauma to that area of my back from the long, high reaches that I could still do in my sleep to the level of those monitor alarms, had finally taken their toll. The same with my neck from constant quick twisting to see the monitors positioned behind me when alarms occurred during patient care. My back and neck blew out at the weakest points during the lifting incident. The weakest points, developed from years of unnecessary cumulative trauma and from being "a good nurse" in a hazardous work environment.

Two years post-injury I finally was allowed to receive an independent ergonomic evaluation of my workstation. I believed that a fair ergonomic evaluation needed to come from someone not employed by the hospital, someone who was not being told by hospital management "what was possible, what was too expensive," etc. One of the independent recommendations was that I should receive more therapy. I was told that I looked like my upper body was frozen into a forward bending-over position.

At first I was reluctant. I had never used a chiropractor and never had one recommended before. My life changed for the better when I was referred to an excellent chiropractor. Using her chiropractic skills along with deep muscle work, I was able to resolve my injury further. She was able to adjust my spine and neck back into place. And she broke up adhesions that guarded my injuries, which had been reluctant to release. I wondered had I seen her earlier, I would have healed quicker. Nonetheless she was a godsend for me.

My workplace eventually remodeled some of our workstations. My manager who had full knowledge of recommendations from several ergonomic evaluations, including mine, did not utilize that knowledge in making the unit ergonomically safe. Once again these improvements would cost little to nothing. Including the adjustable equipment that is designed to be ergonomically sound was not correctly adjusted — even after nurses communicated that there were problems in the new workstations! I was especially upset about this, concerned about the other staff who continue to be at risk for injury. I would even see and hear about male staff being asked to move the very same monitors that injured me.

Another hazard in NICUs is the use of pulse oximeter monitors, many times referred to as oxygen saturation monitors. The type of these monitors commonly used in NICUs ring off false alarms with patient movement. (What infant is ever still unless they are really sick, and then there are other alarms to ring off!) These constant "motion-detector" false alarms put the premature infant's hearing at risk. The alarm sound needs to be loud enough to get the caregiver's attention, which then could be too loud for the delicate, even underdeveloped auditory sensory organs of the premature patients. Improved pulse oximeter monitors do exist, but often are not even considered for use in NICUs, many times due to hospital purchasing contracts. These monitors would help eliminate the false alarms that can impact neonatal hearing and rest, as well as the need for frequent reaching from neonatal caregivers to silence them. (For more information about hospital purchasing groups and their impact on blocking innovative pulse oximetry technology from use in NICUs, read the *New York Times* article: "Two Powerful Groups Hold Sway Over Buying At Many Hospitals," March 4, 2002.)

Changes that could protect nurses would include placing monitors within view while providing care to patients. Remote control of alarms would eliminate many long frequent reaches. Other suggestions include monitors that have screens that are easy to read from around the area, not just straight in front of the screen, because nurses aren't always directly in front of the monitor when alarms ring. Monitors should have alarm configurations that are easier to read— such as large out of parameter values.

Nurses should be able to gauge their responses to the true patient condition, rather than reacting to a machine that has limitations on how accurately it can evaluate and then report a patient's condition. We need machines that aren't so needy to be touched! We are busy enough with patients that need to be touched. Machines should not become a burden but an ally. When split-second response to alarms is "the name of the game," nurses shouldn't be twisting their necks quickly and frequently to see monitors. To do so only increases the chances of neck injuries. Another suggestion is to have monitors mounted on equipment that allows them to swivel and to be brought within reach. This equipment is not expensive and it also can make these monitors safe from falling on patients or staff. The area then becomes safer during earthquakes, another commonly overlooked safety issue in many healthcare settings.

Ergonomics of workstations should prioritize the humans, not the machines. What nurses use constantly should be located right next to them. What they use frequently should be located just a little farther out from their hands, requiring barely a reach. What nurses use infrequently can then be placed a little further out from them. The idea is to eliminate long and frequent reaches. Keep things close. Protect shoulders and backs from awkward postures and long/high/low reaches. Avoid twisting one's torso and long periods of being bent over without a pause. Chairs should be readily available to allow nurses to sit when they get a chance. Monitor controls, infusion pumps, respiratory equipment and other medical machinery should be easy to reach and located at one's waist-level, never below one's knees or above one's shoulders. In units like the one where I was injured, it cost nothing to pull things within reach and down to waist-level.

General ergonomic suggestions that make good sense include padded flooring and adequate/adjustable task lighting. Nurses should have breaks they are entitled to, so they can keep hydrated, very difficult to do in the best of circumstances as nurses could attest to the dry environment of hospitals. Even short breaks help — something that has been proven in other industries: breaks actually improve workers' production! Assignments must be made that take into account breaks, otherwise they never seem to happen for nurses. Heavy patient workloads drive nurses out of healthcare, making the shortage even worse. Long shifts do the same. Nurses should be able to work shifts that fit their physical needs and lifestyle, including part-time positions.

In talking to nurses from other areas of the hospital, I learned that lift equipment was slow to become available for adult/pediatric nurses. Gastric bypass surgeries were being increasingly performed at our hospital, without the proper equipment and furniture for these patients who can be larger than 500 pounds. Our flight nurses would fly long hours, hunched over in small jet airplanes, because their request for adequate seating was denied. They would have frequent back problems from this and from lifting people and heavy equipment during their rescue flights. The OR nurses were being injured at high rates as well. One problem was OR nurses holding the heart above the patient's chest, during open-heart surgery. These OR nurses were tasked with holding hearts that have been iced down for the procedure completely still, up to 15 minutes at a time. While they hold these hearts, they must back out of the surgical field with the rest of their body and face to avoid obstructing the work that must be done. The rehabilitation unit's personnel are frequently hurt because in rehabilitation you don't use lifts on patients. This is the area in healthcare where you push people to do for themselves. You test patients' physical limits. If patients fall, they may even fall on or grab nurses on their way down. When it comes to musculoskeletal injuries in healthcare, the examples are endless.

Ergonomic education is another strategy to prevent musculoskeletal injuries in healthcare. After such education, those who work at the workstations could be the best resources for finding solutions for eliminating and reducing their ergonomic hazards. Stretches and other physical activities could help minimize the impact of the work. This education must include identification of hazards and strategies to minimize and eliminate them, signs and symptoms of physical problems, benefits of breaks, and use of physical stretches/exercises. Turning the tide on these injuries takes management buy-in on the importance of education, prevention, and also accurate record-keeping. Management needs to actively engage in hazards analysis and educating and encouraging workers to report the subtle signs and symptoms of the onset of musculoskeletal injuries.

Recently a lot of attention has been paid to the national/global nursing shortage. The consequences of such a shortage are grave and are of concern for those who give care, as well as those who receive care. However, little regard has been given to one of the origins of this shortage: the high illness and injury rate of the healthcare workforce. It is unethical to continue to recruit healthcare workers, only to injure or infect them. It is unethical to sacrifice one person's health for another. It is unethical to treat anyone as expendable. It is unethical to treat people as less valuable than machines. It is unethical to continue to destroy people, especially when there are well-known solutions and fixes available.

Injured nurses, as well as all injured workers, need to make their voices heard. Our stories must be told. These types of injuries will only continue if we keep our silence. In closing, I would like to say what I have said before, when telling my story:

Work shouldn't hurt.

A job should not rob you of your health.

A job should not rob a child of a healthy parent, nor a marriage of a healthy partner.

Workers need to be viewed as being more valuable than machines, more important than profits.
All workers deserve a workplace free of recognized hazards.
Workers' compensation is never a better option than prevention.
Because when a worker develops a musculoskeletal injury,
It is not just another lost work day.
It is not just another grim statistic.
It can be a life lost forever to disability and pain.

Injured Nurse Story #19: The Victoria, Australia Story

by Elizabeth Y. Langford

ELIZABETH LANGFORD'S STORY

I am the coordinator of the Injured Nurses' Support Group (INSG) in Victoria, Australia. I am a registered general nurse, midwife, and researcher and have also worked in various, other general medical and surgical nursing settings, including orthopedics and pediatrics in the U.K. and Australia.

Before my injury I had a pre-existing spondilolisthesis at L5/S1, which was aggravated by 18 years of bedside nursing. My first surgery was performed when I was 36. I eventually had to have bone graft spinal fusion surgery of L5/S1 in October 1988. This was later followed by another graft fusion at L5/S1 in October 1989 and in November 1996 a metal fixation of L4/5. Due to accelerated degeneration, because of lower spinal fusion, I now have multi-level cervical and thoracic spinal degeneration with a bony spur on C6 and prolapsed disc in the same area. There is also some leg involvement. I made several attempts at returning to work including light nursing duties as a school nurse and sexually transmitted disease clinic nurse. In 1991 I undertook a postgraduate degree in Nursing and graduated with a diploma in Advanced Nursing (Education) in 1992. However, despite light duty and continuing education, my condition forced me to leave the workforce entirely in October 1993 at age 43.

In 1988 I joined Injured Nurses' Support Group (INSG) as a founding member. I had become very angry at the terrible stories I heard of how injured nurses were dealt with in the so-called "health system" and decided to do something about it.

After many frustrating years of writing articles for journals (e.g., the *Australian Nursing Journal*) and newspapers, then lobbying politicians, doctors, universities, government bodies, hospitals and other health agencies and various nursing bodies, I decided that major research was required. There was a lack of interest and integrity about the health issues nurses face due to their poor working conditions. I realized that nurses needed a dramatic cultural shift in their thinking, in order for them to see that they were working in settings with poor health and safety standards. I underpinned my research with: "the fact that nurse's health and safety is as important as that of their patients and clients."

In nursing literature, nurses' health has come in at a very poor second throughout history. Patients and doctors were always seen as more important. This comes from our profession being steeped in religious and military thinking, going back centuries. This has led to nurses' perceptions of their own welfare as low on the hierarchy. My premise is that patients, nurses, and other parties in the health scene could all be well cared for if we did things differently. We are all equally important to our patients and clients.

Guinnette Davies, of the Work Injured Nurses' Group in the U.K., visited Victoria in 1995 and gave me some useful ideas through her work in Britain with the RCN Work Injured Nurses' Group (WING) and eventually, after continued lobbying, I received assistance from Belinda Morrieson, secretary of the Australian Nursing Federation, Victorian Branch. I developed a questionnaire with some assistance from Gwyneth Evans and Jeanette Sdrinis, and the questionnaire was published in the *Australian Nursing Journal* (Victorian section) supplement. I received

1-56670-631-9/03/$0.00+$1.50

an adequate number of replies from injured nurses for the study all with similar issues and responses about their injury experiences. I performed a quantitative and qualitative research study, which revealed that nurses have the highest rate of injuries in the female work force in Victoria and found this to be similar both nationally and internationally. The resulting work was named *Buried But Not Dead* due to how nurses were and are dealt with. That is: use them up, spit them out, bury the problem and then bring in the next unsuspecting group of people with healthy bodies to start the whole process over again.

After undertaking my research, I then lobbied newspapers and national radio and television broadcasters in Australia with very good results and managed to have the problems injured nurses encounter examined on prime time national television. I knew that I would have to use the media, as the nursing fraternity and government bodies were "sitting on their hands." If I had not used the media, I knew the study would just be another dusty report sitting on a shelf.

Findings

The major findings of this research can be summarized as follows:

- The injuries nurses sustain are long-term, with 92% of the respondents showing long-term effects.
- Back injuries were the predominant injury sustained (73%), with the lumbar spine being the main site of back injury (70%). Of these back injuries, 57% were intervertebral disc injuries.
- Shoulder/arm injuries accounted for 9% of all injuries.
- Manual handling was the main causal factor of injury at 74%, with 83% of these injuries occurring in the ward.
- Psychological stress-related illnesses accounted for 61% of all reported illnesses, followed by illnesses caused by chemical exposure (11%). The main chemical irritant was glutaraldehyde.
- 39% of respondents stated that they felt they had been unfairly treated by their employer.
- 47% of respondents felt they had been unfairly dealt with by the insurance companies who administer WorkCover claims.
- Only 46% of respondents regarded the insuring agents' examining doctors to be competent and professional.
- The effect on the respondents' financial and professional life was dramatic, with 41% requiring months off work.
- Once able to return to some form of work, 46% were receiving lower wages.
- Only 48% were able to work at their pre-injury job or hours, and the social effects on these nurses were profound.
- Rehabilitation for injured nurses was also seen as inadequate. There were many comments by young injured nurses stating that they had useless degrees because their injuries were such that they could not return to their normal work. (The Victorian WorkCover Authority (VWA) on average last year only spent $52.50 per head for all health industry injured workers.)

Source: Elizabeth Langford, *Buried But Not Dead: A Survey of Occupational Illness and Injury Incurred by Nurses in the Victorian Health Service Industry*, The Australian Nursing Federation (Victorian Branch) Injured Nurses Support Group. 1997, p. 10.

After the television and radio exposure, I then approached our Health Minister at the time, Rob Knowles, and asked for a meeting to discuss the occupational health and safety issues nurses faced. By this stage, the Australian Nursing Federation, Victorian Branch (ANF, Vic Branch) became fully involved and published the report in book form.

At this time a nurse by the name of Louise O'Shea contacted me after she heard a national radio interview I was on. She specialized in a quality "no lifting" program. The ANF then ask Louise to come to Victoria and talk to our nurses about the "no lifting" system she had developed. This was performed in a series of "Expos" that included information and demonstration of the latest equipment.

FIGURE 1 Injured Nurses' Support Group Coordinator Elizabeth Langford, Injured Nurses Support Group Patron Judith Durham, Health Minister John Thwaites, Department of Human Services Nurse Policy Branch Director and Principal Nurse Advisor Belinda Moyes, and Caulfield General Medical Center Director of Nursing and Residential Services Alan Lilly at the Victorian Nurses Back Injury Prevention Project Evaluation Report launch.

From here on, the whole project, the "Victorian Nurses Back Injury Prevention Project," took off with top level government involvement due to my own, and the ANF's (Vic Branch) involvement and, also, because injured nurses were willing to tell their stories.

INJURED NURSES' SUPPORT GROUP

I would now like to say a few words about the Injured Nurses' Support Group (Figure 1), which has played a prominent part in this story. It was started in Victoria, Australia, in 1988 by a union official, but is a self-governing body run by injured nurses. This is an autonomous self-help group. All members are injured or ill due mostly to work-related factors. Meetings are monthly and there is no cost for membership. The group provides a supportive environment geared to assisting members to obtain the help they need in relation to obtaining benefits, rehabilitation, medical and legal assistance, and support with interpersonal problems. The group sponsors occasional workshops and offers a phone support service. Members assist each other through their experiences. The group also tries to raise awareness in the general community and offers assistance to other work groups when needed.

The group is fortunate in having a well-know patron, Judith Durham (OAM), the lead singer of the internationally renowned group "The Seekers." Judith assists us by raising awareness of the issues injured and ill nurses face and the No Lifting campaign through media outlets and attending various functions.

THE AUSTRALIAN NURSING FEDERATION, VICTORIAN BRANCH

The other organization that has played a vital role in this story is the Australian Nursing Federation (ANF) Victorian Branch, which is the main registered nurses' trade union in the state of Victoria, Australia. The ANF has done some great work in initiating the No Lifting policy. It was the first to have No Lifting Expos; formulated the Victorian No Lifting policy; has lobbied ministers and other government officials; won awards for its work, etc. The ANF Vic Branch is now recognized as a leader in their support of No Lift policies and their work has brought interest from overseas.

FIGURE 2 Elizabeth Langford and Jeanette Sdrinis with the "WorkSafe" Occupational Health and Safety Award for Outstanding Leadership in Health and Safety 2001. The award, which pays tribute to pioneering and sustained efforts made in occupational health and safety, was awarded to the Australian Nursing Federation (Victorian Branch) in recognition of development of a "No Lifting" policy involving the elimination of manual lifting for transferring and moving patients. Benefits of the policy include significant reduction in nurse injury rates from handling patients, improved WorkCover claims costs, reduced premium costs, and significant cultural change within the industry.
http://www.workcover.vic.gov.au/dir090/vwa/home.nsf/pages/so_worksafeawards_2001?OpenDocument&ExpandSection=6#_Section6

A Japanese professor of nursing has recently visited. The Japanese hope to initiate a No Lift program in their health system. The professor interviewed the ANF Vic Branch Occupational Health and Safety (OHS) officer, Jeanette Sdrinis and myself, to obtain an overview of how our project started and runs. The professor also visited a few health institutions that have the No Lift system running well.

Jeanette has assisted the ANF Vic Branch to become a leader in occupational health and safety for nurses. The ANF has won the government's prestigious OHS award, which covers every occupation in the state (Figure 2).

We have gone from the "bottom of the pile" in OHS to leaders, and it is now seen as prestigious to belong to this movement. *We still have a very long way to go.*

Jeanette and myself have worked together on the government's "Victorian Nurses Back Injury Prevention Project" since its inception. Jeanette has been a valuable asset to the ANF (Vic Branch) OHS staff.

INFORMATION AND FURTHER READING

For further information on the reports from Victoria, Australia the following list of material is supplied.

1. *Buried But Not Dead: A Survey of Occupational Illness and Injury Incurred by Nurses in the Victorian Health Service Industry.* Author: Elizabeth Langford. The Australian Nursing Federation (Victorian Branch) Injured Nurses' Support Group.
   Introduction: The dubious honor of having the highest injury rate in the female workforce has rested with nurses ever since WorkCare came into being in Victoria and records were kept. The WorkCover Authority's statistics for 1994/95 showed 1028 new injuries were accepted for compensation. In 1995/96, another 1060 claims were accepted. Importantly, this represents only those claims that were successful and, as anecdotal

evidence presented to the Australian Nursing Federation's Victorian Branch has demonstrated, this is only the "tip of the iceberg."

Through my role as contact for the ANF (Vic Branch) Injured Nurses' Support Group, I received a constant stream of information from nurses with injuries about the problems they endured in relation to injury management, treatment, rehabilitation, medico-legal issues, and workplace return to work programs. It became clear that this anecdotal information needed to be collated and critically examined so that the extent of the problem could be accurately assessed, recommendations made, and programs designed and implemented to relieve and prevent this suffering.

This survey of work related injury/illness experiences is the first step in this process.

To obtain *Buried But Not Dead*, go to: www.anfvic.asn.au/services_pubs.htm

2. *Victorian Nurses Back Injury Prevention Project Evaluation Report 2002.*

   Foreword: "The Victorian Nurses Back Injury Prevention Project (VNBIPP) was established in October 1998 to provide funding for healthcare organizations to assist them to implement programs to prevent back injuries amongst nurses.

   The project was established in response to growing concern amongst nurses and the industry regarding the unacceptably high rate of back injuries in the nursing profession and the enormous financial and human costs associated with such injuries. When the VNBIPP was initiated, nurses accounted for more than 54 per cent of compensation claims by health industry workers.

   This Government acknowledges the outstanding success of the VNBIPP that has been demonstrated in this Report. The evaluation indicates a reduction in WorkCover claims for injuries sustained by nurses by 48 per cent, a reduction in days lost due to injury by 74 per cent, and a reduction in the cost of claims by 54 per cent. These statistics represent an outstanding achievement in back injury prevention..." Hon. John Thwaites, MP, Minister for Health.

   To download *Victorian Nurses Back Injury Prevention Project Evaluation Report 2002*:

   Go to: www.nursing.vic.gov.au

   Select Nursing in Victoria — The Nursing Workplace which goes to...

   Victorian Nurses Back Injury Prevention Project Evaluation Report 2002.

   http://www.nursing.vic.gov.au/vnbippreport.pdf

   Also, in the site: *The Victorian Nurses Back Injury Prevention Project Bulletin 2001.*

   http://www.nursing.vic.gov.au/back_1201.pdf

3. *Transferring People Safely — A Practical Guide to Managing Risk: Handling Patients, Residents and Clients in Health, Aged Care, Rehabilitation and Disability Services.*

   "Every year in Victoria thousands of people working in health, aged care, rehabilitation and disability services injure themselves moving patients, residents or clients.

   For workers, this can mean personal pain and discomfort which sometimes lasts for years, affecting not only their work but their everyday lives, families and relationships. For employers, this type of workplace injury may lead to WorkCover claims and increased premiums, other costs and may affect morale....

   Transferring People Safely: A Practical Guide to Managing Risk gives Victorian employers and managers useful tools to examine the risk factors associated with moving or handling people and implement positive changes that eliminate or minimize those risks." Bob Cameron, MP, Minister for WorkCover.

   "This guide has been produced specifically for Victorian employers to assist in reducing the incidence and the severity of injuries to their staff resulting from manual handling risks when transferring patients/residents."

   To download *Transferring People Safely: A Practical Guide to Managing Risk*:

   http://www.workcover.vic.gov.au/vwa/publica.nsf/InterPubDocsA/C4DD9672BD12B51D4A256BDC0082A619/$File/Trans_People_Safely.pdf

4. *Designing Workplaces for Safer Handling of Patients/Residents: Guidelines for the Design of Health and Aged Care Facilities.*

"There is a link between the layout and design of a workplace and the risk of musculoskeletal disorders. These guidelines have been developed by industry for industry and they represent an important contribution to our knowledge about reducing risks through good design practices. Renovations or the building of new facilities provide an ideal time to incorporate occupational health and safety considerations into the planning process." Hon. Bob Cameron, MLA, Minister for WorkCover.

Introduction: "These guidelines are one step towards reducing the continuing high prevalence of musculoskeletal strain injuries among staff who handle patients or residents in Victorian acute and aged care facilities…

Research shows that patient/resident handling is the most frequent cause of back pain and injury to nurses with body stressing and lifting/handling injuries accounting for over half of these. Poor work place design is a major contributing cause of these injuries…"

"Our Guidelines for the Design of Health and Aged Care Facilities will be of assistance to planners, facility managers and direct care staff."

To download *Designing Workplaces for Safer Handling of Patients/Residents*: http://www.workcover.vic.gov.au/vwa/home.nsf/pages/so_aged/$File/age_care.pdf.

To obtain by email: publications@workcover.vic.gov.au.

Appendix A: Ergonomics for the Prevention of Musculoskeletal Disorders: Guidelines for Nursing Homes*

DISCLAIMER

OSHA's Guidelines for Nursing Homes provides information to help employers identify ergonomic stressors (physical demands that have been associated with certain musculoskeletal disorders) in their workplaces and implement practical measures to control such stressors. To develop these guidelines, OSHA reviewed existing ergonomic practices and programs in the nursing home industry, as well as available scientific information regarding ergonomic stressors and control methods. OSHA also conducted one-on-one meetings with major stakeholder groups to gather the best available information on the stressors that are present in typical operations and on practices, programs, and processes that have been successfully used in the nursing home industry.

The Occupational Safety and Health Act of 1970 (OSH Act) mandates that, in addition to compliance with hazard-specific standards, all employers have a general duty to provide their employees with a workplace free from recognized hazards likely to cause death or serious physical harm. These guidelines are advisory in nature and informational in content. They are not a new standard or regulation and impose no new legal requirements. An employer's failure to implement these guidelines is not a violation of the OSH Act.

TABLE OF CONTENTS

* Taken from the U.S. Department of Labor, Occupational Safety and Health Administration.

INTRODUCTION

NURSING HOMES: A DEMANDING WORK ENVIRONMENT

Despite the efforts of nursing home employers and employees in recent years, workers in nursing homes are more than twice as likely as other workers to be injured on the job. According to the Bureau of Labor Statistics, employees in nursing and personal care facilities suffer over 200,000 work-related injuries and illnesses a year (1). Many of these are serious injuries—more than half require time away from work. Workers' compensation costs for the industry now amount to nearly $1 billion per year (10).

Hazards encountered by nursing home workers may include exposure to bloodborne pathogens and other infectious agents, unsafe walking surfaces, hazardous chemicals, and the risk of workplace violence from combative residents. Perhaps the greatest factor contributing to the high number of injuries in nursing homes, however, is the physically demanding nature of nursing home work. Manual lifting, transferring, and repositioning of residents (referred to in this document as resident lifting or resident handling) are tasks that have been associated with an increased risk of pain and injury to caregivers, particularly to the back (3). These tasks can entail high physical demands due to the large amount of weight involved, awkward postures that may result from leaning over a bed or working in a confined area, shifting of weight or unexpected loading that may occur if a resident loses balance or strength while moving, and many other factors. These activities have been reported to account for the majority of injuries, lost and restricted workdays, and workers' compensation costs in nursing homes. (2).

Nursing home residents often require assistance to walk, bathe, or perform other normal daily activities; in some cases residents are totally dependent upon caregivers for mobility. Assistance is commonly provided by nursing assistants (e.g., nurse's aides, certified nursing assistants, and other employees who perform resident-handling tasks), but nurses and other staff members may be involved as well. Considerable physical demands may also be placed on other nursing home staff. Workers who receive, store, and distribute supplies, gather and process laundry, and provide food service to residents, for example, may also be exposed to various ergonomic stressors. The ergonomic stressors that workers in nursing homes face include:

- Force - the amount of physical effort required to perform a task (such as heavy lifting) or to maintain control of equipment or tools,
- Repetition - performing the same motion or series of motions continually or frequently,
- Awkward postures - assuming positions that place stress on the body, such as reaching above shoulder height, kneeling, squatting, leaning over a bed, or twisting the torso while lifting,
- Vibration - rapid oscillation of the body or part of the body, often caused by use of powered hand tools or equipment,
- Contact stress - pressing the body or part of the body against a hard or sharp edge, such as using the hand as a hammer.

Exposure to these stressors in the workplace can result in a variety of disorders in affected workers, including muscle strains and tears, ligament sprains, joint and tendon inflammation, pinched nerves, herniated spinal discs, and other conditions (4, 5, 6). These conditions, collectively referred to as musculoskeletal disorders (MSDs), may develop gradually over time or may result from instantaneous events such as a single heavy lift. Pain, loss of work, and disability may result (4, 6). Although these conditions may be classified as either injuries or illnesses for Occupational Safety and Health Administration (OSHA) recordkeeping purposes, they are referred to as injuries in this document.

Not all MSDs are related to work activities. Many MSDs are related to non-work activities, genetic causes, age, and other factors. MSDs may also result from accidents such as trips or falls. Finally, there is evidence that MSDs may result from certain psychosocial factors such as job dissatisfaction, monotonous work, and limited job control. However, these guidelines address only physical factors in the workplace; they do not specifically address slips, trips, falls, and other similar accidents, workplace violence, or psychosocial factors.

Ergonomics: An Effective Approach to Reducing the Number and Severity of Work-Related Injuries

The number and severity of injuries resulting from physical demands in nursing homes— and associated costs—can be dramatically reduced. Ergonomics, the practice of designing equipment and work tasks to conform to the capability of the worker, provides a means for adjusting the work environment and work practices to prevent injuries before they occur.

OSHA recommends that manual lifting of residents should be minimized in all cases and eliminated when possible. Minimizing and, where possible, eliminating resident lifting is the primary goal of the ergonomics process in the nursing home setting and of these guidelines. However, the approach to controlling the hazards associated with manual handling of residents should take into account other factors such as:

- Resident rehabilitation needs
- The need to restore functional abilities
- Medical contraindications
- Emergency situations
- Resident dignity and rights

Addressing work-related pain and disability is one of the greatest challenges facing nursing homes today. Successfully meeting this challenge may also serve to alleviate another difficulty confronted by nursing homes—the recruitment and retention of qualified staff members. A recent national survey of nursing home employers found double-digit vacancy rates for staff registered nurses, licensed practical nurses, and certified nurse assistants. Each of these groups also had annual turnover rates of more than 50% (7). Complaints of back pain are relatively common in nursing staff. Some evidence indicates that many trained, experienced staff members who leave their profession do so because of back pain (9).

Facilities that have implemented ergonomics-based injury prevention programs using effective engineering and work practice controls have achieved considerable success in reducing work-related injuries and workers' compensation costs. In addition, some institutions have experienced additional benefits, including reduced staff turnover and associated training and administrative costs, reduced absenteeism, increased productivity, improved morale, reduced resident injury, and increased resident comfort. Many examples exist of effective ergonomics programs in nursing homes. The following examples highlight some of the key aspects of successful ergonomics programs.

- Wyandot County Nursing Home in Upper Sandusky, Ohio, reported that staff had suffered back injuries, including a single injury that resulted in workers' compensation costs of $240,000. The facility acquired 18 ceiling lifts, as well as portable total lifts, sit-to-stand lifts, a lift walker, and 58 electrically adjustable beds at a cost of approximately $150,000. Since Wyandot implemented a policy of performing all assisted resident transfers with mechanical lifts or gait belts, back injuries from resident lifting have been eliminated. Increased efficiency has allowed staff members to spend more time with residents, and caregivers' attitudes and energy levels have reportedly improved. In addition, residents no longer complain of shoulder pain and bruises that had previously been associated with manual resident handling (12).
- At Citizens Memorial Health Care Facility in Bolivar, Missouri, concern about the number of injuries related to lifting and their economic impact led to the establishment

of an ergonomics component in the existing safety and health program. The facility emphasized education and the use of assistive devices. In each of the four years after the program was established, the number of OSHA-recordable lifting-related injuries declined by at least 45% over previous levels, and the number of associated lost workdays declined by at least 55%. These reductions contributed to a direct savings of approximately $150,000 in workers' compensation costs over a five year period (25).

- The Sisters of Charity Health System in Lewiston, Maine, established an ergonomics program including staff involvement through a number of committees and an ergonomics task force, purchase of 15 mechanical lifts, specialized ergonomics training, and provisions for medical management when injuries occur. Two years later, workers' compensation costs related to MSDs had declined by approximately 35%. The ergonomics program was also reported to have contributed to reduced turnover and absenteeism, increased efficiency and effectiveness, and improved morale among employees (92).

The nursing home industry has made substantial efforts in recent years to address work-related musculoskeletal disorders, and the results achieved by these facilities and others demonstrate that methods are available to better protect workers in nursing homes from the risk of injury. These guidelines provide practical suggestions for employers to reduce the number and severity of workplace injuries in their facilities by identifying, evaluating, and controlling hazards using methods that have been found to be successful in the nursing home environment. The recommendations made here are based on discussions with and information from a number of sources, including trade and professional associations, labor organizations, the medical community, individual firms, published documents, OSHA records, and OSHA settlement agreements with employers. These voluntary guidelines are intended for nursing and personal care facilities only. Other employers with similar work environments, such as hospitals and home healthcare providers, may find the information provided useful. Care should be taken, however, to ensure that distinctive circumstances found in different work environments are taken into account in developing ergonomic solutions for specific workplaces.

These *Guidelines for Nursing Homes* provide information and recommendations to address hazards related to forceful exertions, awkward postures, and other ergonomic stressors commonly found in nursing home settings. The information presented is organized into three sections. First, management commitment and employee participation, ergonomics training, occupational health management of MSDs, and methods of evaluating an ergonomics program are discussed in a *Management Practices* section. Second, methods of identifying and evaluating ergonomic stressors are discussed in a *Worksite Analysis* section. The third section, *Control Methods*, presents methods of controlling exposure to ergonomic stressors and gives many examples of controls that can be used in nursing home settings.

MANAGEMENT PRACTICES

Importance of Management Commitment and Employee Participation

Management commitment and employee participation form the foundation for any effective ergonomics process. Addressing ergonomics requires a sustained effort, allocation of resources and frequent follow-up that can only be achieved through management commitment. Management should attempt to tailor the ergonomics process to their specific workplace conditions and assign responsibilities to individuals throughout the organization to effectively implement the program.

Employee involvement is also critical to success. In the nursing home environment where workers must work independently, employees must "buy into" the program to achieve results. Employee participation is important to ensure that workers have input into the decisions regarding the equipment they will use. Employee input will certainly increase worker acceptance of the ergonomics process and specific control methods. In addition, many workers have learned proper body mechanics as the preferred method for resident handling. Unfortunately, even when proper body mechanics are used, employees are exposed to ergonomic stressors. Proper body mechanics are certainly needed skills in many situations but should not be relied upon for routine resident handling operations. Shifting the emphasis from proper body mechanics to minimizing or eliminating manual lifting of residents may be difficult and is best accomplished when employees are involved in the ergonomics process, including worksite analysis, selection of equipment, and program evaluation.

Ergonomics Training

To adequately identify and control ergonomic stressors and minimize the severity of workplace injuries, employees should receive training. Ergonomics training should enable employees at all levels of the organization—administrators, charge nurses, nursing assistants, maintenance workers, and equipment purchasers—to further the ergonomics program. People who work in nursing homes should be able to: (1) recognize the signs and symptoms of MSDs so that they can report them early and respond to them appropriately; (2) identify those jobs or tasks that have ergonomic stressors capable of causing MSDs; and (3) know how to control ergonomic stressors. Success of the ergonomics program depends to a great extent on the effectiveness of ergonomics training.

To be effective, training should be provided to all individuals in a nursing home who are at risk of incurring MSD injuries, as well as all employees who have responsibility for implementing ergonomics policies and procedures. Nursing home employers should consider training the following employees:

- Nursing assistants and other workers exposed to MSD stressors
- Ergonomics program administrators
- Charge nurses, supervisors, and those involved in receiving employee reports of injuries or symptoms
- Other healthcare providers who are on staff or otherwise work at the nursing home
- Maintenance and purchasing personnel (persons involved in selecting and procuring equipment)
- Management and human resources personnel

Many sources of ergonomics training materials exist. OSHA intends to offer advice and training on the guidelines it develops through the Agency's website as well as through the 12 OSHA Education Centers nationwide. In addition, a free consultation service largely funded by OSHA provides assistance to small businesses in identifying available training resources, as well as with other ergonomics-related topics. The content of the training should be targeted to each specific employee group, as illustrated by the following examples.

Nursing Assistants and Other Workers Exposed to MSD Stressors

Nursing assistants, housekeepers, dietary workers, and other employees who are exposed to ergonomic stressors should receive ergonomics awareness training that enables them to identify, control, and report problems. These employees should know how to identify MSD symptoms, ergonomic stressors, and control measures. Learning objectives for this group should include:

- The nursing home's ergonomics program and employees' role in the program
- The signs and symptoms of MSDs
- The procedures for reporting potential problems
- Existing ergonomic stressors and methods of control
- The use of engineering, administrative, and work practice controls, particularly safe resident handling techniques
- Techniques for informing a resident about the benefits of a lift assist

Initial ergonomics training should be provided in several short sessions or included as part of routine safety and health training. All new employees should receive this training before they begin working with residents.

Training sessions should include case studies or demonstrations based on examples from the nursing home, with ample time for questions (13). Training should be provided in a manner that all employees can understand. The training should be updated and presented to employees as changes occur in the workplace, equipment, facilities, procedures, or policies. Employees should be able to demonstrate an understanding of the stated objectives.

Ergonomics Program Administrators

Every nursing home should have an individual or a group that is responsible for implementing, promoting, and maintaining the ergonomics process (14). This individual or group should receive training on the following:

- Fundamental ergonomic principles
- Quantitative methods for task evaluation
- Problem-solving
- Control alternatives, including administrative controls
- How to implement the ergonomics program
- Signs and symptoms of MSDs
- MSD reporting and responsibilities

The training should consist of lectures and learning activities, such as examples of job evaluation protocols and how to use them, brainstorming, and case studies. Ample time should be provided for employee questions. Follow-up training through conferences and courses should be provided to keep the ergonomics program administrator current on ergonomic issues and tools.

Charge Nurses and Supervisors

Charge nurses and supervisors are often the first to receive reports of injuries, and thus should be able to recognize early signs and symptoms of MSDs and know the procedures for responding to them. Nursing home administrators should hold supervisors accountable for the identification of ergonomic stressors and control of ergonomic hazards in their respective areas. Supervisors' training objectives should include:

- Understanding MSD reporting procedures
- Understanding the ergonomics program, especially how to respond to reported MSDs
- Recognizing MSD signs and symptoms
- Identifying ergonomic stressors and control measures

Training should include discussion of ergonomic problems in the nursing home and possible controls. Nursing home administrators should participate in this training to show their support of the ergonomics program.

Other Healthcare Providers and Maintenance and Purchasing Personnel

Similar training should be provided for on-staff healthcare providers and maintenance and purchasing personnel, who are in a position to identify ergonomic stressors before an injury occurs. These employees are also often involved in suggesting controls. In addition, during their work in the nursing home they may use certain ergonomic controls. Training objectives for this group of employees include:

- Understanding the nursing home's ergonomics program
- Identifying ergonomic stressors and appropriate controls
- Becoming familiar with the use of controls

Training should address nursing home policy, ergonomic stressors, and available controls.

Management and Human Resources Personnel

In large nursing homes and organizations with multiple facilities, management should understand ergonomics issues to demonstrate support for the ergonomics program and encourage employee participation. Human resources personnel should be aware that an ergonomics program likely will increase reporting of injuries and signs and symptoms, as well as the number of workers' compensation cases in the short term. Cases will probably be less severe, however, and the number of workers' compensation cases should ultimately decrease. Training objectives for this group include:

- Understanding the nursing home's ergonomics program and its impact on the business
- Appreciating the importance of management leadership and support for the ergonomics program
- Knowing how to respond properly when reports of MSD signs and symptoms or problems are received

These objectives can be met by addressing the role of management in a successful ergonomics program and how that program relates to business objectives.

The training matrix contained in Table 1 provides a summary of the training topics for the various groups of nursing home employees. This matrix illustrates the common topics among the various groups, which suggests that some of the topics can be efficiently presented to more than one group at a time. The ergonomics program administrator will have additional responsibilities that will require a more in-depth knowledge of hazard evaluation techniques, control options, program implementation methods and problem solving methodologies. The ergonomics program administrators may be qualified, following their in-depth training, to train to other employee groups.

Occupational Health Management of MSDs

Even after an effective ergonomics program is implemented, work-related musculoskeletal injuries can still occur. Thus, an ergonomics program should include a process for addressing work-related injuries. Nursing home employers should develop an occupational health management process that ensures that workers receive prompt access to care for evaluation, treatment, and follow-up of MSD injuries related to workplace ergonomic stressors. Establishing such a process helps to identify problems and prevent the recurrence of MSDs.

Benefits of occupational health management include preventing injuries, minimizing time away from work, reducing severity of an injury, and decreasing associated medical costs. The following major components of occupational health management should be included in a facility's policies and procedures (13, 14, 114):

- A mechanism for employees to report MSDs and MSD signs and symptoms early
- A procedure that allows trained healthcare professionals to address reports of MSDs effectively
- Clear procedures for returning employees to work
- Accurate recordkeeping and documentation suitable for evaluation purposes

TABLE 1
Recommended Training Matrix

Employee Groups	Training Topics								
	Ergonomics program and employees role	Signs and symptoms of MSDs	Reporting procedures for ergonomic stressors and sign/ symptoms	Existing ergonomic stressors and controls	The use of different types of controls and safe resident handling	Resident interaction	Additional ergonomic specific task (problem solving, controls, task evaluation)	MSD reporting and responsibilities	Leadership, business impact
Nursing assistants and other exposed employees	X	X	X	X	X	X			
Ergonomics program administrator[1]	X	X	X	X	X	X	X	X	
Charge nurses/ supervisors	X	X	X	X	X	X	X	X	
Other healthcare providers	X	X	X	X	X	X			
Maintenance, purchasing, dietary, laundry	X	X	X	X					
Management and Human resources	X	X	X	X	X	X	X	X	X

[1] The training for ergonomics program administrators is a detailed, intense program because of the broad range of responsibilities necessitates a more in-depth knowledge of hazard evaluation techniques and control options. The ergonomics program administrators also may serve as the trainer for other employees.

Employers should assign a person to be responsible for occupational health management and this person should have the proper training, authority, resources, and experience to carry it out effectively.

A system should be in place so an employee can report to the employer work-related signs or symptoms (e.g., pain, numbness) and injuries. Early reporting is essential to effective management of MSDs, allowing for timely and appropriate treatment and therefore should be encouraged (13, 14, 15). For example, at the Veterans Health Administration Hospital in Tampa, Florida, employees report MSDs or MSD signs or symptoms to their supervisors, who then refer the employees to the occupational health services. The supervisor retains a copy of the MSD report and maintains statistics for the specific unit (16). [Note: Employers should be aware of OSHA's recordkeeping rule (29 CFR 1904). Under this rule employers must record certain injuries including those related to workplace ergonomic stressors.]

In addition to establishing a good reporting system, employers should create a procedure that provides employees with prompt and appropriate assessment, treatment, and follow-up of an MSD by an appropriately trained and licensed healthcare professional (13, 14, 17). The healthcare professional should be familiar with the nursing home facility and the employee's job duties and hazards, work history, and personal factors that may contribute to the problem (15, 17). Often such healthcare professionals can help identify any work-related factors that may contribute to a reported injury.

The healthcare professional, in consultation with the employer and the employee, should prepare return-to-work procedures that ensure that assigned restricted duties do not further injure the employee. When making recommendations to return an employee to work, the healthcare professional should first determine that the employee's physical capabilities match the work requirements. The employer should make sure the job assigned to the employee is appropriate considering the work restrictions. The employer and employee should follow any work restrictions designated by the healthcare professional.

Maintaining documentation of injury reports related to ergonomic stressors is critical to preventing their recurrence. In this process, the information in individual employees' health records should be kept confidential (8).

Information gained from work-related MSD reports can be used to make the workplace safer. An analysis of such reports can identify areas of the nursing home that need focused worksite analysis and prevention. Reports can also help determine if injury trends are emerging that should be addressed.

Ergonomics Program Evaluation

To be effective, an ergonomics program should have procedures and mechanisms in place to evaluate the program's implementation and monitor its progress. These procedures should ensure that the principles outlined in these guidelines are followed and the program is accomplishing measurable goals. Evaluating a program involves more than inspecting or auditing a workplace; employers should determine whether the management system it establishes adequately addresses ergonomic stressors.

Leading and Trailing Indicators

Effective evaluation requires the use of both leading and trailing indicators. Leading indicators seek to measure processes or events that can prevent injuries, and accidents from occurring.

Leading indicators measure how well an employer is following the guidelines. Leading measures can include the number and quality of worksite analyses; frequency of ergonomic interventions; extent of management commitment to reducing injuries related to ergonomic stressors; adherence to reporting mechanisms; the effectiveness of training programs; and the extent of employee involvement. Although leading indicators can often be difficult to measure, employers should recognize the benefits of leading indicators and use them when appropriate.

Trailing indicators measure historical results (e.g., number of injuries that occurred). The most commonly used trailing indicators are OSHA recordable injury rates, the total incidence rate, the lost-time incidence rate, and the workers' compensation experience modifier rate. Large worksites may find the analyses to be more useful than smaller worksites.

The manner in which these indicators are examined is at least as important as the indicators themselves. While the specific programs will vary from site to site, the basic tools used for evaluation should remain consistent. These tools include document reviews, employee interviews (including interviews of managers, supervisors, and front-line employees), employee surveys, and reviews and observations of workplace conditions.

Adherence to the Principles of the Ergonomic Guidelines for Nursing Homes

The employer, with employee assistance, should periodically evaluate whether its system for identifying problems related to exposure to ergonomic stressors is functioning effectively. While trailing indicators such as injury rates provide useful information toward accomplishing this goal, the inquiry should not stop there. The employer should ensure that reports of problems are being filed, their contents evaluated, and decisions made regarding their disposition.

Once control measures are introduced as the result of worksite analysis, the employer should ensure that the controls are effective. Trailing indicators can provide useful empirical data at this stage, as can other techniques such as employee interviews. For example, after introducing a new lift at a nursing home, the employer should follow-up by talking with employees to ensure that the problem has been adequately addressed. In addition, interviews provide a mechanism for ensuring that the control is not only in place, but is being used properly. Employers should combine their worksite assessments with their evaluation of control methods to achieve the most comprehensive picture of how the program is functioning.

In examining the occupational health management program, employers should ensure that a system exists for reporting injuries related to ergonomic stressors and responding to employee medical needs. Reviewing OSHA logs or other empirical data may provide an incomplete picture of the facility's occupational health management program. Thus, employees should be interviewed, or surveyed anonymously, to ensure that they are reporting injuries and receiving follow-up care.

An effective training program should have clear goals for developing employees' ergonomic awareness. When evaluating a training program, the employer should examine the content and frequency of training sessions. Does the training curriculum reflect the needs of the facility's ergonomics program? Does the frequency of training sessions allow for prompt training of new employees and follow-up training for veteran staff? Also, the training is effective only if the material is understood and integrated into practice. Thus when conducting an evaluation, employers should observe trained employees to ensure that they are adhering to the work practices and ergonomic principles discussed in training sessions. Employers should ensure that employees follow the program through such actions as:

- Reviewing the program and training to make sure that employees are fully informed of their responsibilities
- Evaluating the workplace culture to ensure that poor performance is not rewarded
- Ensuring that employees have the needed tools and equipment to accomplish their jobs safely
- Establishing policies that address employee compliance with program procedures

Finally, an ergonomics program that exists only on paper is unlikely to realize any health and safety benefits. Determining whether employees are accepting the program is often a difficult outcome to measure. Anonymous surveys of employees, as well as employee interviews, can be effective tools for gauging employee buy-in to the ergonomics program.

Evaluating Injury Data

While it should not be the sole indicator of the effectiveness of an ergonomics program, the role of empirical data should not be understated. OSHA 300 logs can be particularly useful in this effort because they include a description of the incident, time and date of incident, unit where the incident occurred, body parts affected, and days of work lost. Incident reports or OSHA 301 forms may provide more detailed information for assessing the effectiveness of an ergonomics program. Other useful records may include reports of workers' compensation claims, accident and near-miss investigation reports, and insurance company reports.

Implementing and Following Up on Evaluations

Program evaluations can be conducted by those responsible for implementing the ergonomics program, but evaluations performed by persons who are not involved in the day-to-day operation of the program are often even more valuable. These employees bring a fresh perspective to the task and can identify program weaknesses that those routinely involved in the program implementation fail to see.

The frequency of evaluations will vary by the size and complexity of the facility, but a nursing home should evaluate its ergonomics program at least every three years, and more frequently if it has reason to believe that the program is ineffective. Management should revise the program promptly in response to any identified deficiencies and communicate the results of the program evaluation and program revisions to the employees.

WORKSITE ANALYSIS

Analysis of ergonomic stressors in a job is the process of identifying where and how job requirements exceed the physical capabilities of workers. Worksite analysis is the centerpiece of any program addressing injuries in the workplace because it identifies the conditions of the job and aspects of work activities that result in increased risk to employees. Without proper analysis and assessment, any corrective actions taken may not only be ineffective, but may worsen existing hazards or introduce new hazards into the workplace.

Worksite analysis involves examining workplace conditions and individual tasks or elements of a job to distinguish factors that may result in an increased risk of injury. Information about the job and any associated problems can be obtained from a variety of sources. Once information is obtained, it can be used to identify and evaluate elements of a job that are associated with problems. Suitable options can then be selected to address any problems found.

Assessment of the worksite begins with the identification of tasks that require analysis. Job activities are candidates for analysis if reason exists to believe that they pose a risk to employees. Review of injury and illness logs, workers' compensation records, reports of problems, discussions with employees, or general observation of the workplace can all be used to identify tasks that warrant analysis.

The assessment of work tasks involves an examination of the duration, frequency, and magnitude of exposure to ergonomic stressors - force, repetition, awkward postures, vibration, and contact stress - to determine if employees are at risk of pain or injury. In many cases the relevant stressors are apparent after observation of the job and discussions with employees. In other cases, causes of pain and injury may not be readily apparent. This is because many activities involve exposure to a combination of stressors. For example, lifting a 30 pound box in front of the body from knuckle height to shoulder height (force) may not create a problem. However, if such a lift is performed frequently during the work shift (repetition), or if the worker must twist the torso while lifting the box (awkward posture), an increased risk of injury may result (19).

A variety of methods can be used to analyze exposure to ergonomic stressors. Procedures may be simple and informal, consisting of observation of the job and discussions with affected employees. A more formalized process, such as a breakdown of the tasks performed in a job, videotaping or photographing employees performing the job, employee questionnaires, checklists, or quantitative measurements may also be used. A number of protocols and checklists have been developed that can be used in performing these evaluations (16, 18, 19, 20, 21, 22, 23, 24).

The process of assessing job requirements in nursing homes is complicated by considerations that should be taken into account when examining resident handling tasks. Resident handling tasks can be variable, dynamic, and unpredictable in nature. In addition, factors such as resident safety and medical contraindications should be taken into account. As a result, different techniques are used for assessing resident handling tasks than are used for other nursing home activities.

Assessment of Resident Handling Tasks

An analysis of any resident handling task involves an assessment of the needs and abilities of the resident involved. This assessment allows staff members to account for these characteristics in determining the safest methods for performing the task, within the context of a care plan that provides for appropriate care and services for the resident, including consideration of resident safety, dignity and other rights, and the need for maintenance or restoration of a resident's functional abilities. The resident assessment should include examination of factors such as:

- The size and weight of the resident
- The level of assistance the resident requires
- The ability and willingness of the resident to cooperate
- Any medical conditions that may influence the choice of methods for transfer or repositioning

These factors are critically important in determining appropriate control measures for resident handling tasks. The size and weight of the patient will, in some situations, determine what equipment is needed and how many caregivers are required to provide assistance. The physical and mental abilities of the resident also play an important role in selecting appropriate control methods. For example, a resident who is able and willing to partially support their own weight may be able to move from their bed to a chair using a standing assist device, while a mechanical sling lift may be more appropriate for those residents who are unable to support their own weight. Other factors related to resident condition may need to be taken into account as well. For instance, a resident who has recently undergone hip replacement surgery may require specialized equipment for assistance in order to avoid placing stress on the affected area.

A number of individuals in nursing homes can contribute to resident assessment and the determination of appropriate methods for assisting in transfer or repositioning. Staff nurses, nursing supervisors, physical therapists, physicians, and the resident or his/her representative may all be involved. Of critical importance is the involvement of employees directly responsible for resident care and assistance, as the needs and abilities of residents may vary considerably over a short period of time, and the employees responsible for providing assistance are in the best position to be aware of and accommodate such changes.

A number of protocols have been developed for systematically examining resident needs and abilities with regard to transfer and repositioning, and for recommending procedures and equipment to be used for performing tasks (13, 14, 16, 17). Figure 1 provides an example of one set of assessment criteria that has been developed. These assessment criteria can assist in planning for safe lifting, transfer, and repositioning of residents and can be useful when incorporated into the assessment process. Figures 2 through 7 show examples of algorithms that can be used as guides when planning lifting, transfer, and repositioning tasks. Algorithms can provide useful guidance in planning resident handling tasks, but do not substitute for professional judgment needed to assure the safety of residents and caregivers.

I. Patient's Level of Assistance:

___ Independent – Patient performs tasks safely, with or without assistive devices.

___ Partial Assist – Patient requires no more help than stand-by, cueing, or coaxing, or no more than 50% physical assistance by the nurse.

___ Dependent – Patient requires more than 50% assistance by nurse, or is unpredictable in the amount of assistance offered.

An assessment should be made prior to each task if the patient has varying level of ability to assist due to medical reasons, fatigue, medications, etc. When in doubt, assume the patient cannot assist in the transfer/repositioning.

II. Can the patient bear weight?

___ Yes, Full
___ Yes, Partial
___ No

III. Does the patient have upper extremity strength needed to support his/her weight during transfers?

___ Yes
___ No

IV. Patient's level of cooperation and comprehension:

___ Cooperative – may need prompting; able to follow simple commands.

___ Unpredictable or varies (patient whose behavior changes frequently should be considered as "unpredictable"), not cooperative, or unable to follow simple commands.

V. Patient's weight: ________ **height:** ________

The presence of the following conditions are likely to affect the transfer/repositioning process and should be considered when identifying equipment and technique needed to move the patient.

VI. Check applicable conditions likely to affect transfer/repositioning techniques.

___ Abdominal Surgery Wounds
___ Bilateral Amputation
___ Colostomy
___ Contractures/Spasms
___ Fractures

___ Hip/Knee Replacements
___ History of Falls
___ Paralysis
___ Presence of Tubes (IV, chest, etc.)
___ Pressure Ulcers

___ Postural Hypotension
___ Severe Osteoporosis
___ Splints/Traction
___ Unstable Spine
___ Other

Comments : ____

VII. Care Pl			
Algorithm #	**Task**	**Equipment/Assistive Devices**	**# Staff**
1	Transfer To and From: Bed-to-Chair, Chair-to-Toilet, Chair-to-Chair, or Car-to-Chair		
2	Lateral Transfer to and from: Bed-to-Stretcher, Trolley.		
3	Transfer to and from: Chair-to-Stretcher, or Chair-to-Exam Table		
4	Reposition in Bed: Side-to-Side, Up in Bed		
5	Reposition in Chair: Wheelchair and Dependency Chair		

Sling Type (circle choices): 1) Standard 2) Amputation 3) Head Support **Sling Size:** _____

Signature: ______________________________ **Date:** ____________________

FIGURE 1 Assessment criteria and care plan for safe patient handling and movement.

The algorithms can be used to identify appropriate options among the measures presented in the Control Section of this document. For example, Figure 2 presents an algorithm that can be used to assess procedures and equipment for transferring a resident from a bed to a chair, a chair to the toilet, or between chairs. The algorithm indicates that when a resident cannot bear weight, and is not cooperative or does not have upper extremity strength, a full body sling lift and two caregivers should be used to perform the transfer. A ceiling-mounted lift device (shown as control method number 9) is an example of a control method appropriate for this situation. If the resident can partially bear weight and is cooperative, the algorithm indicates that either a stand and pivot technique using a gait/transfer belt or a powered standing assist lift (in each case with the assistance of one caregiver) would be appropriate. An example of an option for this situation would be control number 14, a portable powered standing assist device.

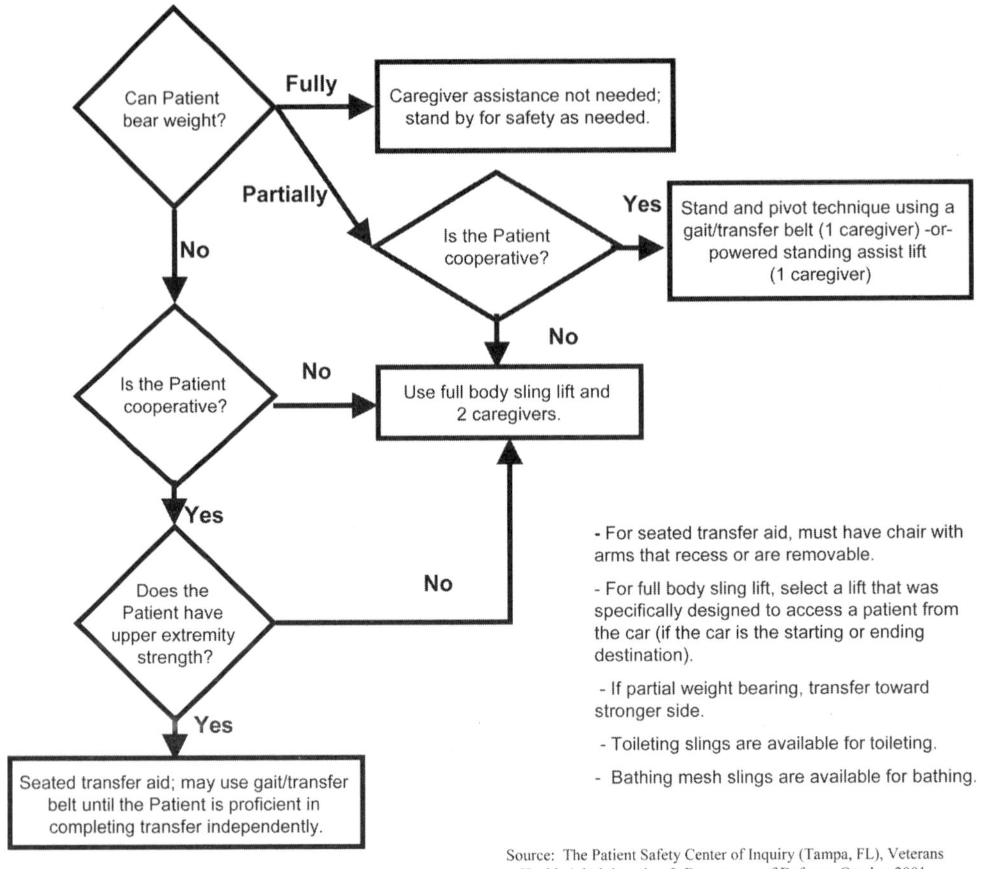

FIGURE 2 Transfer to and from: bed to chair; chair to toilet, chair to chair, or car to chair.

In many cases a number of control options will be available. For instance, Figure 3 presents an algorithm that can be used when performing lateral transfers between beds, stretchers or trolleys. If the resident in unable or only partially able to assist in the transfer, and the weight of the resident is less than 100 pounds, a lateral sliding aid with the assistance of two caregivers is recommended. Many alternative lateral sliding aids are available. The Control Section shows gurneys with transfer devices (control method number 1), free-standing mechanical devices control method number 3), draw sheets (control method number 4), vinyl-covered boards and rollers (control method number 5), inflatable mattresses (control method number 6), and transfer cots with handles (control method number 7) that could be used in this situation. The employer should determine which device or devices are most appropriate to protect workers given the particular circumstances found in the workplace.

In some cases the algorithms indicate not only a choice in the particular type of control measure, but in the method used as well. For example, Figure 3 indicates several methods are appropriate for performing a transfer when a resident is unable or only partially able to assist in the transfer, and the resident weighs in excess of 200 pounds. Use of a lateral sliding aid with three caregivers, use of a friction-reducing device or lateral transfer device with two caregivers, or use of a mechanical lateral transfer device would all be considered acceptable. A range of options for each of these methods is presented in the Control Section. Again, the employer should determine which method or methods are most appropriate to protect workers given the particular circumstances found in the workplace. For instance, if staff availability precludes assigning three caregivers to assist with such a transfer, an employer may choose to adopt methods that can be performed safely with fewer staff members.

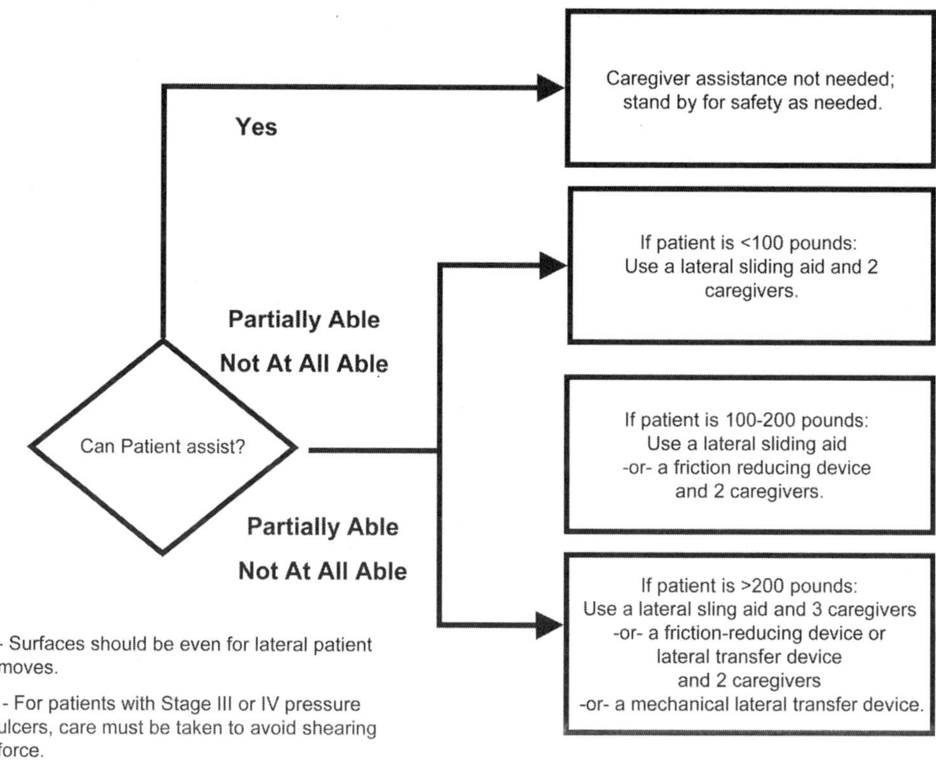

FIGURE 3 Lateral transfer to and from: bed to stretcher, trolley.

An assessment of resident handling activities should also take into account other considerations that may impact the task being performed. Considerations may include the presence of apparatuses such as oxygen tanks or intravenous connections, restrictions related to the environment such as limited room for equipment use in a bathroom, the availability of sufficient numbers of staff to perform assigned tasks properly, and safety concerns such as a wet floor or the potential for a chair or bed to move unexpectedly while a task is being performed. The special needs of bariatric residents may also need further examination. Assistive devices must be capable of handling the heavier weight involved, and modification of work practices may be necessary.

Assessment of Activities Other than Resident Handling

Although the majority of work-related MSDs in nursing homes are associated with the manual transfer and repositioning of residents, a number of other activities may also involve increased risk to workers. The assessment process for these activities involves examination of the physical demands of work, layout and condition of the workplace or workstation, and characteristics of any objects handled. Tasks that may require assessment include:

- Cleaning rooms
- Bending to make a bed or feed a resident
- Lifting food trays above shoulder level or below knee level
- Waste collection

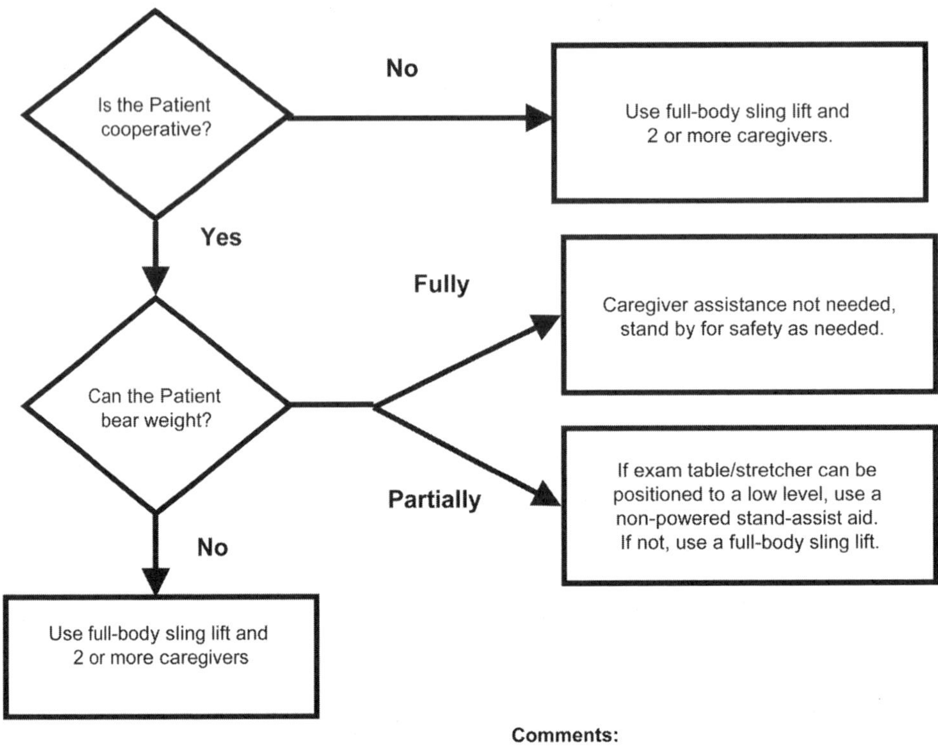

FIGURE 4 Transfer to and from: chair to stretcher.

- Pushing heavy carts
- Bending to remove items from a deep cart
- Lifting and carrying when receiving and stocking supplies
- Bending and manually cranking an adjustable bed
- Removing laundry from washing machines and dryers

The assessment includes an evaluation of the duration, frequency, and magnitude of exposure to ergonomic stressors, as discussed at the beginning of this section. In the vast majority of cases, job assessments can be accomplished by observing employees performing the task, and by discussing with employees the activities and conditions that they associate with difficulties. Observation provides general information about the workstation layout, tools, equipment, and general environmental conditions in the workplace. Discussing tasks with employees helps to ensure that a complete picture of the process is obtained. Employees who perform a given task are also often the best sources for identifying the cause of a problem, and developing the most practical and effective solutions. Once information is obtained and problems identified, suitable improvements can be implemented.

CONTROL METHODS

This section on control methods for ergonomic stressors presents changes to equipment, work practices, and work methods that can reduce injury levels among staff and residents, help control

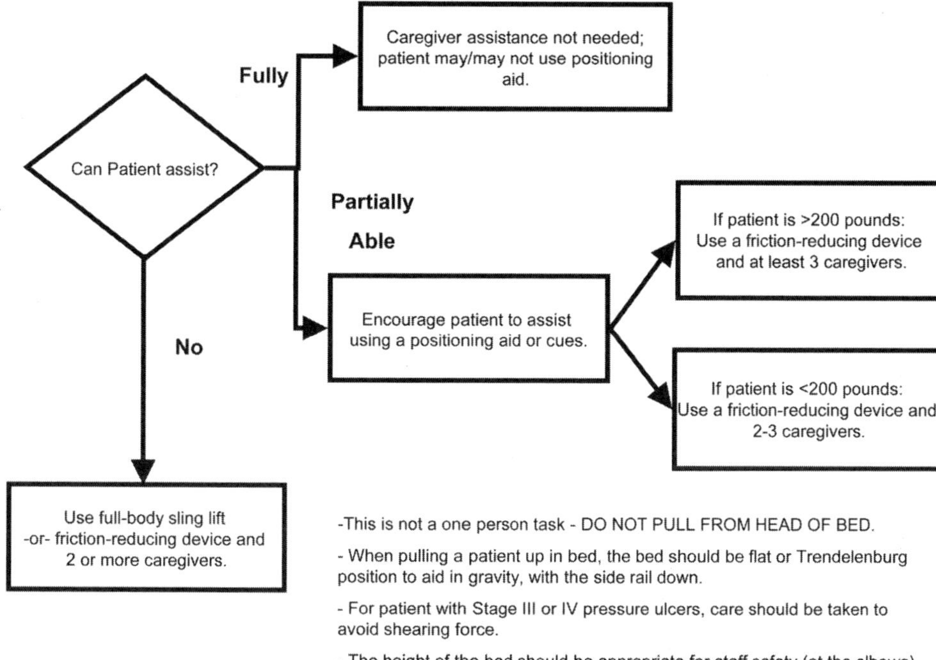

FIGURE 5 Repositioning in bed: side-to-side, up in bed.

costs, and make employee recruitment less difficult. The selection of appropriate controls should be based on worksite analysis described in the previous section. But other factors should also be considered. For example, what types of controls are possible for conditions in an individual facility. This decision can be based on a number of factors, including the number of residents, the current state of equipment used, the availability of caregivers, and other on-going or prospective projects that will compete for scarce resources.

The integration of ergonomic controls into the nursing home is a strategic decision that should be carefully planned and executed and that will pay long term benefits. The prudent administrator will gain staff support and commitment by sharing information and soliciting input from those affected by these decisions—other managers, the caregivers, the residents, and the residents' families. Keeping everyone informed of the possible changes and obtaining input and acceptance on the use of the controls, will smooth the transition to a safer workplace.

Forty-nine control approaches are presented in the following pages. The controls are not intended to be an exhaustive list, nor does OSHA expect that all the presented controls will be used in any given facility. The information represents a range of options available that a facility can consider using. Many of the control approaches are simple, common sense modifications to equipment or procedures that require little time or few resources to implement. Other control approaches may require more substantial efforts.

As stated earlier, OSHA recommends that manual lifting of residents should be minimized in all cases and eliminated when possible. Administrators should also be cognizant of several factors that might restrict the application of certain controls, such as residents' rehabilitation needs, the need for restoration of functional abilities, other medical contraindications, emergency conditions, and patients' dignity and rights.

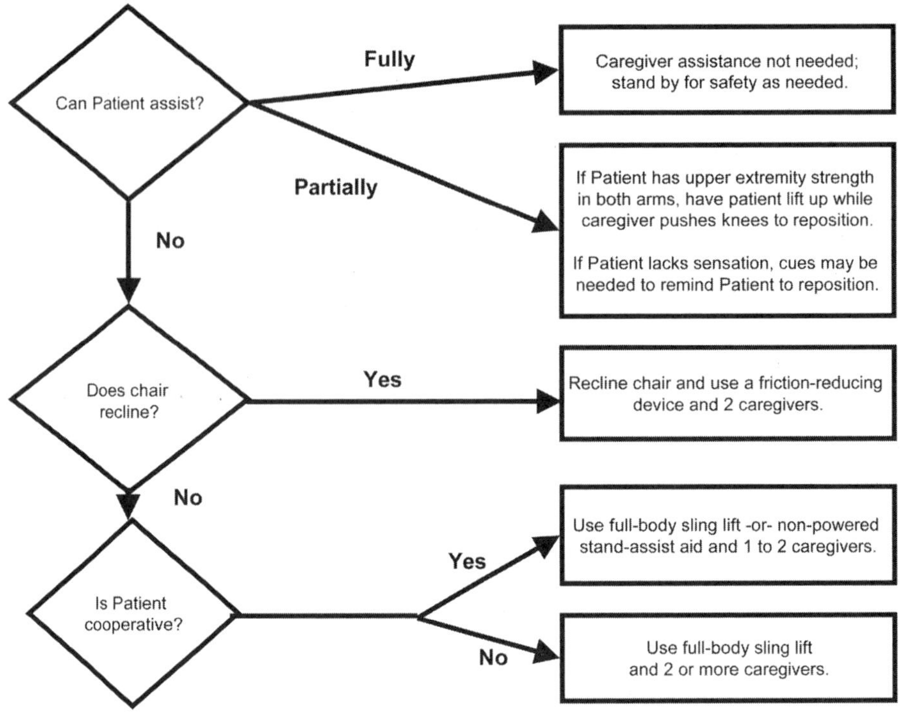

FIGURE 6 Reposition in chair: wheelchair and dependency chair.

Many controls are applicable to several conditions in a typical nursing home. Table 2 provides an index to the control descriptions for easy access. Each of the control descriptions begins with a title and a representative picture and includes the following information:

- Activity briefly describes the general work category being addressed. More than one activity may be listed here and some controls will have many activities listed.
- Description explains the types of equipment or handling methods used.
- When To Use describes the conditions under which it is appropriate to use the control.
- Points to Remember provides reminders about using the control, especially regarding safe and effective use.
- References refers to the source of the information used (see the Reference List at the end of this document).

One important consideration on the procurement of control equipment is the selection of the equipment supplier. Employers should establish a relationship with equipment suppliers. Such a relationship is critical because it facilitates obtaining training for employees in how to use equipment, modifying the equipment for special circumstances, and procuring parts and service when needed. The following questions are designed to aid in the selection of the supplier that best meets the needs of an individual nursing home.

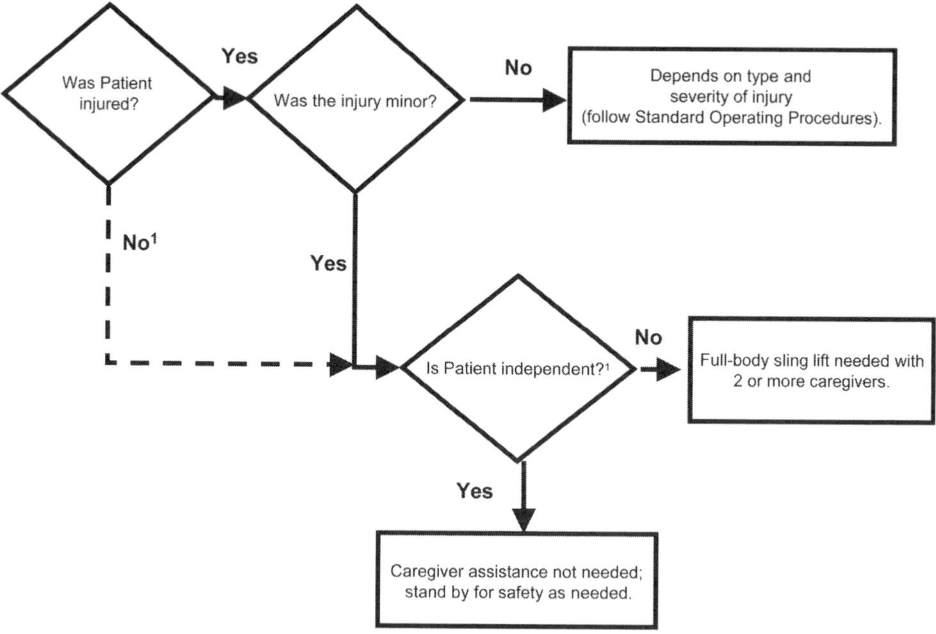

FIGURE 7 Transfer a patient up from the floor.

- Availability of technical service – Is over-the-phone assistance, as well as onsite assistance, for repairs and service of the lift available?
- Availability of parts – Which parts will be in stock and available in a short time frame and how soon can they be shipped to your location?
- Storage ability of the mechanical lift – Is it too big for your facility? Can it be stored in close proximity to the area(s) where it is used?
- If needed, is a charging unit and back up battery included?
- What is the simplicity of the charging unit and space required for a battery charger if one is needed?
- If the lift has a self-contained charging unit, what is the amount of space necessary for charging and what electrical receptacles are required?
- What is the minimum charging time of a battery?
- How high is the base of the lift and will it fit under the bed and various other pieces of furniture?
- How wide is the base of the lift or is it adjustable to a wider and lockable position? How many people are required to operate the lift for lifting of a typical 200-pound person?
- Does the lift activation device (pendant) have remote capabilities?
- How many sizes and types of slings are available?
- What type of sling is available for optimum infection control?
- Is the device versatile? Can it be a sit-to-stand lift, as well as a lift device? Can it be a sit-to-stand lift and an ambulation-assist device?
- What is the speed and noise level of the device?
- Will the lift go to floor level? How high will it go?

TABLE 2
Index of Control Methods

Topics	See These Controls
Resident Handling	
Transfers	1, 2, 3, 4, 5, 6, 7, 11, 12, 13, 14, 15, 16, 17, 18, 29
Repositioning	4, 6, 11, 19, 20, 21, 22, 23, 36
Lifting	2, 8, 9, 10, 14, 15, 27, 34, 35
Ambulation	17, 24
Activities of Daily Living	25, 26, 27, 28, 29, 30, 31, 32, 33, 37
Non-Resident Handling	
Dispense of Medication	38
Moving Equipment	39, 43
Dietary	40, 41, 43, 45
Housekeeping/Laundry	40, 41, 42, 43, 44, 45, 46, 47, 48, 49
Maintenance	41, 43, 44, 45

Based on the characteristics of the resident population and the layout of the facility, employers should determine how many devices are needed and where to locate the devices so they are accessible to workers. If resident lifting equipment and controls are not accessible when they are needed, it is likely that all other aspects of the ergonomic process will be ineffective. Employers should also establish routine maintenance schedules to ensure that the equipment is in good working order.

1

Lateral Transfer

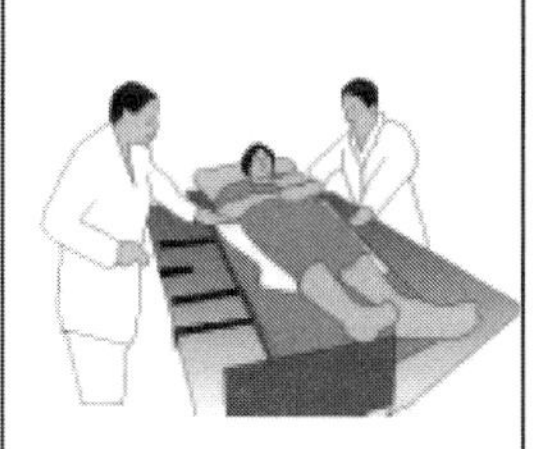

Activity:
Resident Handling

Description:
Gurneys with transfer devices

When to Use: Transferring a partial- or non-weight-bearing resident between 2 horizontal surfaces, such as a bed to a stretcher or gurney in supine position.

Points to Remember: Two caregivers are required to perform this type of transfer. Additional assistance may be needed depending on resident status, e.g., for heavier or non-cooperative residents. Motorized height-adjustable gurneys with built-in slide boards are preferred to those adjusted by crank mechanism to minimize physical exertion required by caregiver. Always ensure that lifting device is in good working order before use and is rated for the load weight to be transferred. Ensure wheels on equipment are locked. Ensure transfer surfaces are same level and at caregiver's waist level to avoid extended reaches and back flexion.

References: 13, 16, 39, 72

2

Lateral Transfer

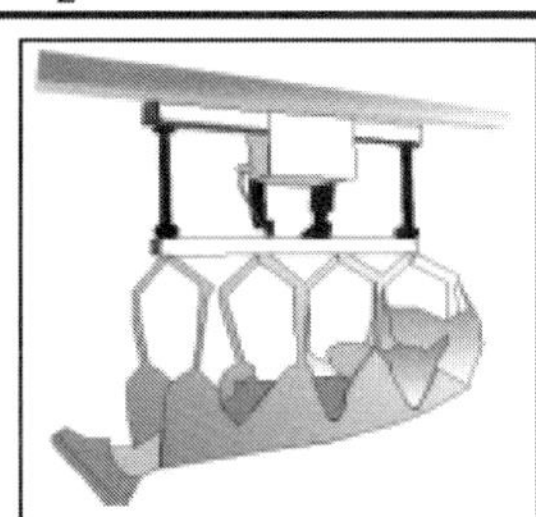

Activity:
Resident Handling

Description:
Ceiling-mounted device with horizontal frame system or litter

When to Use: Transferring residents who are totally dependent, non-weight bearing, have other physical limitations, or are very heavy and cannot be safely transferred by staff between 2 horizontal surfaces, such as a bed to a stretcher or gurney in supine position.

Points to Remember: Motors can be fixed or portable (lightweight). Device can be operated by hand-held control attached to unit or by infrared remote control. Always ensure lifting device is in good working order before use and is rated for the load weight to be lifted.

References: 40, 47, 55, 88

3

Lateral Transfer

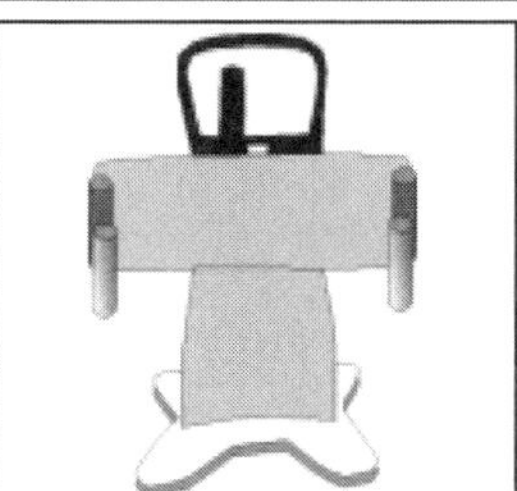

Activity:
Resident Handling

Description:
Free-standing lateral transfer devices used *with* height-adjustable stretcher or gurney

When to Use: Transferring a partial- or non-weight-bearing resident between 2 horizontal surfaces, such as a bed to a stretcher or gurney in supine position.

Points to Remember: Two caregivers are required to perform this type of transfer. Additional assistance may be needed depending on resident status, e.g., for heavier or non-cooperative residents. Always ensure that lifting device is in good working order before use and rated for the load weight to be transferred. Ensure wheels on equipment are locked. Ensure transfer surfaces are at same level and at a height that allows caregivers to work at waist level to avoid extended reaches and back flexion.

References: 16, 54

4

Lateral Transfer; Repositioning

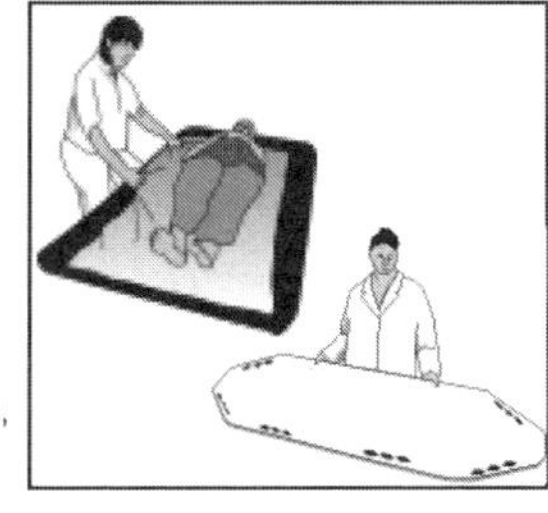

Activity:
Resident Handling

Description:
Draw sheet to be used in combination with friction-reducing devices such as slippery sheets, plastic bags, low friction mattress covers, or slide boards

When To Use: Transferring a partial- or non-weight bearing resident between 2 horizontal surfaces such as a bed to a stretcher or gurney in supine position or when repositioning resident in bed.

Points to Remember: Two caregivers are required to perform this type of transfer. Additional assistance may be needed depending upon resident status, e.g., for heavier or non-cooperative residents. May not be suitable for bariatric residents. Use a good hand-hold by rolling up draw sheets or use other friction-reducing devices with handles such as slippery sheets. Narrower slippery sheets with webbing handles positioned on the long edge of the sheet may be easier to use than wider sheets. Look for a combination of devices that will minimize risk of skin trauma. Ensure wheels on equipment are locked. Ensure transfer surfaces are at same level and at a height that allows caregivers to work at waist level to avoid extended reaches and back flexion. Count down and synchronize the transfer motion between caregivers.

References: 13, 16, 27, 29, 30, 39, 56, 70, 76, 79, 89, 97, 113

5

Lateral Transfer

Activity:
Resident Handling

Description:
Boards or mats with vinyl coverings and rollers

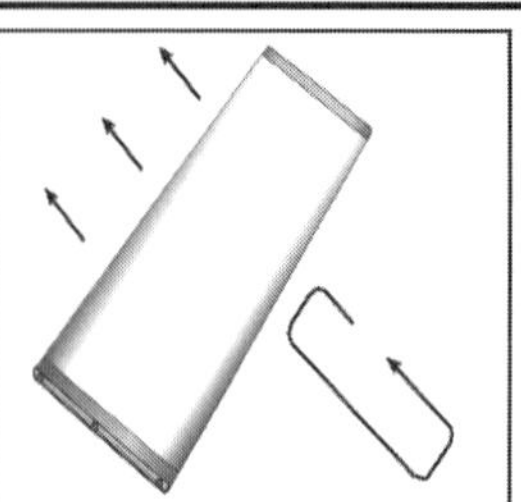

When to Use: Transferring a partial- or non-weight bearing resident between 2 horizontal surfaces, such as a bed to a stretcher or gurney in supine position.

Points to Remember: Two caregivers are required to perform this type of transfer. Additional assistance may be needed depending on resident status, e.g., for heavier or non-cooperative residents. Move resident to new surface using gentle push and pull motion. Device may not be suitable for bariatric residents. Ensure equipment wheels are locked. Ensure transfer surfaces are at same level and at a height that allows caregivers to work at waist level to avoid extended reaches and back flexion. Count down and synchronize the transfer motion between caregivers.

References: 13, 16, 90, 97

6

Lateral Transfer; Repositioning

Activity:
Resident Handling

Description:
Air-assist lateral sliding aid; flexible mattress inflated by portable air supply

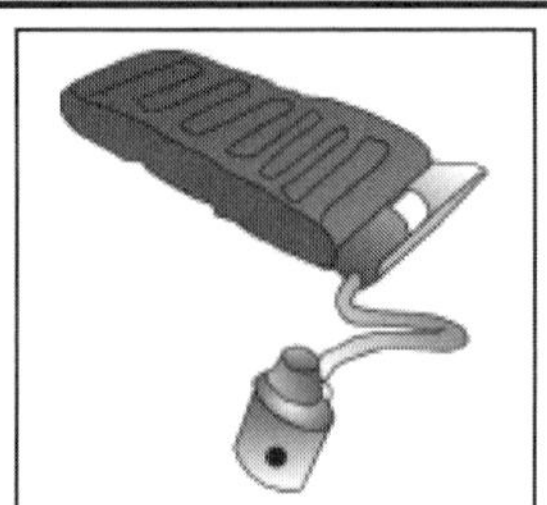

When to Use: Transferring a partial- or non-weight bearing resident between two horizontal surfaces such as a bed to a stretcher or gurney in supine position. Can also be used for repositioning a resident in bed. Increases resident's comfort and reduces risk of tissue damage during transfer.

Points to Remember: Two caregivers are required to perform this type of transfer. Additional assistance may be needed depending on resident status, e.g., for heavier or non-cooperative residents. Ensure wheels on equipment are locked. Ensure transfer surfaces are at level and at a height that allows caregivers to work at waist level to avoid extended reaches and back flexion. Count down and synchronize the transfer motion between caregivers.

References: 16, 39

7

Lateral Transfer

Activity:
Resident Handling

Description:
Transfer cots with handles

When to Use: Transferring a partial- or non-weight-bearing resident between 2 horizontal surfaces, such as a bed to a stretcher or gurney in supine position.

Points to Remember: Two caregivers are required to perform this type of transfer. Additional assistance may be needed depending on resident status, e.g., for heavier or non-cooperative residents. Technique may not be suitable for bariatric residents. Use in combination with friction-reducing devices such as slide boards or slippery sheets. Ensure wheels on equipment are locked. Ensure transfer surfaces are at same level and at a height that allows caregivers to work at waist level to avoid extended reaches and back flexion. Count down and synchronize the transfer motion between caregivers.

References: 13, 90

8

Resident Lifting

Activity
Resident Handling

Description:
Portable lift device (sling type); can be a universal/ hammock sling or a band/leg sling

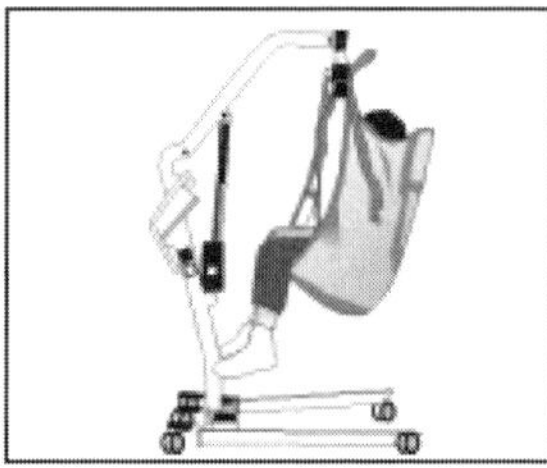

When to Use: Lifting residents who are totally dependent, are partial- or non-weight bearing, are very heavy, or have other physical limitations. Transfers from bed to chair (wheel chair, Geri or cardiac chair), chair or floor to bed, lateral transfers, or for bathing and toileting. Enhances resident safety and comfort.

Points to Remember: May require 2 or more caregivers. Look for a device with a variety of slings, lift-height range, battery portability, hand-held control, emergency shut-off, manual override, and boom pressure sensitive switch. Having multiple slings allows one of them to remain in place while resident is in bed or chair for only a short period, reducing the number of times the caregiver lifts and positions resident. Ensure lifting device is in good working order before use and is rated for the load weight to be lifted. Electric/ battery powered lifts are preferred to crank or pump type devices to allow a smoother movement for the resident, and less physical exertion and risk of musculoskeletal injury to the caregiver.

References: 3, 13, 14, 16, 27, 29, 33, 38, 41, 43, 46, 50, 51, 52, 57, 65, 66, 69, 72, 73, 74, 75, 76, 77, 86, 90, 93, 94, 97, 98, 108, 110, 113

9

Resident Lifting

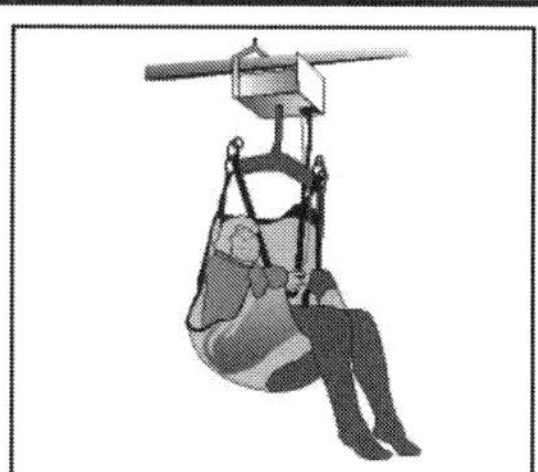

Activity
Resident Handling

Description:
Ceiling-mounted lift device

When to Use: Lifting residents who are totally dependent, are partial- or non-weight bearing, very heavy, or have other physical limitations. Transfers from bed to chair (wheel chair, Geri or cardiac chair), chair or floor to bed, or for bathing and toileting. Increases residents' safety and comfort during transfer.

Points to Remember: May require 2 or more caregivers. Some residents can use the device without assistance. May be quicker to use than portable device. Motors can be fixed or portable (lightweight). Device can be operated by hand-held control attached to unit or by infrared remote control. Ensure lifting device is in good working order before use and is rated for the load weight to be lifted.

References: 13, 16, 39, 40, 47, 51, 74, 75, 78, 85, 88, 90, 94

10

Resident Lifting

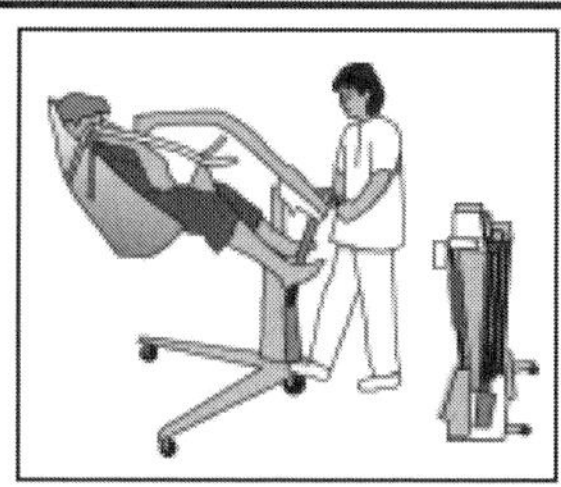

Activity:
Resident Handling

Description:
Portable compact lifts

When to Use: Lifting residents who are totally dependent, partial- or non-weight bearing, or have other physical limitations. Transfers from bed to chair (wheelchair, Geri or cardiac chair), chair or floor to bed, chair to car, lateral transfers, or for bathing and toileting. Increases resident safety and comfort during transfer. Can be used where space or storage is limited.

Points to Remember: May require 2 or more caregivers. Look for a device with a variety of slings, lift height range, battery portability, hand-held control, emergency shut-off, manual override, and boom pressure-sensitive switch. Having multiple slings allows one of them to remain in place while resident is in bed or chair for only a short period reducing the number of times a caregiver lifts and positions the resident. Always ensure lifting device is in good working order before use and rated for the load weight to be lifted. Electric/battery powered lifts are preferred to crank or pump type devices to allow a smoother movement for the resident and less physical exertion and risk of musculoskeletal injury to the caregiver.

References: 13, 16, 51, 69, 76

11

Lateral Transfer; Repositioning

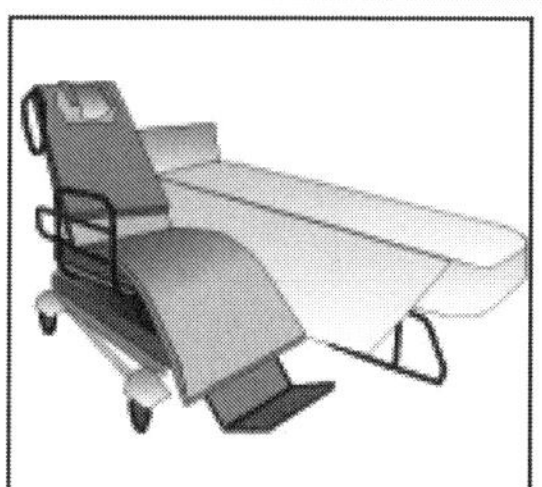

Activity:
Resident Handling

Description:
Convertible wheelchair or Geri chair to stretcher

When to Use: For lateral transfer of residents who are partial- or non-weight bearing. Eliminates the need to perform lift transfer in and out of wheelchairs.

Points to Remember: Two caregivers are required to perform lateral transfer. Additional assistance for lateral transfer may be needed depending on residents status, e.g., for heavier or non-cooperative residents. Wheelchairs that convert to stretchers may also have a mechanical transfer aid built in for bed-to-stretcher or stretcher-to-bed type transfers. Motorized height-adjustable devices are preferred to those adjusted by crank mechanism to minimize physical exertion and risk of musculoskeletal injury to the caregiver. Always ensure device is in good working order before use and is rated for the load weight to be transferred. Ensure wheels on equipment are locked. Ensure transfer surfaces are at same level and at a height that allows caregivers to work at waist level to avoid extended reaches and back flexion.

References: 16, 26, 42

12

Lateral Transfer in Sitting Position

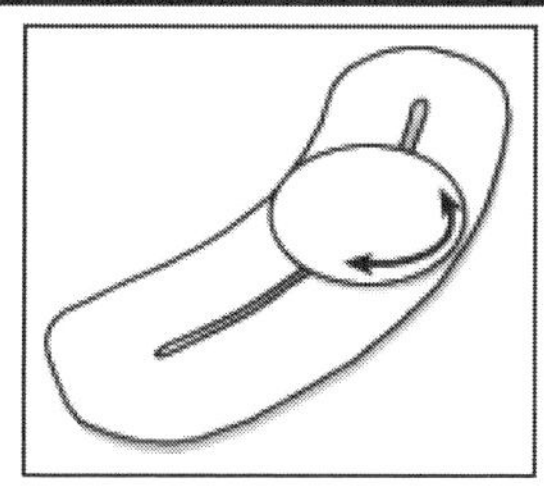

Activity:
Resident Handling

Description:
Transfer boards –
wood or plastic
(some with movable seat)

When to Use: Transferring (sliding) residents who have good sitting balance and are cooperative from one level surface to another, e.g., bed to wheelchair, wheelchair to car seat or toilet. Can also be used by residents who require limited assistance but need additional safety and support. Movable seats increase resident comfort and reduce incidence of tissue damage during transfer.

Points to Remember: Ensure clothing is present between the resident's skin and the transfer device. The seat may be cushioned with a small towel for comfort. May be uncomfortable for larger residents. Usually used in conjunction with gait belts for safety depending on resident status. Ensure boards have tapered ends, rounded edges, and appropriate weight capacity. Ensure wheels on equipment are locked and transfer surfaces are at same level. Remove lower bedrails from bed and remove arms and foot rests from chairs as appropriate.

References: 13, 16, 57, 86, 90, 94, 97, 113

13

Lateral Transfer in Sitting Position

Activity:
Resident Handling

Description:
Transfer slings

When to Use: Transferring residents who are partially dependent, cooperative, have some weight-bearing capacity, can sit up unaided and are able to bend hips, knees, and ankles. Transfers from bed to chair, or chair to chair, chair or toilet.

Points to Remember: Can be used by 1 or 2 caregivers with pivot or transfer disc to assist transfer as necessary. Place sling under resident's hips. Caregiver(s) *support* resident during the transfer. Use for short distance transfers only. Sling should not be used to lift residents. Position sling so as not to dig into or slip off of resident. Ensure sling is in good condition before use and resident is wearing non-slip clothing. Adjust bed so resident's feet are flat on floor. Use smooth motion by moving feet while pivoting.

References: 13, 90, 97

14

Transfer from Sitting to Standing Position

Activity:
Resident Handling

Description:
Powered sit-to-stand or standing-assist devices.

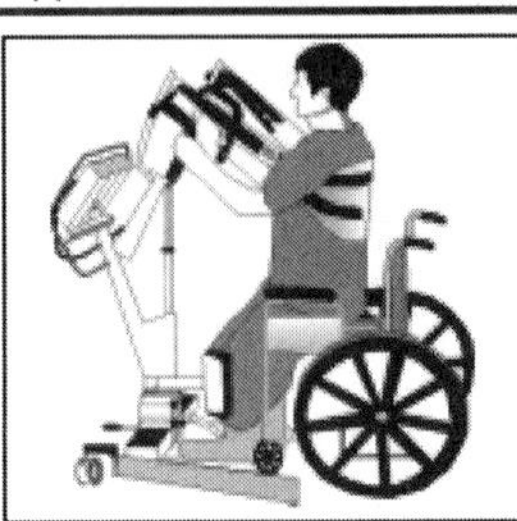

When to Use: Lifting residents who are partially dependent, have some weight-bearing capacity, are cooperative, can sit up on the edge of the bed with or without assistance, and are able to bend hips, knees, and ankles. Transfers from bed to chair (wheel chair, Geri or cardiac chair), or chair to bed, or for bathing and toileting. Can be used for repositioning where space or storage is limited.

Points to Remember: Usually requires 1 caregiver. Look for a device that has a variety of sling sizes, lift-height range, battery portability, hand-held control, emergency shut-off, and manual override. Ensure lifting device is in good working order before use and is rated for the load weight to be lifted. Electric/battery powered lifts are preferred to crank or pump type devices to allow a smoother movement for the resident, and less physical exertion and risk of musculoskeletal injury to the caregiver.

References: 13, 16, 33, 38, 43, 46, 48, 86, 90, 94, 113

15

Transfer from Sitting to Standing Position

Activity:
Resident Handling

Description:
Lift cushions and lift chairs

When to Use: Transferring residents who are weight-bearing and cooperative but need assistance when standing and ambulating. Can be used for independent residents who need an extra boost to stand. Can aid resident independence.

Points to Remember: Lift cushions use a lever that activates a spring action to assist residents to rise up. Lift cushions may not be appropriate for heavier residents. Lift chairs are operated via a hand-held control that tilts forward slowly, raising the resident. Residents need to have physical and cognitive capacity to be able to operate lever or controls. Always ensure device is in good working order before use and is rated for the load weight to be lifted.

References: 13

16

Transfer from Sitting to Standing Position

Activity:
Resident Handling

Description:
Stand-assist devices; can be fixed to bed or chair or be free-standing

When to Use: Transferring residents who are weight-bearing and cooperative and can pull themselves up from sitting to standing position. Can be used for independent residents who need extra boost to stand. Can aid resident independence.

Points to Remember: Check that device is stable before use and is rated for resident weight to be supported. Ensure frame is firmly attached to bed, or if relies on mattress support that mattress is heavy enough to hold the frame.

References: 13, 16, 90, 113

17

Transfer from Sitting to Standing Position; Ambulation

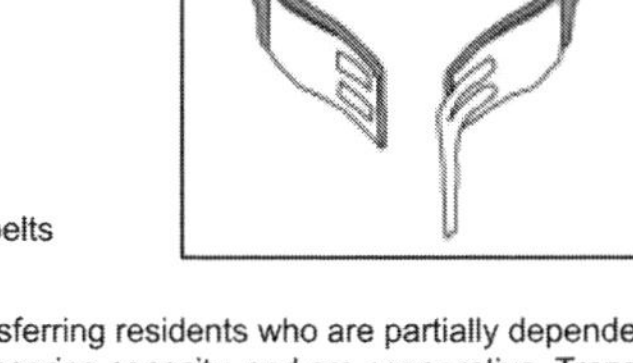

Activity:
Resident Handling

Description:
Gait belts/transfer belts with handles

When to Use: Transferring residents who are partially dependent, have some weight-bearing capacity, and are cooperative. Transfers such as bed to chair, chair to chair, or chair to car; when repositioning residents in chairs; supporting residents during ambulation; and in some cases when guiding and controlling falls.

Points to Remember: May require 1 or 2 caregivers. Belts with padded handles are easier to grip and increase security and control. Always transfer to resident's strongest side. Use rocking and pulling motion rather than lifting when using a belt. Belts may not be suitable for ambulation of heavy or non-weight bearing residents or residents with recent abdominal or back surgery, abdominal aneurysm, etc. Should not be used for lifting residents. Ensure belt is securely fastened and cannot be easily undone by the resident during transfer (i.e., Velcro fasteners). Ensure a layer of clothing is between residents' skin and the belt to avoid abrasion. Keep resident as close as possible to caregiver during transfer. Lower bedrails, remove arms and foot rests from chairs, and other items that may obstruct the transfer.

References: 3, 13, 16, 27, 29, 38, 41, 52, 57, 76, 87, 90, 91, 97, 113

18

Transfer from Sitting to Standing Position

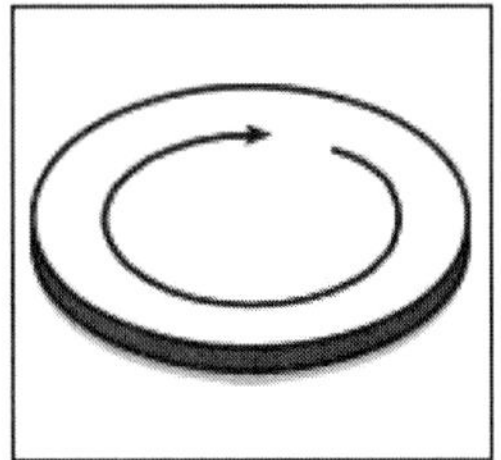

Activity:
Resident Handling

Description:
Pivot discs or boards; some discs have a stand-assist device attached for independent residents

When to Use: Transferring residents who are partially dependent, have some weight-bearing capacity, and are cooperative. Transfers such as bed to chair, chair to chair, or chair to car. Helps caregivers perform transfers without twisting.

Points to Remember: May require 1 or 2 caregivers. The disc is placed on the floor and used to rotate the resident 90 degrees to a bed or chair. Lower bed so that resident's feet are supported on the floor before standing. Resident's feet should be in the center of the disc, not touching the outer rim. A transfer or gait belt can be used in addition to the disc. May not be appropriate for heavier residents. Ensure disc has non-slip material on either side. Ensure wheels on equipment are locked and transfer surfaces are at same level. Lower bedrails, remove arms and foot rests from chairs and other items that may obstruct the transfer.

References: 13, 29, 41, 90, 97

19

Repositioning

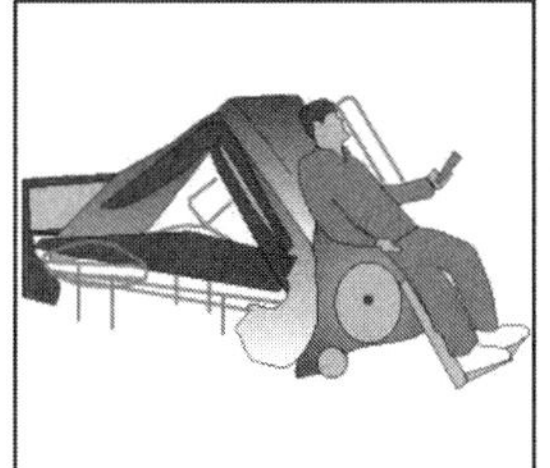

Activity:
Resident Handling

Description:
Beds that convert to chairs

When to Use: Repositioning residents who are totally dependent, non-weight bearing, very heavy, or have other physical limitations. Can also be used to assist residents who are partially weight bearing from a sit-to-stand position.

Points to Remember: Additional friction-reducing devices may be required to reposition resident. Heavy duty beds are available for bariatric residents. Device should have easy-to-use controls located within easy reach of the caregiver, sufficient foot clearance, and wide range of adjustment. Electric motor operation is preferred to reduce physical exertion and risk of musculoskeletal injury to caregiver and to facilitate positioning of resident.

References: 16, 39, 54

20

Repositioning

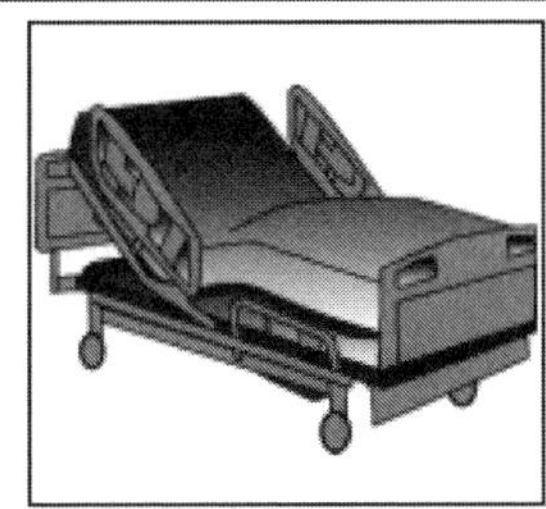

Activity:
Resident Handling

Description:
Electric powered height adjustable bed

When To Use: For all activities involving resident care, transfer, repositioning in bed, etc., to reduce caregiver trunk flexion when interacting with resident.

Points to Remember: Device should have easy-to-use controls located within easy reach of the caregiver, sufficient foot clearance, and wide range of adjustment. Heavy duty beds are available for bariatric residents. Beds raised and lowered with an electric motor are preferred over crank-adjust beds which require trunk flexion, force, and repetitive motion to use.

References: 29, 57, 67, 68, 70, 73, 79, 80, 94, 107

21

Repositioning

Activity:
Resident Handling

Description:
Trapeze bar

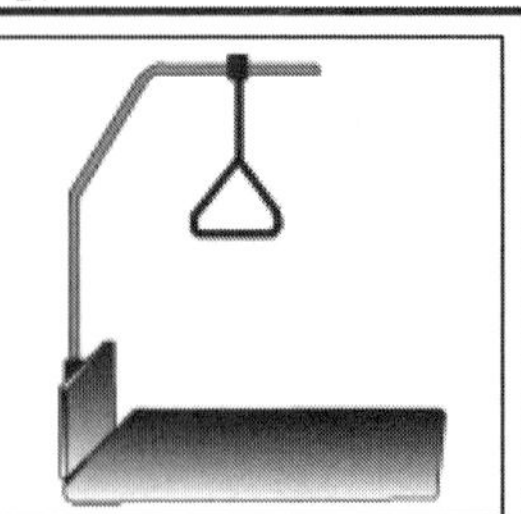

When to Use: Reposition residents that have the ability to assist the caregiver during the activity, i.e., residents with upper body strength and use of extremities, who are cooperative and can follow instructions.

Points to Remember: Residents use device by grasping bar suspended from an overhead frame to raise themselves up and reposition themselves in a bed. Heavy duty trapeze frames are available for bariatric residents. Ensure that bed wheels are locked, bedrails are lowered and bed is adjusted to caregiver's waist height to reduce back flexion.

References: 13, 75, 90, 113

22

Repositioning

Activity:
Resident Handling

Description:
Hand blocks and push up bars.

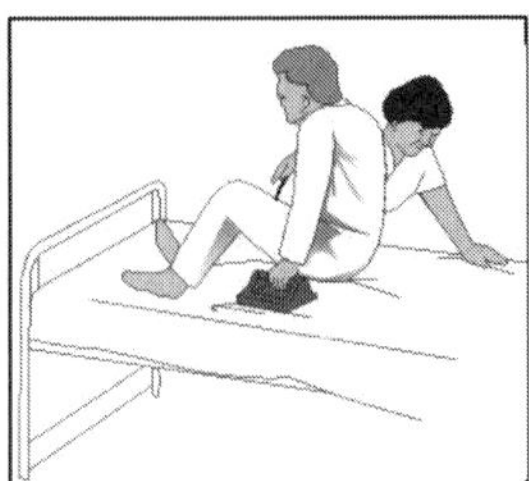

When to Use: Repositioning residents that have the ability to assist the caregiver during the activity, i.e., residents with upper body strength and use of extremities, who are cooperative and can follow instructions.

Points to Remember: Blocks also enable residents to raise themselves up and reposition themselves in bed. Bars attached to the bed frame serve the same purpose. May not be suitable for heavier residents. Ensure that bed wheels are locked, bedrails are lowered, and bed is adjusted to caregiver's waist height to reduce back flexion.

References: 13, 57

23

Repositioning

Activity:
Resident Handling

Description: Pelvic lift devices (hip lifters)

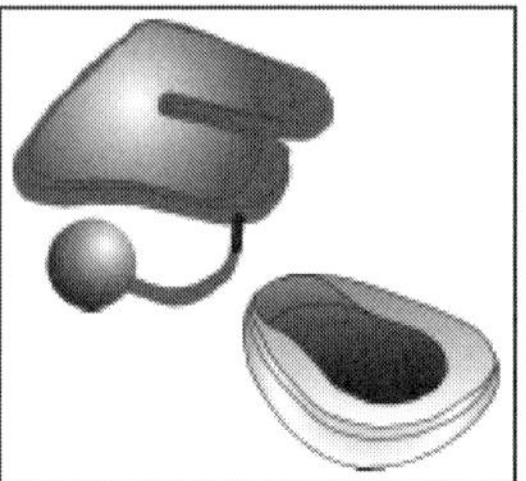

When to Use: To assist residents who are cooperative and can sit up with or without assistance, to position on a special bed pan. May reduce need for extra toileting.

Points to Remember: Device is positioned under hips and inflates like a pillow to lift hips. Use correct body mechanics, lower bedrails and adjust bed to waist height to reduce back flexion.

References: 13, 27

24

Ambulation

Activity:
Resident Handling

Description:
Ambulation assist device

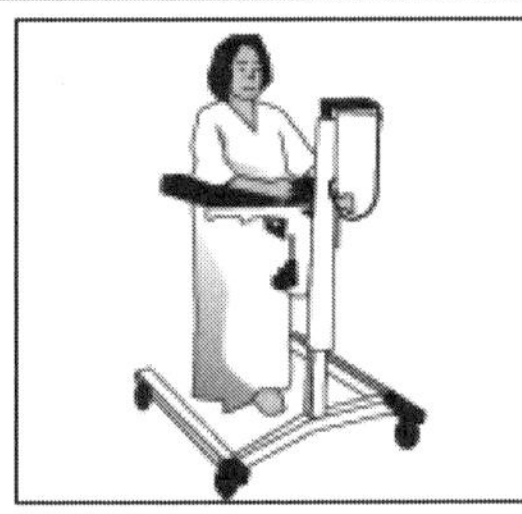

When to Use: For residents who are weight bearing and cooperative and who need extra security and assistance when ambulating. Increases resident safety during ambulation and reduces risk of falls.

Points to Remember: Usually requires one caregiver. The device supports residents as they walk and push it along during ambulation. Ensure height adjustment is correct for resident before ambulation. Ensure device is in good working order before use and rated for the load weight to be lifted. Apply brakes before positioning resident in or releasing resident from device.

References: 13, 16, 28

25

Bathtub, Shower, and Toileting Activities

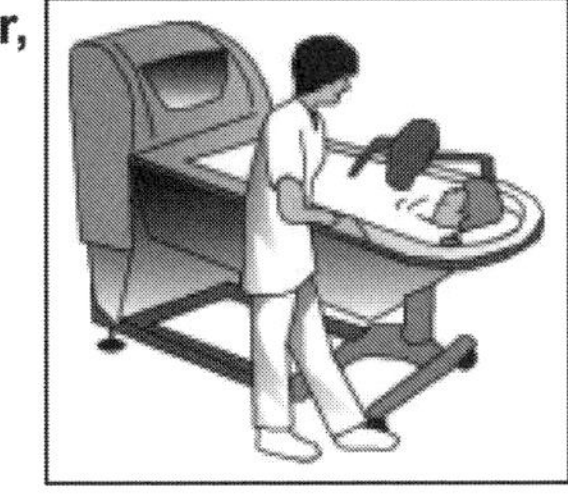

Activity:
Resident Handling

Description:
Height-adjustable bathtub and easy-entry bath tubs

When to Use: Bathing residents who sit directly in the bathtub, or to assist ambulatory residents climb more easily into a low tub, or easy-access tub. Bathing residents in portable-powered or ceiling-mounted lift device using appropriate bathing sling. Reduces risk of back and shoulder injuries to caregiver and to those who clean the tub after use. Increases resident safety and comfort.

Points to Remember: The tub can be raised to eliminate bending and reaching for the caregiver. Use correct body mechanics, and adjust the tub to waist height to reduce back flexion, when performing hygiene activities.

References: 13, 33, 94

26

Bathtub, Shower, and Toileting Activities

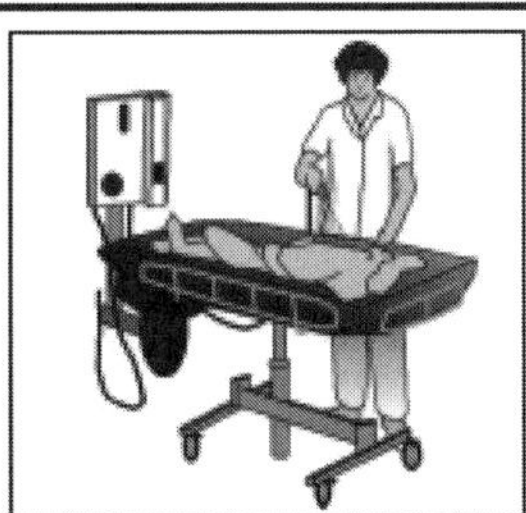

Activity:
Resident Handling

Description:
Height-adjustable shower gurney or lift bath cart with waterproof top

When to Use: For bathing non-weight bearing residents who are unable to sit up. Transfer resident to cart with lateral transfer boards or other friction-reducing devices.

Points to Remember: The cart can be raised to eliminate bending and reaching to the caregiver. Feet and head supports are available for resident comfort. May not be suitable for bariatric residents. Look for carts that are power-driven to reduce force required to move and position device.

References: 13, 29, 98

27

Bathtub, Shower, and Toileting Activities

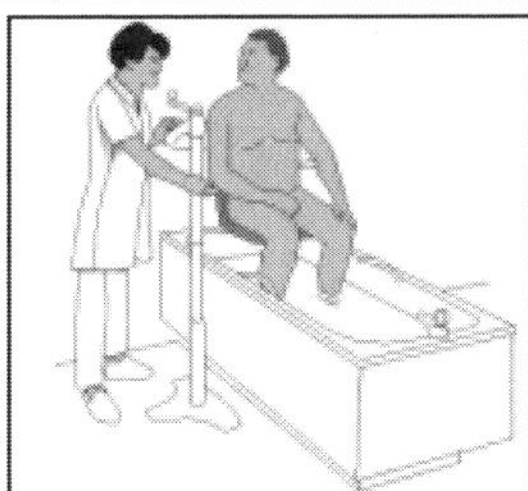

Activity:
Resident Handling

Description:
Built-in or fixed bath lifts

When to Use: Bathing residents who are partially weight bearing, have good sitting balance, can use upper extremities (have upper body strength), are cooperative, and can follow instructions. Useful in small bathrooms where space is limited.

Points to Remember: Ensure that seat raises so resident's feet clear tub, easily rotates, and lowers resident into water. May not be suitable for heavy residents. Always ensure lifting device is in good working order before use and rated for the load weight to be lifted. Choose device with lift mechanism that minimizes physical effort by caregiver when raising and lowering device.

References: 28, 90, 108, 113

28

Bathtub, Shower, and Toileting Activities

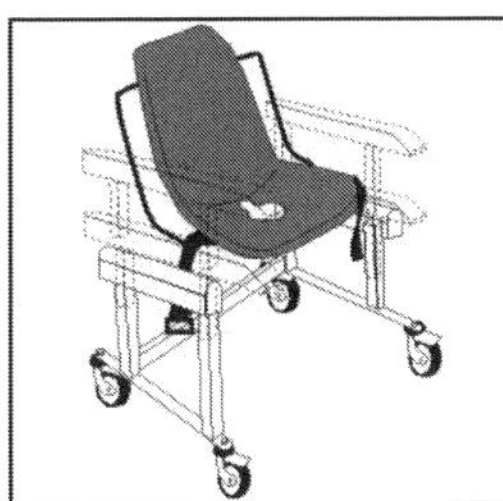

Activity:
Resident Handling

Description:
Shower and toileting chairs

When to Use: Showering and toileting residents who are partially dependent, have some weight bearing capacity, can sit up unaided, and are able to bend hips, knees, and ankles.

Points to Remember: Ensure that wheels move easily and smoothly; chair is high enough to fit over toilet; chair has removable arms, adjustable footrests, safety belts, and is heavy enough to be stable, and that the seat is comfortable, accommodates larger residents, and has a removable commode bucket for toileting. Ensure that brakes lock and hold effectively on at least two wheels and that weight capacity is sufficient for resident handling.

References: 13, 29, 43, 90, 108

29

Bathtub, Shower, and Toileting Activities

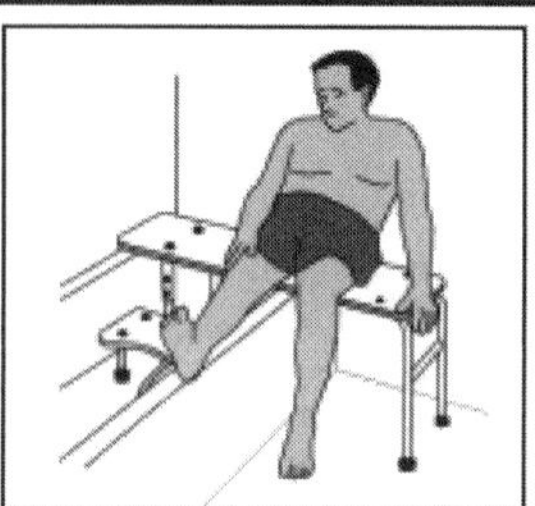

Activity:
Resident Handling

Description:
Bath boards and transfer benches

When to Use: Bathing residents who are partially weight bearing, have good sitting balance, can use upper extremities (have upper body strength), are cooperative, and can follow instructions. Independent residents can also use these devices.

Points to Remember: To reduce friction and possible skin tears, use clothing or material between the resident's skin and the board. Can be used with a gait or transfer belt and/or grab bars to aid transfer. Back support and vinyl padded seats add to bathing comfort. Look for devices that allow for water drainage and have height-adjustable legs. May not be suitable for heavy residents. Ensure wheels on equipment are locked, transfer surfaces at same level, and device is securely in place and rated for load to be transferred. Remove arms and foot rests from chairs as appropriate and ensure that floor is dry.

References: 13, 29, 90, 108

30

Bathtub, Shower, and Toileting Activities

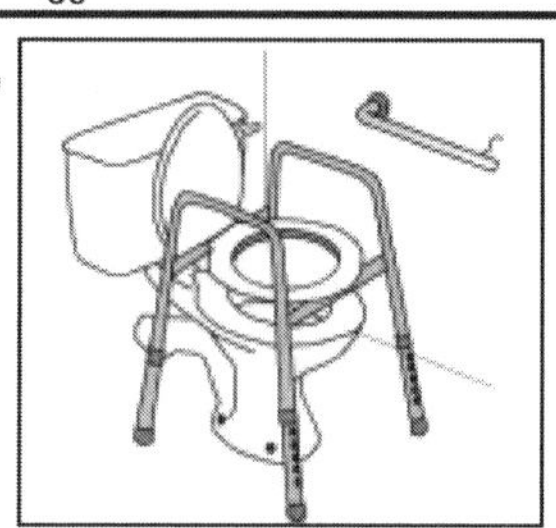

Activity:
Resident Handling

Description:
Toilet seat risers

When to Use: For toileting partially weight-bearing residents who can sit up unaided, use upper extremities (have upper body strength), are able to bend hips, knees, and ankles, and are cooperative. Independent residents can also use these devices.

Points to Remember: Risers decrease the distance and amount of effort required to lower and raise residents. Grab bars and height-adjustable legs add safety and versatility to the device. Ensure device is stable and can accommodate resident's weight and size.

References: 13, 80, 113

31

Bathtub, Shower, and Toileting Activities

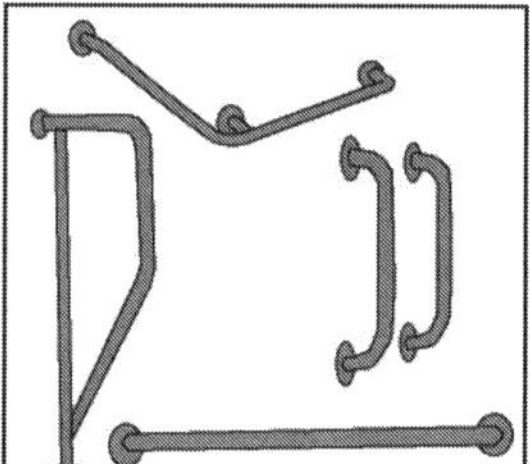

Activity:
Resident Handling

Description
Grab bars and stand assists; can be fixed or mobile

When to Use: When toileting, bathing, and/or showering residents who need extra support and security. Residents must be partially weight bearing, able to use upper extremities (have upper body strength), and be cooperative. Independent residents can also use these devices.

Points to Remember: Movable grab bars on toilets minimize workplace congestion. Ensure bars are securely fastened to wall before use.

References: 108, 113

32

Bathtub, Shower, and Toileting Activities

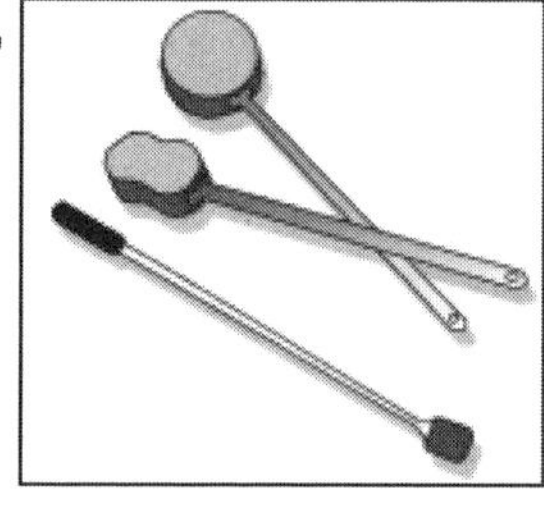

Activity:
Resident Handling

Description:
Long-handled extension tools on hand-held shower heads, wash or scrub brushes

When to Use: When bathing or showering residents.

Points to Remember: These devices reduce the amount of bending, reaching, and twisting required when washing feet, legs, and trunk of residents. Residents who are independent can also use these devices to facilitate personal hygiene activities.

References: 13, 29, 44, 105

33

Weighing

Activity:
Resident Handling

Description:
Scales with ramp to accommodate wheelchairs; portable-powered lift devices with built-in scales; beds with built-in scales.

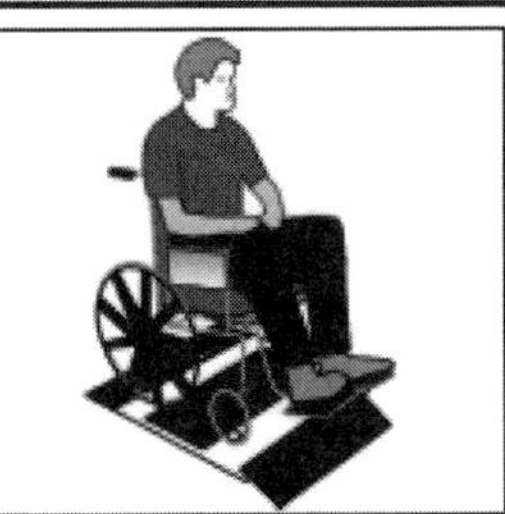

When to Use: To reduce the need for additional transfer of partial- or non-weight-bearing or totally dependent residents to weighing device.

Points to Remember: Some wheelchair scales can accommodate larger wheelchairs. Built-in bed scales may increase weight of the bed and prevent it from lowering to appropriate work heights.

References: 13, 43, 57, 71, 94

34

Guiding and Slowing Falls

Activity:
Resident Handling

Description:
Method for guiding and slowing falls

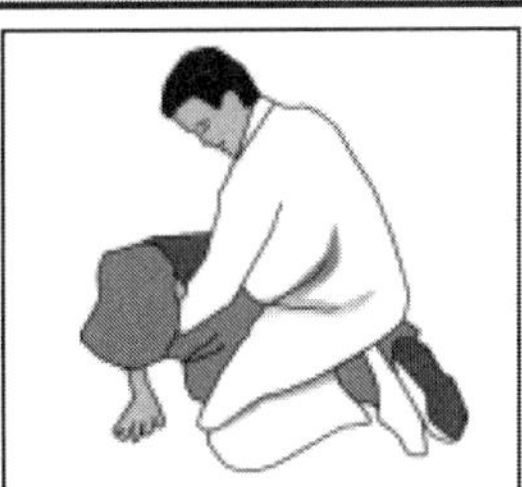

When To Use: When resident is falling.

Points to Remember: The use of transfer or gait belts may assist the caregiver in guiding the fall. Hold onto the belt/handles and slowly lower the resident to the floor using correct body mechanics. Reviewing resident assessments and watching for signs of weakness are effective ways of preventing falls. Keep back straight, tighten abdominal muscles, bend legs, and stay close to resident if safe to do so. Do not attempt to stop the fall abruptly as this may contribute to caregiver injury.

References: 13, 56, 76, 113

35

Lifting from the Floor

Activity:
Resident Handling

Description:
Methods to lift residents from floor

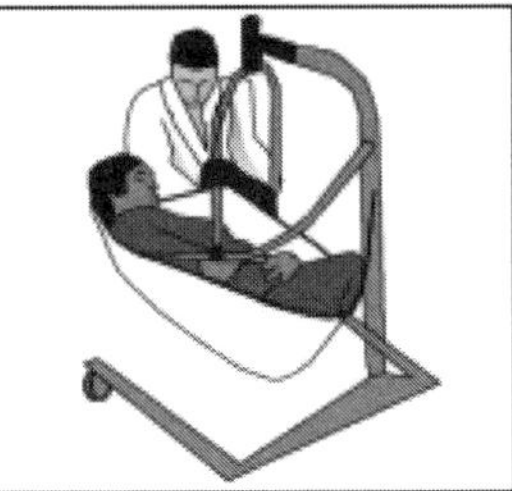

When To Use: After a resident fall.

Points to Remember: Assess resident for injury prior to lifting. If resident cannot stand with minimal assistance, use a powered portable or ceiling-mounted lift device to move resident. If resident can regain standing position with minimal assistance, use gait or transfer belt with handles to aid resident. If manual assistance is required insure adequate number of caregivers are available to provide needed assistance. Use 2 or more caregivers when assisting larger residents. Keep back straight, bend legs, and stay as close to resident as possible.

References: 13, 16, 27, 29

36

Repositioning in Chair

Activity:
Resident Handling

Description:
Variable position Geri chairs

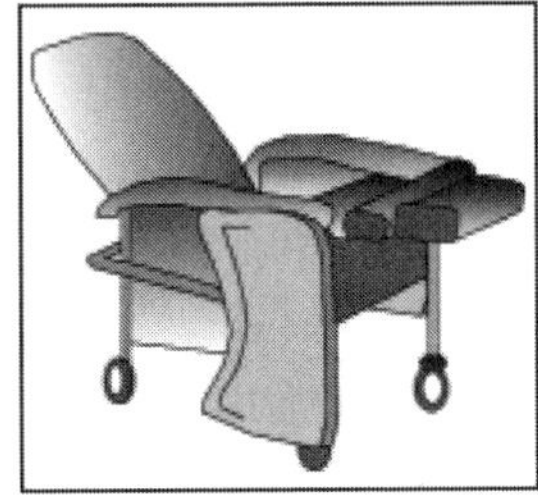

When to Use: Repositioning partial- or non-weight-bearing residents who are cooperative.

Points to Remember: One caregiver can assist if resident has upper extremity strength in both arms. If resident cannot assist to reposition self in chair, use at least 2 caregivers and friction-reducing device. Wheels on chair add versatility. Ensure that chair is easy to adjust, move, and steer. Lock wheels on chair before repositioning. Remove trays, foot rests, and seat belts where appropriate. Ensure device is rated for the resident weight.

References: 16, 31, 94

37

Various Activities of Daily Living and Bedside Assistance

Activity Description: Various Activities of Daily Living

Description: Work practices for feeding, dressing, and grooming

When to Use: During feeding, dressing, personal hygiene tasks, vital sign assessment, and other bedside assistance.

Bedside Assistance: Lower bed rails, position resident as close to edge as safely possible. Sit or stand as close as possible to resident's side and face resident. Adjust tables and electric beds to a height to allow caregivers to work at waist level and place supplies close by to avoid back flexion and twisting. Gather supplies in advance and place them on a table that is positioned perpendicular to the resident. Avoid reaching across resident; rather, walk to other side. Carry objects close to the body. Provide adaptive equipment for resident use when appropriate to increase independence and reduce assistance from caregiver.

Feeding: Cut food before placing in front of resident.

Dressing and grooming: Ensure that resident's feet are flat on the floor or a stool for balance when sitting. Place weaker limb in pant or sleeve first. Use appropriate adaptive equipment for dressing, grooming and oral hygiene.

References: 56, 99, 100, 101, 102, 103, 104

38

Activities of Daily Living

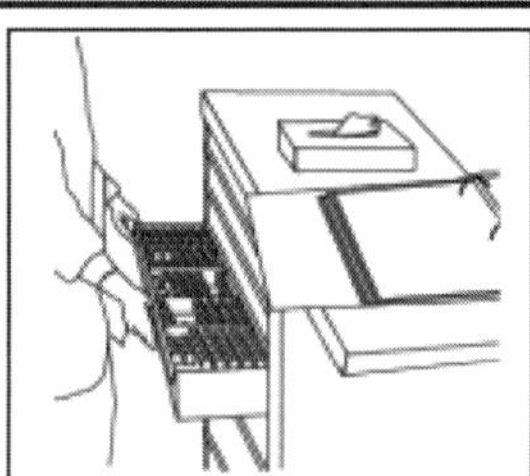

Activity: Dispensing medications

Description: Low profile medication cart and cartridge pill dispenser

When to Use: Dispensing medications. The cart increases accuracy and reduces time required to perform task.

Medications cart: Sort medications by day and time. Low profile carts with easy-open side drawers are recommended to accommodate hand height of shorter nurses.

Cartridge dispenser: Use cartridges with a "flip top" to store medications until dispensed rather than foil wrapped doses or small bottles. Individually wrapped medications in foil and paper may require high finger forces and a sharp object to break the seal.

References: 43, 56, 92, 106

39

Activities of Daily Living

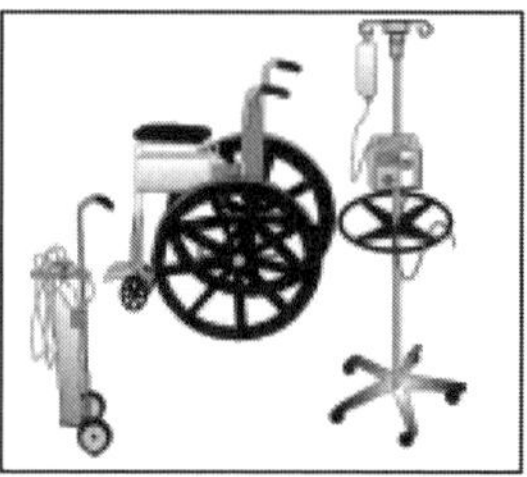

Activity: Transporting Equipment

Description: Work methods and tools to transport equipment

When to Use: When transporting assistive devices and other equipment

Oxygen tanks: Use small cylinders with handles to reduce weight and allow for easier gripping. Secure oxygen tanks to transport device.

Medication pumps: Use stands on wheels.

Transporting equipment: Push equipment, rather than pull, when possible. Keep arms close to the body and push with whole body and not just arms. Remove unnecessary objects to minimize weight. Avoid obstacles that could cause abrupt stops. Place equipment on a rolling device if possible. Take defective equipment out of service. Perform routine maintenance on all equipment.

References: 16, 29, 43, 82

40

Pouring Liquids

Activity: Dietary and Housekeeping

Description: Pouring containers that tilt

When to Use: In dietary and housekeeping areas when pouring soups or other liquid foods that are heavy, and in housekeeping areas when emptying buckets with floor drain arrangements. Reduces risk of spills and burns, speeds process, and reduces waste.

Points To Remember: Tilt handle or mechanism should allow the worker to reach it without bending and be positioned to assist in controlling the weight of the container and liquid. If the worker stands for more than 2 hours per day, shock-absorbing floors or insoles will minimize back and leg strain. With hot liquids, ensure a splash guard is included.

References: 43

41

Hand Tools

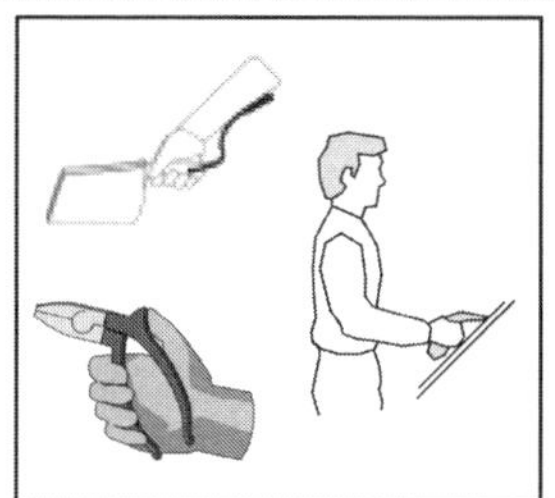

Activity:
Dietary, Laundry, Housekeeping and Maintenance

Description:
Select and use properly designed tools

When to Use: When selecting tools for the kitchen, housekeeping, laundry and maintenance areas. Enhances tool safety, speeds process, and reduces waste.

Points To Remember: Handles should fit the grip size of the user. If tool must be used with a bent wrist, purchase bent-handled tools. Minimize tool weight. Select tools that have minimal vibration or vibration damping devices. Implement a regular maintenance program for tools to keep blades sharp and edges and handles intact. Always wear the appropriate personal protective equipment.

References: 34

42

Linen Carts

Activity:
Laundry

Description:
Spring loaded carts that automatically bring linen within easy reach

When to Use: Moving or storing linen. Speeds process for handling linen, and reduces wear on linen due to excessive pulling.

Points to Remember: Select a spring tension that is appropriate for the weight of the load. Carts should have wheel locks and height-appropriate handles that can swing out of the way. Heavy carts should have brakes.

References: 14, 35, 43, 56, 59, 64, 79, 92, 94

43

Storage and Transfer of Food and Supplies

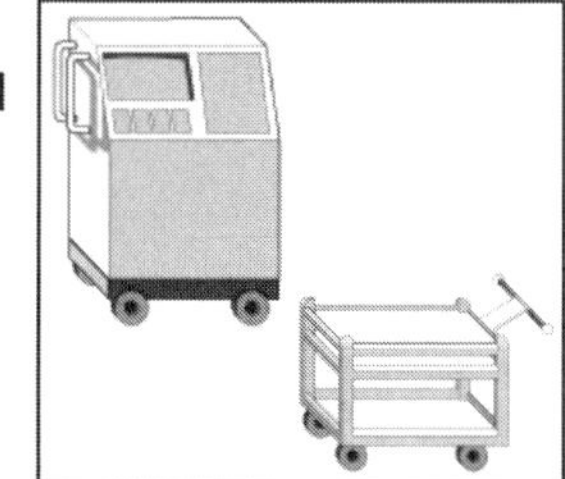

Activity:
Dietary, Laundry, Housekeeping and Maintenance

Description:
Use of carts

When to Use: When moving food trays, cleaning supplies, equipment, and maintenance tools. Speeds process for accessing and storing items.

Points to Remember: Placement of items on the cart should keep the most frequently used and heavy items within easy reach between hip and shoulder height. Carts should have full-bearing wheels of a material designed for the floor surface in your facility. Cart handles that are vertical, with some horizontal adjustability will allow all employees to push at elbow height and shoulder width. Carts should have wheel locks. Handles that can swing out of the way may be useful. Heavy carts should have brakes. Balance loads and keep loads under cart weight restrictions. Ensure stack height does not block vision.

References: 15, 35, 60, 84

44

Handling Bags

Activity:
Laundry, Housekeeping, and Maintenance

Description:
Equipment and practices for handling bags

When to Use: When handling laundry, trash and other bags. Reduces risk of items being dropped, and speeds process for removing and disposing of items.

Points to Remember: Receptacles that hold bags of laundry or trash should have side openings that keep the bags within easy reach and allow employees to slide the bag off the cart without lifting. Minimize the size and weight of bags and provide handles to decrease the strain of handling. Chutes and dumpsters should be at or below grade level. Provide automatic opening or hardware to keep doors open to eliminate twisting and awkward handling.

References: 11, 26, 35, 53, 56, 62, 63

45

Reaching into Sink

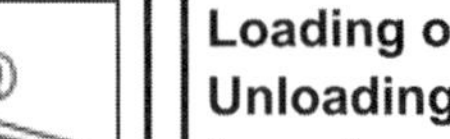

Activity:
Dietary, Housekeeping, and Maintenance

Description:
Tools used to modify a deep sink for cleaning small objects

When to Use: Cleaning small objects in a deep sink.

Points to Remember: Place an object such as a plastic basin in the bottom of the sink to raise the work surface and reduce back flexion sink bottom. An alternative is to use a smaller porous container to hold small objects for soaking, transfer to an adjacent countertop for aggressive cleaning, and then transfer back to the sink for final rinsing. Store inserts and containers in a convenient location to encourage consistent use.

References: 15, 36, 56, 109

46

Loading or Unloading Laundry

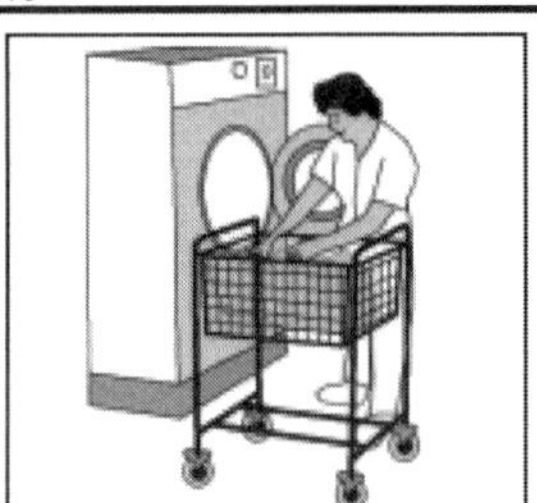

Activity:
Laundry

Description:
Front-loaded washers and dryers

When to Use: When loading or unloading laundry from washers, dryers and other laundry equipment. Speeds process for retrieving and placing items, and minimizes wear-and-tear on linen.

Points to Remember: Washers with tumbling cycles separate clothes, making removal easier. For deep tubs, a rake with long or extendable handle can be used to pull linen closer to the door opening. Raise machines so that opening is between hip and elbow height of employees.

References: 15, 43, 45, 81

47

Cleaning Rooms (Dry Method)

Activity:
Housekeeping

Description:
Work methods or tools to clean resident rooms without water and chemical products

When to Use: When cleaning rooms, beds, counters, walls, and furniture; sweeping and dusting floors.

For hand tool use: Alternate leading hand; avoid tight and static grip; and use padded non-slip handles.

For cleaning: Raise beds to waist level; use knee pads when kneeling; use tools with long handles, step stools, or ladders to avoid overhead reaching; alternate tasks frequently; and use carts to transport cleaning supplies.

Sweeping and dusting: Use flat head duster or "doodle bugs" and push with leading edge; avoid lifting leading edge of duster; sweep all areas into one pile for pick up; and vacuum if possible.

References: 11, 26, 37, 50, 56, 60, 61, 63, 83

48

Cleaning Rooms (Wet Method)

Activity:
Housekeeping

Description:
Work methods and tools to clean resident rooms with water and chemical products

When to Use: When cleaning with water and chemical products; mopping; and using spray bottles.
For hand tool use: Alternate leading hand; avoid tight; static grip and use padded non-slip handles.
For all cleaning: Use chemical cleaners and abrasive sponges to minimize scrubbing force. Use kneepads when kneeling. Avoid bending and twisting. Use extension handles, step stools, or ladders for overhead needs. Use carts to transport supplies. Carry only small quantities and weights of supplies. Maintain all equipment regularly. Ventilation of rooms may be necessary when chemicals are used.
Mopping: Alternate mopping styles frequently (e.g. push/pull, figure 8 and rocking side to side). Use rubber-soled shoes in wet areas to prevent slipping. Wheeled buckets should have functional brakes. Use dry mop vs. wet mop to minimize back strain.
Cleaning wheelchairs: Push wheelchair up a ramped platform to perform work at waist height.
Spray bottles: Use trigger handles long enough for the index and middle fingers. Avoid using the ring and little fingers.
References: 11, 26, 37, 60, 61, 63, 83, 111, 112

49

Cleaning Rooms (Electrical)

Activity:
Housekeeping

Description:
Work methods and tools to vacuum and buff floors

When to Use: Vacuuming and buffing floors.

Buffing: Buffers should have lightweight construction, triggers long enough to accommodate at least the index and middle fingers, adjustable handle height, and easy to reach controls.

Vacuuming: Vacuums should have lightweight construction, adjustable handle height, and controls that are easily accessed. Avoid short strokes by walking with the machine. Alternate leading hand and avoid tight grip. Use telescoping or extension handles for areas that are high, low, or far away. Remove vacuum bag when 1/2 to 3/4 full. Vacuums and powered devices are preferred over manual devices for moderate-to-long duration use. Heavy rolling canisters should have brakes.

Emptying trash: limit size of container to limit weight of load; dump carts of trash into receptacle at lower level for gravity assistance; use frame vs. solid can to prevent plastic bag from sticking to inside of container; avoid tying bag to frame; and place receptacles in unobstructed and easy to reach places.

References: 11, 26, 56, 62, 63, 109

REFERENCE LIST

1. Bureau of Labor Statistics. 2000. Number of nonfatal occupational injuries and illnesses by industry and selected case types.
2. Garg A. 1995. Effectiveness of Ergonomics Interventions at United Health, Inc. Preliminary Report, Industrial and Manufacturing Engineering, University of Wisconsin-Milwaukee, Milwaukee, WI.
3. Garg A. 1999. Long-Term Effectiveness of "Zero-Lift Program" in Seven Nursing Homes and One Hospital. U.S. Department of Health and Human Services, Centers for Disease Control and Prevention, National Institute for Occupational Safety and Health (NIOSH), Cincinnati, OH. August. Contract No. U60/CCU512089-02.
4. National Institute for Occupational Safety and Health (NIOSH). 1997. Musculoskeletal Disorders and Workplace Factors – A Critical Review of Epidemiologic Evidence for Work-Related Musculoskeletal Disorders of the Neck, Upper Extremity, and Low Back. July.
5. National Research Council. 1999. Work-Related Musculoskeletal Disorders. National Academy of Sciences. Washington, DC: National Academy Press.
6. National Research Council and Institute of Medicine. 2001. Musculoskeletal Disorders and the Workplace – Low Back and Upper Extremities. National Academy of Sciences. Washington, DC: National Academy Press.
7. American Health Care Association. 2002. Results of the 2001 AHCA nursing position vacancy and turnover survey. Health Services Research and Evaluation, American Health Care Association.
8. Cohen A.L., Gjessing C.G., Fine L.J., Bernard B.P., McGlothlin J.D. March 1997. Elements of Ergonomics Programs: A Primer Based on Workplace Evaluations of Musculoskeletal Disorders. US Department of Health and Human Services.
9. Pheasant S., Stubbs D. 1992. Back pain in nurses: Epidemiology and risk assessment. Applied Ergonomics. 23(4):226-232.
10. Service Employees International Union. 1997. Caring Till It Hurts.
11. Lawrence Livermore National Laboratory. n.d. Custodial Cleaning Manual.
12. Documents submitted to OSHA by Wyandot County Nursing Home.
13. Feletto M., Graze W. 1997. A Back Injury Prevention Guide for Health Care Providers. November. Cal/OSHA Consultation Service.
14. SEIU Education and Support Fund. 1996. Back Facts: A Training Workbook To Prevent Back Injuries in Nursing Homes.
15. New Zealand Department of Labor. 1993. Back in Care: Preventing Musculoskeletal Injuries in Staff in Hospitals and Residential Care Facilities. Occupational Safety & Health Service, Dept. of Labour, Wellington, New Zealand. ISBN0-477-03521-3.
16. Veterans Administration Hospital, Tampa, Florida. 2001. Patient Care Ergonomic Resource Guide: Safe Patient Handling and Movement. Patient Safety Center of Inquiry, Veterans Health Administration, and Department of Defense. November.
17. Veterans Health Administration, Patient Safety Center of Inquiry. n.d. Draft: Safe Patient Handling and Movement Policy.
18. Moore J., Garg A. 1995. The strain index: a proposed method to analyze jobs for risk of distal upper extremity disorders. AIHA Journal 56(5): 443-458.
19. Waters T., Putz-Anderson V., Garg A. 1994. Applications Manuals for the Revised NIOSH Lifting Equation. National Institute for Occupational Safety and Health. DHHS, NIOSH Publication No. 94-110. January.
20. Snook S., Ciriello V. 1991. The design of manual handling tasks: revised tables of maximum acceptable weights and forces. Ergonomics 34(9): 1197-1213.
21. McAtamney L., Corlett E. 1993. RULA: a survey method for the investigation of work-related upper limb disorders. Applied Ergonomics 24(2): 91-99.
22. Hignett S., McAtamney L. 2000. Rapid Entire Body Assessment (REBA). Applied Ergonomics 31: 201-205.
23. American Conference of Governmental Industrial Hygienists. 1998. 1998 Threshold Limit Values for physical agents in the work environment. In: 1998 TLVs and BEIs Threshold Limit Values for Chemical Substances and Physical Agents Biological Exposure Indices, pp. 109-131.

24. United Auto Workers-General Motors Center for Human Resources, Health and Safety Center. 1998. UAW-GM Ergonomic Risk Factor Checklist RFC2.
25. Documents submitted to OSHA by Citizens Memorial.
26. Worker's Compensation Board of British Columbia. 2002. MSI Prevention Bulletin 4: Room Cleaning in Healthcare.
27. National Institute for Occupational Safety and Health. 1997. Ergonomics: Effective Workplace Practice and Programs.
28. National Institute for Occupational Safety and Health. 1997. Health Hazard Evaluation Report, University of Cincinnati Hospital, Cincinnati, OH.
29. University of California at San Diego. 1998. UCSD Healthcare Ergonomic Guidelines. Univ. of California San Diego Medical Center. http://www-ehs.ucsd.edu/MCERGO.HTML
30. U.S. Department of Veterans Affairs. 2001. Veteran Health Administration: Use of Physical Restraint for Veterans at Risk of Falling. January.
31. National Institute for Occupational Safety and Health. 1988. Guidelines for Protecting the Safety and Health of Health Care Workers. DHHS (NIOSH) Publication No. 88-119. U.S. Government Printing Office. September.
32. (Reserved)
33. Lincoln Lutheran Minimal Lift CQI Team. 1996. The Minimal Lift Program for Maximum Safety. October.
34. Worker's Compensation Board of British Columbia. n.d. Draft ergonomics regulations: ergonomics code of practice part 3. Canada.
35. Tessler J. 1999. Backs for the Future - an Ergonomics Training Program for the SUNY Health Science Center Workforce. Public Employees Federation AFL-CIO Health and Safety Dept. Albany, N.Y.
36. New Zealand Department of Labor. n.d. Back in Care - Preventing Back Pain and Back Injuries in Caregivers.
37. Glan Hafren National Health Service Trust. n.d. Minimal Manual Handling Policy. NHS Trust, UK.
38. Yip Y. 2001. A study of work stress, patient handling activities and the risk of low back pain among nurses in Hong Kong. Journal of Advanced Nursing 36(6):794-804.
39. Shepherd C. 2001. Dimensions of care—ergonomics for the hospital setting. Occupational Health Tracker, Summer, SYSTOC, pp.8-9.
40. Ronald L., Yassi, A., Spiegel J., Tate R., Tait D., Mozel, M. 2002. Effectiveness of installing overhead ceiling lifts: reducing musculoskeletal injuries in an extended care hospital unit. AAOHN Journal 50(3):120-127.
41. Chin-Cheng Pan, Andris Frelvalds. 2000. Ergonomic evaluation of a new patient handling device. In: Proceedings of the IEA 2000/HFES 2000 Congress. Human Factors and Ergonomics Society.
42. Fragala G. 1996. Ergonomic Solutions for Preventing Patient Care Worker Back Injury. Wy'East Medical, Clackamas, OR.
43. Washington Department of Labor and Industries. 2001. Ergonomics Demonstration Project: Skilled Nursing Facility. October. <www.lin.wa.gov>
44. Proteau R. 2000. Ergonomics in home care. In: Proceedings of the IEA 2000/HFES 2000 Congress, Vol. 4, p. 4-275.
45. Rollins G. 2000. Ergonomics solutions in health care. NSC Safety and Health, November, pp. 26- 31.
46. Roth P., Ciecka J., Wood E., Taylor R. 1993. Evaluation of a unique mechanical client lift. AAOHN Journal. Vol. 41, No. 5:229-234.
47. Occupational Health and Safety Agency for Healthcare, British Columbia. 2000. Evaluation of Resident Lifting System Project, St. Joseph's Hospital, Comox, BC.
48. Washington Department of Labor and Industries. 1999. Frequently Asked Questions About Sit-Stand Patient-Resident Devices. May. <www.lin.wa.gov>
49. Marras W., Davis K., Krikling B., Bertsche P. 1998. Low back disorder and spinal loading during patient transfer. In: Proceedings of the Human Factors and Ergonomics Society 42nd Annual Meeting, Vol 2, pp. 901-905.
50. Kothiyal K., Yuen T. 2000. Manual handling in nursing jobs: an ergonomic study of a patient transferring aid. In: Proceedings of the IEA 2000/HFES 2000 Congress Vol. 5, p. 5-610.

51. Engkvist I., Wigaeus Hjelm E., Hagberg M. 2000. Patient transfers and the preventative effects for over exertion back injury of training and use of transfer devices among nursing personnel. In: Proceedings of the IEA 2000/HFES 2000 Congress, Vol. 5, pp. 5-427 – 5-429.
52. Caillard J., Iwatsubo Y. 2000. Prevention of musculoskeletal disorders among health care workers. In: Proceedings of the IEA 2000/HFES 2000 Congress, Vol. 5, pp. 5-781 – 5-784.
53. Washington Department of Labor and Industries. 2002. Simple Engineering Controls Can Improve Workflow and Reduce Twisting and Bending.
54. Hill-Rom Services, Inc. 1998. Total Care Bed System. Hill-Rom A HillenBrand Industry. Pub. No. CTG001 RB2.00. <http://www.hill-rom.com/prodbins/prods/brochures/f_totcr_b01_brc.pdf>
55. Guldmann. 2001. Various articles about ceiling hoists that demonstrate Guldmann's successes globally. <www.guldmann.com>
56. Vredenburgh A. 2000. Worker related MSDs: ergonomic risk to health care workers. Applied Ergonomics. Chapter 17, pp. 146-155.
57. Rhodes L., Rhodes D. 2001. Safety and Health in Nursing Homes - A Pennsylvania Perspective. American Society of Safety Engineers.
58. (Reserved)
59. Steinbrecher S. 1994. The revised NIOSH lifting guidelines—application in a hospital—linen handling. AAOHN Journal 42(2) pp 62-66.
60. Louhevaara V. 2000. Cardiorespiratory strain during floor mopping with different methods. In: Proceedings of the IEA 2000/HFES 2000 Congress, Vol.5 pp. 518- 520.
61. Hopsu L., Toivonen R. Louhevaara, Sjogaard K. 2000. Muscular strain during floor mopping with different cleaning methods (housekeeping). In: Proceedings of the IEA 2000/HFES 2000 Congress, Vol.5 pp. 521- 524. HFES.
62. Loussenhop S., Krueger D., Huth, E. 2000. Musculoskeletal disorders in cleaning personnel - interventions for prevention and rehabilitation. In: Proceedings of the IEA 2000/HFES 2000 Congress, Vol.5 pp. 747- 750. HFES.
63. Woods V., Buckle P. 2000. Recommendations for reducing musculoskeletal health problems among cleaners. In: Proceedings of the IEA 2000/HFES 2000 Congress, Vol. 5, pp. 510-513. HFES.
64. Intilli H. 1999. The effects of converting wheels on housekeeping carts in a large urban hospital. AAOHN Journal 47(10):466-69.
65. Garg A., Wen B., Beller D., Banaag D. 1991. A biomechanical and ergonomic evaluation of patient transferring tasks: bed to wheelchair and wheelchair to bed. Ergonomics 34(3):289-312.
66. Yassi A., Cooper J., Tate R., Gerlach S., Muir M., Trottier J., Massey K. 2001. A randomized controlled trial to prevent patient lift and transfer injuries of health care workers. Spine 26(16):1739-46.
67. de Looze M., Zinzen E., Caboor D., Heyblom P., van Bree E., van Roy P., Toussaint H, Clarijs J. 1994. Effect of individually chosen bed-height adjustments on the low-back stress of nurses. Scandinavian Journal of Work, Environment and Health 120(6):427-34.
68. Caboor D., Verlinden M., Zinzen E., Van Roy P., Van Riel M., Clarys J. 2000. Implications of an adjustable bed height during standard nursing tasks on spinal motion, perceived exertion and muscular activity. Ergonomics 43(10):1771-8.
69. Retsas A., Pinikahana J. 2000. Manual handling activities and injuries among nurses: an Australian hospital study. Journal of Advanced Nursing 31(4):875-83.
70. Owen B. 2000. Preventing injuries using an ergonomic approach. AORN Journal 72(6):1031-6.
71. Owen B., Garg A. 1994. Reducing back stress through an ergonomic approach: weighing a patient. International Journal of Nursing Studies 31(6):511-9.
72. Brophy, M., Achimore, L., Moore-Dawson J. 2001. Reducing incidence of low-back injuries reduces cost. AIHA Journal 62(4):508-11.
73. Tessler J. 1999. Back Injury Prevention Strategies for the Nursing Home Manager: A Guide to an Ergonomic Approach to Back Injury Prevention in Nursing Homes. Prepared for the Public Policy and Education Fund of New York, NYSDOL contract # C008311.
74. Kraker K., Vajdik C. 1997. Designing the environment to make bathing pleasant in nursing homes. Journal of Gerontological Nursing 23(5):50-1.
75. White C. 1998. How to prevent injury caused by moving and handling. Nursing Times 94(34):58- 62.
76. Lloyd P. 1997. Moving patients. Community Nurse 3(8):25-6.
77. Sykes K. 1998. Sweden: preventing health workers' back injuries. Public Health Reports 113(6):559.

78. Holiday P., Femie G., Plowman, S. 1994. The impact of new lifting technology in long-term care. AAOHN Vol. 42; No. 12: 582-589.
79. Warthington K. 2000. Watch your back. American Journal of Nursing 100(9):96.
80. Fern Gold M. 1994. The ergonomic workplace. Charting a course for long term care. Provider 2/94.
81. Pugliese G. 1993. Error-free linen handling. Materials Management in Health Care 2(9):36-9.
82. Griffin W. 2001. All in the timing: scheduling is key to floor care success. Health Facilities Management. 14(11):33-35.
83. Finestone H., Helfenstein S. 1994. Spray bottle epicondylitis: diagnosing and treating workers in pain. Canadian Family Physician 40:336-7.
84. Huang J., Ono Y., Shibata E., Takeuchi Y., Hisanaga, N. 1988. Occupational musculoskeletal disorders in lunch centre workers. Ergonomics 31(1):65-75.
85. Guldmann. 2001. Success With Ceiling Hoist in Canada. Canadian ASSTSAS. <www.guldmann.com>.
86. Schiff L. 2001. Lift and transfer devices—market choices. RN Vol. 64 No 8: 61-62.
87. Muir III T. n.d. Back injury prevention in health care requires training techniques, exercise. Occupational Health and Safety. Stevens Publishing.
88. Holiday P., Femi G., Plowman S. 1994. The impact of new lifting technology in long term care. Journal 42(12):582-589.
89. Spiegel J.,Yassi A., Ronald L., Tate F., Hacking P., Colby, T. 2002. Implementing a resident lifting system in an extended care hospital: demonstrating cost-benefit. AAOHN Vol. 50, No. 3, pp 128-134.
90. Health Care Health and Safety Association of Ontario. 2000. Transfers and Lifts for Caregivers.
91. Health Care Health and Safety Association of Ontario. 2001. Ergonomics for Health Care. The Ergonomic Resource Guide for Organizations in Health and Community Care: ERGO. Health Care Health & Safety Association of Ontario.
92. U.S. General Accounting Office. 1997. Worker Protection – Private Sector Ergonomics Programs Yield Positive Results. August. GAO/HEHS-97-163.
93. Washington Department of Labor and Industries. 1999. Frequently Asked Questions About Portable Total Body Patient-Resident Lifts. <www.lin.wa.gov>
94. Ohio Bureau of Workers' Compensation, Division of Safety and Hygiene. 2002. Draft: Extended Care Facilities Ergonomics Best Practices.
95. (Reserved)
96. (Reserved)
97. Medical Devices Agency. 1993-1997. Disability Equipment Assessment. Documents A3, 10, 19, and 23.
98. Medical Devices Agency. 1998/2000. Moving and Handling: Portable bath Lifts – A comparative evaluation. MH1, June, 1998. Also Moving and Handling: Mobile electric hoists – An evaluation. MH2, July 2000. Medical Devices Agency, UK.
99. Mills M. 1983. A gooseneck feeding device. The American Journal of Occupational Therapy 37(2):112.
100. Ahasan R., Campbell D., Salmoni A., Lewko J. 2001. HFs/ergonomics of assistive technology. Journal of Physiological Anthropology and Applied Human Science 20(3):187-97.
101. Yates J. Whitehead G. 1986. Aids to feeding. Nursing 3(7):244-8.
102. Shinnar S.. 1983. Use of adaptive equipment in feeding. Journal of the American Dietetic Association 83(3):321-2.
103. Marshall S. 1968. Bowl stabilizer. American Journal of Occupational Therapy 22(1):38-9.
104. Weiss D., Weiss L. 1976. The sandwich holder. American Journal of Occupational Therapy 30(6):384.
105. Paul S., Baron T. 1988. Toileting device for patients with decreased hand function. Archives of Physical Medicine and Rehabilitation 69(2):142-3.
106. Huffman M., Cummins J. 1986. Providing 24-hour pharmaceutical services with mobile medication carts. American Journal of Hospital Pharmacy 43(6):1504-6.
107. Medical Devices Agency. 2002. Entrapment Risk in Electric Operated Adjustable Variable Height Beds.
108. Center for Assistive Technology. 2002. RERC Aging—Safety in the Bathroom—Products To Assist With Bathing. University of Buffalo. <http://cat.buffalo.edu/rerc-aging/rerca-benches.php>
109. Manual Handling in the Health Industry: Support Services Identification and Solutions.

110. Bogue B. 2001. Focus on caregiving: a look at no-manual-lift programs. Provider 43-5.
111. Taylor and Francis. 1988. Redesigning Tools, Workstations and Jobs. Chapter 9 in: Cumulative Trauma Disorders: A Manual for MSDs of the Upper Limb. Putz-Anderson V, Ed., p. 107.
112. Rodgers S. 1986. Design and Selection of Containers, Hand Carts, and Hand Trucks. Chapter 21 in: Ergonomic Design for People at Work, Rodgers S, Ed., p. 375. Volume 2. Van Nostrand Reinhold Co.
113. Occupational Health and Safety Agency for Healthcare in British Columbia. 2002. Safe Patient and Resident Handling.
114. Royal College of Nursing. n.d. RCN Code of Practice for Patient Handling.

Appendix B: Frequently Asked Questions about Portable Total Body Patient/Resident Lifts

What Is a Portable Total Body Lift?

Portable total body lifts are used to transfer patients/residents to or from a bed to a wheelchair, toilet chair, bathtub, commode, etc. (These devices used to be called HoyeFM lifts because Hoyer was one of the first companies to make them.) A total body lift is typically used with patients/residents who can't bear weight, have physical limitations (quadriplegia, amputee), or are very heavy and can't be safely transferred manually by staff. A portable total lift supports the entire weight of the resident with a sling attached to a stand on wheels that can be freely moved or positioned to allow a transfer to a different surface.

How Is a Lift Designed to Handle a Patient/Resident Safely?

Most modern total body lifts have several common features as seen in the illustration to the right. With a wide adjustable base, extremely strong sling materials, steel or aluminum mast and boom, and electrical motorized lifting mechanisms, most newer lifts can lift residents up to 400 pounds. Some lifts have even been designed to handle residents up to 600 or even 1000 lbs!

What Are the Benefits of Using a Total Body Lift?

The physical demands required to transfer a resident using a total lift are significantly less than manually lifting a resident. This means less risk of a back or shoulder overexertion injury, which are two of the most common injuries in direct care staff. The less physically demanding the task, the less fatigued the caregiver is over the course of a shift, which also means less risk of injury.

Common Features of a Portable Total Body Lift

Not only does the caregiver benefit from using a lift, but common features of a portable total body lift the patient/resident also benefits. Using a total body lift is much safer for the patient/resident since a lift is specifically designed to handle the weight of a patient/resident. With a manual lift, the caregiver must rely on their own physical strength to perform the transfer; when handling the total weight of a resident, this often means working beyond their physical capabilities. This in turn means greater risk of dropping or mishandling the resident during a manual lift. Most modern lifts are very stable, require very little force to push or pull even with a heavy person in the sling, and are designed with slings that reduce the risk of skin tear or abrasion. Also, with plenty of practice, most lifts can be used by one caregiver, freeing up additional staff to focus on patient/resident care.

Should a Total Body Lift Be Used for all Patient/Resident Transfers?

No. For those patients/residents that can bear some body weight, a total body transfer may not be appropriate. Instead, a sit-stand lift can be used that requires the patient/resident to bear some weight as they are transferred from a sitting to standing posture. A sit-stand lift is similar in design to a total body lift except a footplate is attached to the mast where the patient/resident stands while they are transferred. In addition, a different type of sling is used. Since the sit-stand lift only needs to support the upper body, a smaller sling is used that attaches around the back and under the arms of the patient/resident. Use of a sit-stand lift also requires the patient/resident to be able to sit up on the edge of the bed with or without assistance, and to be able to bend their hips, knees, and ankles. Further information on sit-stand lifts will be provided in a subsequent document from Labor and Industries that evaluates sit-stand lifts.

Why Is the Department of Labor and Industries Interested in Total Lifts?

As mentioned before, total lifts significantly reduce the physical effort transferring patients/residents, which results in less risk of shoulder and back overexertion injuries. According to Labor and Industries workers' compensation data and the Bureau of Labor Statistics on workplace injuries, the nursing home industry has one of the highest rates of overexertion injuries in the State of Washington. This is primarily due to shoulder and back injuries from transferring residents. Labor and Industries is interested in devices that could prevent workers from being hurt on the job and reduce industrial insurance costs.

Having one or several total body lifts in your facility does not ensure that employees will use them. Research on the use of lifts in healthcare settings has found considerable reluctance from caregivers to use a mechanical lift when performing a transfer. Therefore it is important that the introduction of total lifts also include the following:

Commitment from management that the use of lifts benefits patients/residents and staff. It is an important part of the facility's overall health and safety program.

Training on the proper use of the lift, which should include ample hands-on practice.

Written policies and procedures that provide guidance on how and when total lifts are to be used.

Consistent and fair plan or process for ensuring compliance with policies and procedures on patient/resident handling.

Education for residents and resident family members on the benefits of a total lift.

How Did Labor and Industries Collect Information on Total Lifts?

In an effort to better educate ourselves and the nursing home industry about the use of these devices, as well as their capabilities and limitations, Labor and Industries evaluated a variety of assistive devices that can be used to help caregivers transfer residents. Total body lifts were included in the first evaluation. Several different total lifts were evaluated in a skilled nursing facility with staff and residents. Eight total lift manufacturers were originally contacted; seven participated. Of the seven, six were evaluated at a skilled nursing facility and one was evaluated at a hospital. The evaluations at the nursing facility were conducted at Liberty Country nursing home in Centralia, Washington with staff and residents on one of the long-term care units. Each evaluation lasted one week. Before staff used the lifts, a representative from the manufacturer conducted approximately a one hour in-service training for staff on the proper use of the lift. Staff then used each lift for one week as part of their normal resident transfer duties.

Does Labor and Industries Endorse any of the Lifts Evaluated?

No. Labor and Industries does not endorse any of the specific manufacturers of lifts that were evaluated. The information on the different manufacturers is provided to assist you in making comparisons between the various brands and to highlight the most important features to consider when purchasing a lift.

What Are the most Important Features to Consider when Purchasing a Total Lift?

After evaluating each lift and talking to nursing home staff and residents who used each lift, it was determined that some features were the most critical to consider before purchasing. This is based on the use of the lift in a long-term care unit in a skilled nursing facility. Below is a description of each critical feature.

1. **Price:** Expect to pay somewhere between $4050 and $6000 for a new total lift with a built in scale (used to weigh patients/residents) or $3200 to $4900 without a scale. The scale is an important feature since the resident can be weighed during a regularly scheduled transfer without having to perform an additional transfer just for weighing.
2. **Weight capacity:** A weight capacity of at least 400 lb. is recommended to accommodate transferring very heavy residents.
3. **Lifting mechanism:** Most new lifts have a hydraulic lifting mechanism that is powered by an electric motor and battery. The electric motor is an important feature since it eliminates the need to pump or crank the lift by hand. The motor is controlled by a hand control with buttons for up and down. The electric motor also makes the lifting and lowering of the resident a smooth, continuous movement without jerky or rapid accelerations that are common with older hand crank or pump lifts. Some manufacturers' lifts also come with a two-speed motor, with slow and fast speeds.
4. **Lift height range:** The lift needs to lower far enough to reach a resident who has a low bed or has fallen to the floor. Caregivers are at an even greater risk of injury if manually lifting a resident in these situations. The lift should also be able to transfer a resident into a high bed such as a Clinitron™.
5. **Various sizes and types of slings:** Obviously not all residents are built the same, and not all sizes of slings will fit all residents. That is why it is important for slings to come in a variety of sizes at least small, medium, large, and extra large. Also important is the availability of slings for special purposes such as toileting and bathing.
6. **Sling position control:** A sling position control means either the sling is made with handles on the outside to help position the resident, or the sling bar has a handle as part of a pivoting frame. This feature is important since the handles or the bar can be used to help position the resident in a more upright posture before they are lowered into a chair or into a recumbent posture as they are lowered into a bed.
7. **Battery portability:** Battery portability is a lift feature that allows a dead battery to be quickly exchanged with a fully charged battery. Some manufacturers use a portable battery system as a standard, whereas others offer it as an option. Those that offer a portable battery as an option use a nonportable system as a standard, which requires the lift to be directly plugged into an outlet to be recharged.
8. **Hand-held control:** A hand held control is typically a push button control used to raise or lower the mast. An important feature is the ability to quickly place the control on the lift during the transfer process. This will free up the caregiver's hands to assist or position the resident. For this reason, a control with a magnetic attachment is preferred over a clip since it allows the control to be placed almost anywhere on the lift, including the boom.

9. **Emergency shut-off control:** This control stops the motor in case of an emergency and is a separate control from the hand-held push button that activates the power. This safety feature serves as a back up to the hand-held control. It could be used in a situation where the resident grabs the hand-held control and the caregiver needs to quickly shut off the power to protect the resident from harm.
10. **Manual override control:** In a situation where the battery loses power during the transfer, it will be important that the resident can be safely lowered using an override control. These controls are usually a manual crank, although one manufacturer provides a bit that can be attached to a powered screwdriver that substitutes as the manual crank.
11. **Boom pressure-sensitive switch:** Another safety feature that protects the resident in case the boom is inadvertently lowered onto them. A pressure-sensitive switch means the lift senses the upward resistance of an object or a person caught underneath the boom and automatically stops the motor. This reduces the risk of a resident being injured by coming in contact with the boom.
12. **Turnaround for replacement parts/country lift made in:** If the lift requires repair or a part needs to be replaced, how soon will the lift be back in service? Most manufacturers can provide replacement parts within one or two days. Since many of the manufacturers' sales representatives also service the lifts, they will have a stock of replacement parts on hand. If they do not, the part may have to be ordered from the manufacturer in which case a U.S. manufacturer can usually provide the parts faster.
13. **Manufacturers' sales representative:** Most sales representatives also service their lifts, so finding a reliable representative that serves your geographic area is an important consideration.

What Are some Additional Features to Consider before Purchasing a Lift?

There are many additional features that may need to be considered before purchasing and may be more important depending on your facility's unique needs. For instance, you may prefer a lift with a sling made out of a particular material or a lift that comes with a battery that can be recharged in three hours or less. A list and description of these additional features is provided below.

Whatever the case, don't just rely on the information in this document before you purchase. *TRY BEFORE YOU BUY!* All of the manufacturers listed here should allow you to try their lift in your facility for a trial period (at least a week) before you purchase. Using the equipment in your facility is the best way to get feedback from employees and residents and find out what works and what doesn't in your particular facility.

14. **Type of base:** All lifts have adjustable bases that allow the legs to fit around chairs, commodes, etc. A "V" base indicates the legs pivot around a central point when spread apart thus forming a "V." A "U" base has legs that slide away from the mast and remain parallel to each other when spread. Power means the base legs are spread in and out through the use of the hand control and the electric motor. With a manual base, the legs are spread with a bar that is moved by hand or a foot control.
15. **Base length:** This is the length of the base from the mast to the end of the legs. This feature may be important if the lift is used in tight spaces such as the bath or shower room.
16. **Base width adjustment range:** All the lifts have an adjustable base where the legs can be moved in and out to allow the lift to be positioned around chairs, commodes, etc. This measurement was taken from the inside of the legs with the base positioned all the way in (the first number) and the spreader all the way out (second number).
17. **Height of base legs:** This height may be important if the lift needs to be positioned under a very low bed or stretcher. This measurement was taken with the corresponding caster diameter listed in feature #28. Some lifts can be lowered further with smaller casters.

18. **Type of sling bar:** The sling bar is the part of the lift where the sling attaches to the boom. A fourpoint sling bar means the sling is attached at four separate points, a three point at three points, etc. Some lifts utilize a pivoting frame that's attached to the sling bar. Attached to the pivoting frame is a handle that can be used to position the resident in a reclined or upright seated posture during the transfer.
19. **Type of sling:** A one-piece sling is standard, however, some lifts do offer a two piece sling for residents with very limited flexibility. Those lifts that use a pivoting frame use a key lock system, which has plastic end-attachments that secure the sling to the sling bar and are used in place of an end loop. Some slings can wrap under or between the resident's legs depending on the how the caregiver wants to position that individual.
20. **Sling color-coded:** Many of the slings are color coded by size (e.g., yellow: small, red: medium, etc.) either by trim on the sling, color-coded sling handles, loops, or tags.
21. **Sling material:** A variety of materials are used, and each may have different washing and drying requirements. Some slings also have plastic stays in the head support, which must be removed before laundering.
22. **Sling laundering:** Depending on the material, the sling may have certain laundering restrictions. For those nursing homes that send their slings to a laundry service this may be an important consideration.
23. **Sling easily removed:** The standard bed-to-wheelchair sling had to be easy to remove from underneath the resident after they had been transferred to the chair. Sling removal involved pulling the sling out from underneath while lifting the resident's thighs and then leaning the resident forward and pulling the sling up and out from behind the resident.
24. **Sling head and neck support:** Adequate head and neck support is especially important for residents who don't have the strength to keep their head upright. Some slings utilize plastic stays that are embedded in pockets to provide additional support.
25. **Sling stretcher adapter:** A stretcher adapter allows the resident who needs to remain horizontal to be transferred.
26. **Peak push force:** A push force gauge was used to measure the force required to push the lift with a 120-lb resident on a tile floor. The peak force is the force required to initially push the lift. Once the lift is moving, even less force is required.
27. **Height of handles:** This is the height from the floor to the bottom and top of the handles used to push/pull the lift. Longer vertical handles (those with the greatest distance between the top and bottom of the handle) will accommodate a larger percentage of shorter and taller caregivers.
28. **Minimum storage space:** This is the minimum amount of space needed to store the lift (with the boom lowered and the base retracted fully). The amount of space needed could be important in facilities with limited storage area.
29. **Diameter of casters:** The diameter of the casters was measured because in general, larger casters require less force to push/pull and maneuver.
30. **Brakes:** Brakes are used to lock the rear wheels in place during certain transfer situations.
31. **Battery type:** Rechargeable lead acid or gel cell batteries are used in all of these lifts.
32. **Battery recharge time:** Depending on the battery, it can take anywhere from one to six hours to fully recharge. Some batteries can also be recharged in 5 to 10 minutes to get a few more lifts if needed.
33. **Battery life:** The number of lifts per charge is highly dependent on the weight of the resident transferred and if a powered base is used (powered bases require additional battery power to adjust the legs in and out). Therefore, the number of lifts per charge is very approximate. The total life of the battery also depends on usage and varies from nine months to three years.

34. **Low battery indicator:** All lifts were equipped with a Iow battery indicator, although the amount of charge left in the battery when the indicator came on varies. In general, the lifts have either a lighted indicator or auditory warning or both.
35. **Battery replacement cost:** This is the cost to purchase a new battery once the original(s) can no longer hold a charge. Ask the vendor if the battery can be purchased at a retail store rather than only from the vendor. If the battery can be purchased at a retail store it is often cheaper.
36. **New lift delivery time:** Depending on where the lift is manufactured and assembled, delivery time can vary from one day to four weeks.
37. **Warranty:** Some manufacturer warranties cover all parts for a standard period, whereas others only cover certain parts for certain periods.
38. **Repair parts from:** Some sales representatives stock their own parts, whereas others rely on the manufacturer to supply parts.
39. **Average repair time:** Average repair time was based on the sales representative's estimate of how long a typical repair takes maintenance staff once the replacement part is delivered.
40. **Loaners available:** Loaners are available in case extensive service or replacement is required.
41. **Zero-lift program offered:** In addition to selling total body lifts, most manufacturers also offer additional assistance to facilities to reach the goal of zero or no lifts. This assistance can be in the form of an audit or evaluation of the facility to determine the ideal quantity of lifts, in-service on the use of the lifts, recommendations on additional needs to reach zero lift, etc.
42. **Leasing:** Leasing is also an option if the up front cost of a lift(s) is prohibitive.
43. **Special features:** Some features could not be easily categorized or only pertain to one particular lift and are mentioned here.
44. **Tested on site:** All of the lifts were evaluated and tested at Liberty Country Nursing Home in Centralia, Washington, except for the Liko lift, which was evaluated at a hospital.

ACKNOWLEDGMENTS

We would like to acknowledge the staff and residents at Liberty Country Home in Centralia, Washington for their cooperation and feedback during the on-site evaluations of each total body lift; Lynn Ford and Pauline McDaniel, administrators at Liberty Country for allowing the Department of Labor and Industries to conduct these evaluations at their facility, and all of the sales representatives for their assistance in supplying their equipment and their willingness to allow Labor and Industries the opportunity to evaluate it. Special thanks go to the Department of Labor and Industries' Nursing Home Zero Lift Initiative Team for its feedback and comments on drafts of this document.

Appendix C: Frequently Asked Questions about Sit-to-Stand Patient/Resident Devices

What Is a Sit-To-Stand Device?

Sit-to-stand devices are used to transfer patients/residents between two seated postures (e.g., seated on the edge of the bed to a wheelchair, or wheelchair to commode or shower cabinet). A sit-to-stand device is designed to support only the upper body of the resident and therefore requires the resident to be able to bear some weight. This is different than a total body lift (also called a Hoyer™ lift, because Hoyer™ was one of the first companies to make them) that is meant to support the entire weight of the resident/patient. A sit-to-stand device is meant to replace the manual stand-and-pivot transfer that is performed frequently by caregivers when transferring a weight-bearing resident/patient from a seated posture to a standing posture or different seated surface.

How Is a Sit-To-Stand Device Designed to Handle a Patient/Resident Safely?

Most modern sit-to-stand devices have several common features as seen in the illustration to the right. With a wide adjustable base, extremely strong sling materials, steel or aluminum mast and boom, and electrical motorized lifting mechanisms, most newer devices can transfer residents up to 300 pounds.

Common Features of a Sit-To-Stand Device

What are the Benefits of using a Sit-to-Stand Device?

The physical demands required to transfer a resident using a sit-stand device are significantly less than manually performing a stand-and-pivot transfer, even if the caregiver is using a transfer belt. This means less risk of a back or shoulder overexertion injury, which are two of the most common injuries in direct care staff. The less physically demanding the task, the less fatigued the caregiver is over the course of a shift, which also means less risk of injury.

Because sit-to-stand devices are designed to quickly transfer a resident between two seated surfaces, caregivers can eliminate two or three manual transfers when toileting or showering a resident. For instance, when toiletlng a resident, instead of manually transferring a resident from the bed to the wheelchair, the wheelchair to the commode, and then the commode back to the wheelchair, a sit-to-stand device can be used instead. Since sit-to-stand devices typically have a shorter base than a total body lift and place the resident in a standing or nearly standing posture, they can more easily fit and maneuver a resident into tight spaces such as bath and shower rooms.

Not only does the caregiver benefit from using a sit-to-stand device, but the patient/resident also benefits. Using a sit-to-stand device is much safer for the patient/resident since the device is specifically designed to handle the weight of a patient/resident. With a manual stand-and-pivot transfer, the caregivers must rely on their own physical strength to perform the transfer, which often means working beyond their physical capabilities. This in turn means greater risk of dropping or mishandling the resident during a manual lift. Since sit-to-stand devices are designed to be used when toileting they also eliminate the manual transfer of the resident off the commode, which is often cited by caregivers as the most difficult transfer. In addition, modern sit-to-stand devices are very stable, require very little force to push or pull even with a heavy resident in the sling, and are designed with slings that reduce the risk of skin tear or abrasion. Also, with plenty of practice, a sit-to-stand device can be used by one caregiver, freeing up additional staff to focus on patient/resident care.

Can a Sit-to-Stand Device Be Used with any Resident/Patient?

No. A sit-to-stand device should only be used with residents/patients that can bear some body weight. Depending on how much weight bearing capacity the resident has, the sit-to-stand device can raise the resident just high enough for short distance transfers such as bed to wheelchair or wheelchair to commode, or to a fully standing posture for longer distance transfers. Use of a sit-to-stand device also requires the patient/resident to be able to sit up on the edge of the bed with or without assistance, and to be able to bend their hips, knees, and ankles.

For those residents that can bear some body weight, a sit-to-stand device can also be a helpful rehabilitation tool. It can be used to promote increased weight bearing by controlling the resident's position. The closer the resident is to upright, the more weight their lower extremities will be bearing.

Why Is the Department of Labor and Industries Interested in Sit-to-Stand Devices?

As mentioned before, sit-to-stand devices significantly reduce the physical effort transferring patients/residents, which results in less risk of shoulder and back overexertion injuries. According to Labor and Industries workers' compensation data and the Bureau of Labor Statistics on workplace injuries, the nursing home industry has one of the highest rates of overexertion injuries in the State of Washington. This is primarily due to shoulder and back injuries from transferring residents. Labor and Industries is interested in devices that could prevent workers from being hurt on the job and reduce industrial insurance costs.

Having one or several sit-to-stand devices in your facility will not ensure employee use. Research on the use of mechanical assistive devices in healthcare settings has found considerable reluctance from caregivers to use a mechanical lift when performing a transfer. Therefore it's important that the introduction of sit-to-stand devices also include the following:

Commitment from management that the use of mechanical assists benefits patients/residents and staff. It is an important part of the facility's overall health and safety program.

Training on the proper use of the device, which should include ample hands-on practice.

Written policies and procedures that provide guidance on how and when sit-to-stands are to be used.

Consistent and fair plan or process for ensuring compliance with policies and procedures on patient/resident handling.

Education for residents and resident family members on the benefits of a sit-to-stand device.

How Did Labor and Industries Collect Information on Sit-to-Stand Devices?

In an effort to better educate ourselves and the nursing home industry about the use of these devices, as well as their capabilities and limitations, Labor and Industries evaluated a variety of assistive devices that can be used to help caregivers transfer residents. Total body lifts were included in the first evaluation and are reported on in the publication "*Frequently Asked Questions about Portable Total Body Patient/Resident Lifts*." The second evaluation involved sit-to-stand devices. Both total lifts and sit-to-stand devices were evaluated in a skilled nursing facility with staff and residents. Nine sit-to-stand device manufacturers were originally contacted; six participated. The evaluations were conducted at Liberty Country nursing home in Centralia, Washington with staff and residents on one of the long term care units. Each evaluation lasted one week. Before staff used the devices, a representative from the manufacturer conducted approximately a one-hour in-service training for staff on the proper use of the device. Staff then used each device for one week as part of their normal resident transfer duties. An ergonomist and physical therapist with Labor and Industries attended each in-service and evaluated the devices for those features in Table 1.

Does Labor and Industries Endorse any of the Sit-to-Stand Devices Evaluated?

No. Labor and Industries does not endorse any of the specific manufacturers of sit-to-stand devices that were evaluated. The information on the different manufacturers is provided to assist you in making comparisons between the various brands and to highlight the most important features to consider when purchasing a sit-to-stand device.

What Are the most Important Features to Consider when Purchasing a Sit-to-Stand Device?

After evaluating each sit-to-stand device and talking to nursing home staff and residents who used each piece of equipment, it was determined that some features were the most critical to consider before purchasing. This is based on the use of the device in a long-term care unit in a skilled nursing facility. Below is a description of each critical feature followed by Table 1, which compares the six devices, based on these features.

1. **Price:** Expect to pay somewhere between \$4000 and \$5500 for a new sit-to-stand device with a built-in scale (used to weigh patients/residents) or \$2900 to \$3700 without a scale. The scale is an important feature since the resident can be weighed during a regularly scheduled transfer without having to perform an additional transfer just for weighing.
2. **Weight capacity:** A weight capacity of at least 300 lb is recommended to accommodate transferring very heavy residents.
3. **Lifting mechanism:** Sit-to-stand devices typically have a hydraulic lifting mechanism that is powered by an electric motor or actuator and battery. The electric motor or actuator is an important feature since it eliminates the need to pump or crank the lift by hand. The motor is controlled by a hand control with buttons for up and down. The electric motor also makes the raising and lowering of the resident a smooth, continuous movement without jerky or rapid accelerations that are common with older hand crank or pump lifts. Some manufacturers' lifts also come with a two-speed motor, with slow and fast speeds.
4. **Battery portability:** Battery portability is a feature that allows a dead battery to be quickly exchanged with a fully charged battery. Some manufacturers use a portable battery system as a standard, whereas others offer it as an option. Those that offer a portable battery as an option use a nonportable system as a standard, which requires the lift to be directly plugged into an outlet to be recharged.

TABLE 1

Manufacturer	Alpha Modalities	Arjo	Hill-Rom	Medcare	Medi-Man	Vanderlift
Model	Elf Stand-up	Sara 2000	Translift - Stand assist	Stand and Weigh	Medi SSL	Vera
1. Price	$4395 w/scale $2895 w/o scale	$5495 w/scale	With scale n/a $3595 w/o scale	$4995 w/scale $3698 w/o scale	With scale n/a $4995 w/o scale	$4000 w/scale $3245 w/o scale
2. Weight Capacity	300 lbs.	400 lbs.	400 lbs.	400 lbs.	350 lbs.	350 lbs.
3. Lifting Mechanism	Electric motor, battery operated, 2 speeds	Electric motor, battery operated, variable speed	Electric motor, battery operated, 1 speed	Electric motor, battery operated, 1 speed	Electric motor, battery operated, 1 speed	Electric motor, battery operated, 1 or 2 speeds available
4. Battery Portability	Yes, standard	Yes, standard	Yes, standard	Yes, standard	Yes, standard	No
5. Hand-held Control	Yes, clip-on	Yes, clip-on	Yes, magnetic	Yes, clip-on	Yes, magnetic	Yes, magnetic
6. Emergency shut Off Control	Yes	Yes	Yes	No	No	Yes, on separate switch
7. Manual Override Control	Yes	Yes	Yes	No	Yes	Yes
8. Turnaround for Replace. Parts/Country Made in	24–48 hours/Australia	24–48 hours/Sweden, England	24–48 hours/Australia	24 hours/USA	24–48 hours/Canada, USA	24 hours/USA
9. Location of Nearest Manufacturer's Sales Rep.	Portland, OR	Seattle, WA or Portland, OR	Puyallup, WA	Tacoma, WA	Beaverton, OR	Mill Creek, WA

5. **Hand-Held control:** A hand held control is typically a push button control used to raise or lower the boom. An important feature is the ability to quickly place the control on the sit-tostand device during the transfer process. This will free up the caregiver's hands to assist or position the resident. For this reason, a control with a magnetic attachment is preferred over a clip since it allows the control to be placed almost anywhere on the lift, including the boom.
6. **Emergency shut-off control:** This control stops the motor in case of an emergency and is a separate control from the hand-held push button that activates the power. This safety feature serves as a back up to the hand-held control. It could be used in a situation where the resident grabs the hand-held control and the caregiver needs to quickly shut off the power to protect the resident from harm.
7. **Manual override control:** In a situation where the battery loses power during the transfer, it will be important that the resident can be safely lowered using an override control. These controls are usually a manual crank, although one manufacturer provides a release lever that can be pulled that Iowers the resident automatically. Another manufacturer provides a bit that can be attached to a powered screwdriver that substitutes as the manual crank.
8. **Turnaround for replacement parts/country lift made in:** If the lift requires repair or a part needs to be replaced, how soon will the lift be back in service? Most manufacturers can provide replacement parts within one or two days. Since many of the manufacturers' sales representatives also service the lifts, they will have a stock of replacement parts on hand. If they do not, the part may have to be ordered from the manufacturer in which case a U.S. manufacturer can usually provide the parts faster.
9. **Manufacturer's sales representative:** Most sales representatives also service their lifts, so finding a reliable representative that serves your geographic area is an important consideration.
10. **What are some additional features to consider before purchasing a sit-to-stand device?** There are many additional features that may need to be considered before purchasing, some may be more important depending on your facility's unique needs. For instance, you may prefer a sit-to-stand device with a sling made out of a particular material or a sit-to-stand that comes with a battery that can be recharged in three hours or less. Whatever the case, do not just rely on the information in this document before you purchase. *TRY BEFORE YOU BUY!* All manufacturers should allow you to try their sit-to-stand in your facility for a trial period (at least a week) before you purchase. Using the equipment in your facility is the best way to get feedback from employees and residents and find out what works and what doesn't in your particular facility.
11. **Type of base:** All sit-to-stand devices have adjustable bases that allow the legs to fit around chairs, commodes, etc. All sit-to-stand devices evaluated utilized a "V" base where the base legs pivot around a central point when spread apart thus forming a "V." Power means the base legs are spread in and out through the use of the hand control and the electric motor. With a manual base, the legs are spread with a bar that's moved by hand or a foot control.
12. **Base length:** This is the length of the base from the mast to the end of the legs. This fcaturc may bc important when the device is used in tight spaces such as the bath or shower room.
13. **Base width adjustment range:** All the devices have an adjustable base where the legs can be moved in and out to allow the lift to be positioned around chairs, commodes, etc. This measurement was taken from the inside of the legs with the base positioned all the way in (the first number) and the spreader all the way out (second number).
14. **Height of base legs:** This height may be important if the sit-to-stand needs to be positioned under a very low bed or stretcher. This measurement was taken with the

corresponding caster diameter listed in feature #24. Some devices can be lowered further with smaller casters.

15. **Lift range:** This is the vertical distance the device moves the resident from a seated to a standing posture. All sit-to-stand devices were capable of transferring a 5 foot 10 inch resident sitting in an 18" high chair to standing or near full standing posture.
16. **Number of different sized slings:** Most of the manufacturers only offer one or two sizes of slings which should fit practically all residents.
17. **Sling attachments:** Most slings use a simple loop as an attachment point from the sling to the boom. The sling loop attaches around a J or C shaped hook on the boom. A few manufacturers use a key lock system, which has plastic end-attachments that snap in place to secure the sling to boom.
18. **Sling color-coded:** Many of the slings are color coded by size (e.g., yellow = small, red = medium, etc.) either by trim on the sling, color-coded sling handles, loops, or tags.
19. **Sling material:** A variety of materials are used, including sheepskin as a cover to help protect the resident from skin abrasions around the arms and axilla. Most of the slings have foam inserts in the back portion of the sling to help provide additional padding against the residents back.
20. **Sling laundering:** Depending on the material, the sling may have certain laundering restrictions. For those nursing homes that send their slings to a laundry service this may be an important consideration.
21. **Adjustable foot plate:** Some manufacturers provide an adjustable foot plate that can be adjusted vertically to accommodate very short residents.
22. **Peak push force:** A push force gauge was used to measure the force required to push the sit-to-stand with a 120 lb. resident on a tile floor. The peak force is the force required to initially push the sit-to-stand. Once the device is moving, even less force is required.
23. **Height of handles:** This is the height from the floor to the bottom and top of the handles used to push/pull the sit-to-stand. Longer vertical handles (those with the greatest distance between the top and bottom of the handle) will accommodate a larger percentage of shorter and taller caregivers.
24. **Minimum storage space:** This is the minimum amount of space needed to store the sit-to-stand (with the boom lowered and the base retracted fully). The amount of space needed could be important in facilities with limited storage area.
25. **Diameter of casters:** The diameter of the casters was measured because in general, larger casters require less force to push/pull and maneuver. It was also noted if there were single or dual wheels.
26. **Brakes:** Brakes are used to lock the rear wheels in place during certain transfer situations.
27. **Battery type:** Rechargeable lead acid or gel cell batteries are used in all of these sit-to-stand devices.
28. **Battery recharge time:** Depending on the battery, it can take anywhere from one to six hours to fully recharge. Some batteries can also be recharged in 5 to 10 minutes to get a few more lifts if needed.
29. **Battery life:** The number of transfers per charge is highly dependent on the weight of the resident transferred and if a powered base is used (powered bases require additional battery power to adjust the legs in and out). Therefore, the number of transfers per charge is very approximate. The total life of the battery also depends on usage and varies from nine months to seven years.
30. **Low battery indicator:** All sit-to-stands were equipped with a Iow battery indicator, although the amount of charge left in the battery when the indicator came on varies. In general, the devices have either a lighted indicator or auditory warning or both.
31. **Battery replacement cost:** This is the cost to purchase a new battery once the original(s) can no longer hold a charge. Ask the vendor if the battery can be purchased at a retail

store rather than only from the vendor. If the battery can be purchased at a retail store it is often cheaper.

32. **New sit-to-stand delivery time:** Depending on where the sit-to-stand is manufactured and assembled, delivery time can vary from one day to six weeks.
33. **Warranty:** Some manufacturer warranties cover all parts for a standard period, whereas others only cover certain parts for certain periods.
34. **Repair parts:** Some sales representatives stock their own parts, whereas others rely on the manufacturer to supply parts.
35. **Average repair time:** Average repair time was based on the sales representative's estimate of how long a typical repair takes maintenance staff once the replacement part is delivered.
36. **Loaners available:** Loaners are available in case extensive service or replacement is required.
37. **Zero-lift program offered:** In addition to selling sit-to-stand devices, most manufacturers also offer additional assistance to facilities to reach the goal of zero or no lifts. This assistance can be in the form of an audit or evaluation of the facility to determine the ideal quantity of devices, in-service on the use of the devices, recommendations on additional needs to reach zero lift, etc.
38. **Leasing:** Leasing is also an option if the up front cost of a sit-to-stand(s) is prohibitive.
39. **Special features:** Some features couldn't be easily categorized or only pertain to one particular sit-to-stand and are mentioned here.

ACKNOWLEDGMENTS

We acknowledge the staff and residents at Liberty Country Home in Centralia, Washington for their cooperation and feedback during the on-site evaluations of each sit-to-stand device; Lynn Ford and Pauline McDaniel, administrators at Liberty Country for allowing the Department of Labor and Industries to conduct these evaluations at their facility, and all of the sales representatives for their assistance in supplying their equipment and their willingness to allow Labor and Industries the opportunity to evaluate it. Special thanks go to the Department of Labor and Industries' Nursing Home Zero Lift Initiative Team for its feedback and comments on drafts of this document.

Appendix D: Equipment Options

LIKO'S MOBILE LIFTS

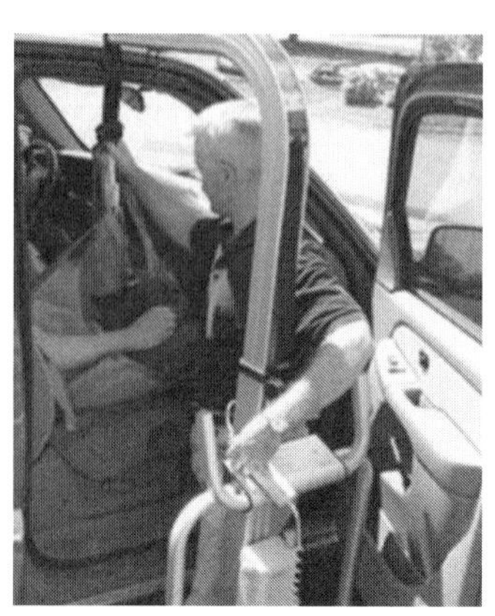

Golvo: Can be used to place a person in a vehicle

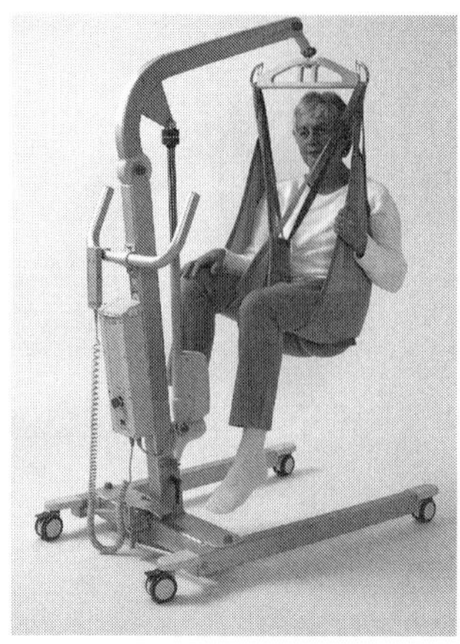

LikoLight: 308-lb lifting capacity

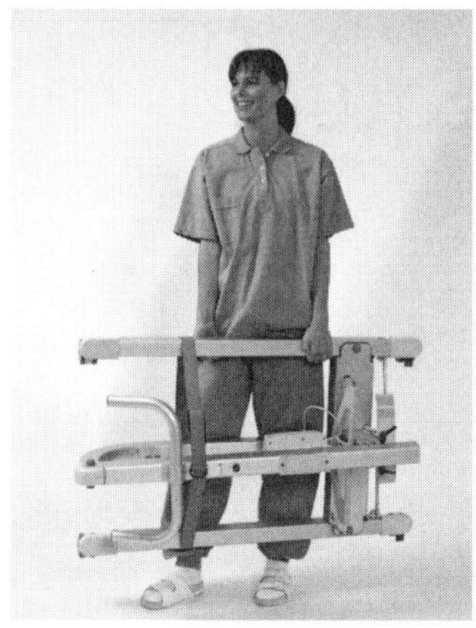

LikoLight: Foldable portable lift

Sabina II: Sit-to-stand lift

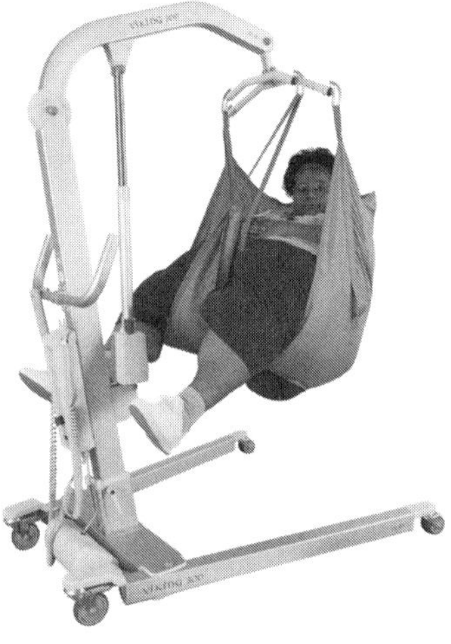

Bariatric lift — Viking 300: 660-lb lifting capacity

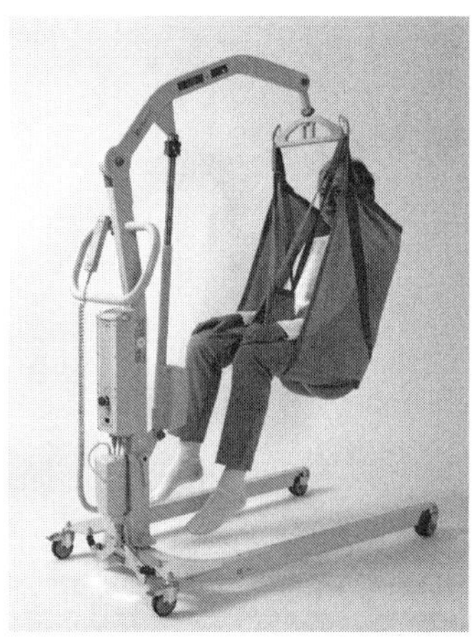

Uno: 385-lb lifting capacity

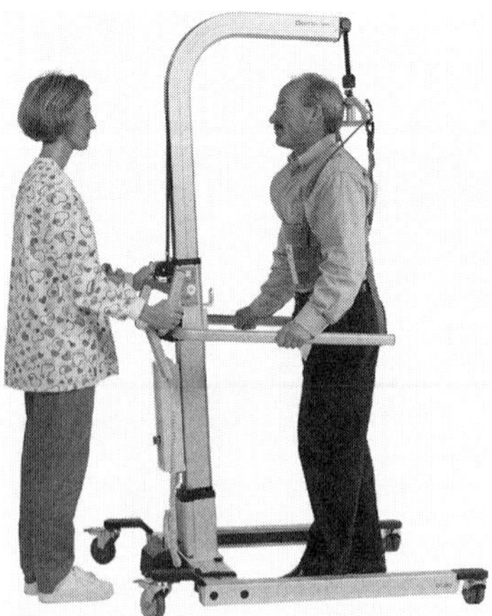

Golvo: Ambulatory gait training

LIKO'S OVERHEAD LIFTS

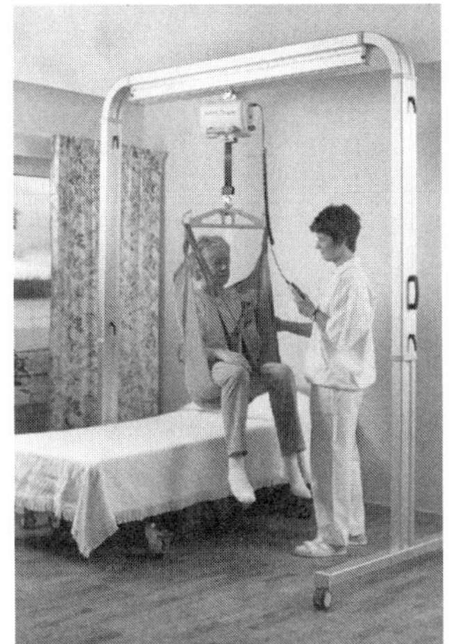

Multirall

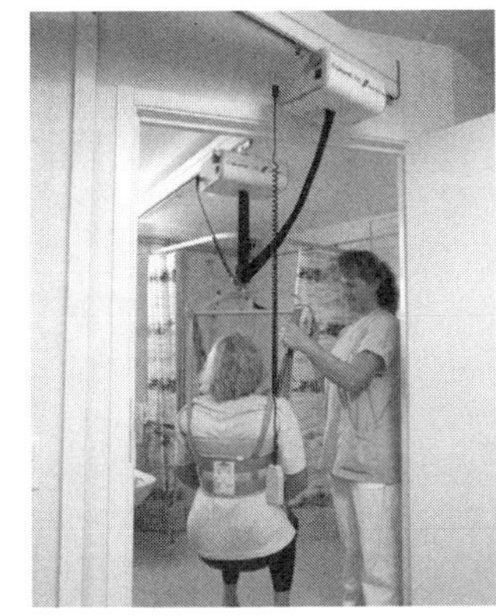

Room-to-Room (Likorall R2R)

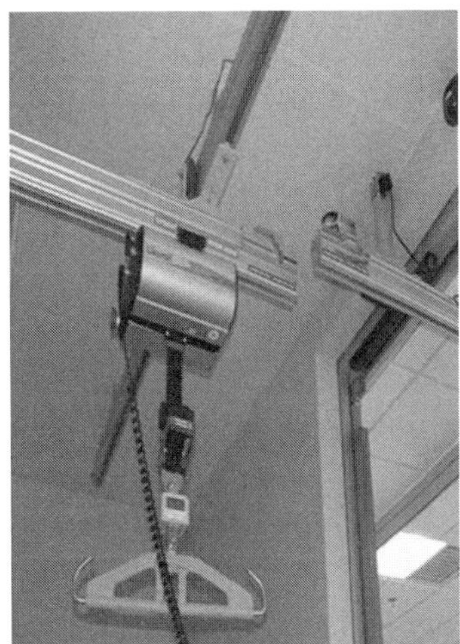

Cross-Over Switch

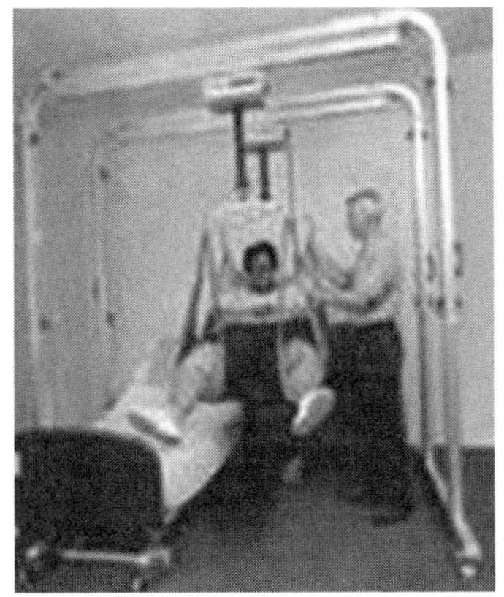

Bariatric Overhead Lift: lifts up to 880 lbs

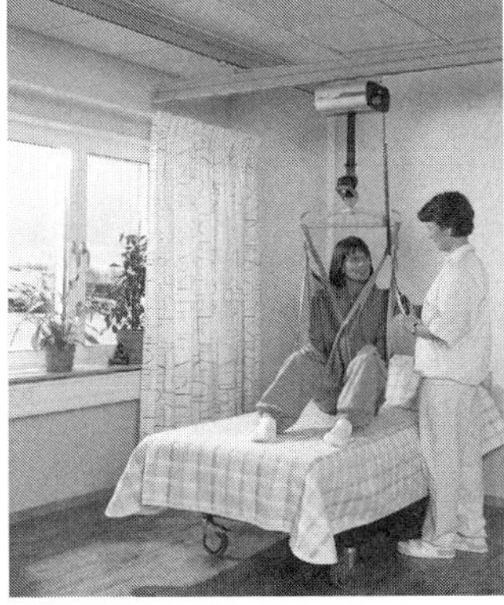

Likorall Overhead Lifting System

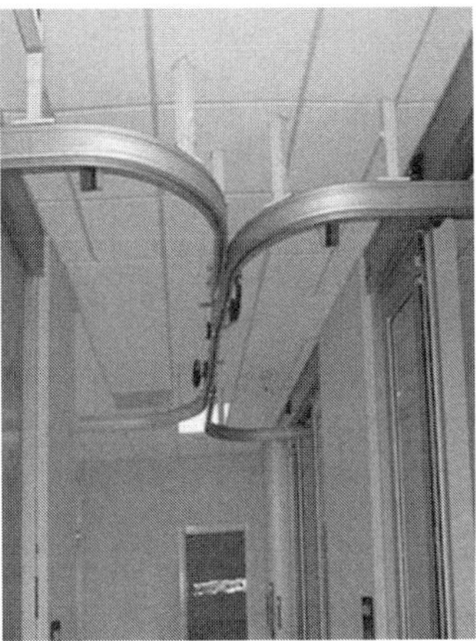

Rail System: installed at the Tampa VA Hospital, Florida

WY'EAST LIFTS

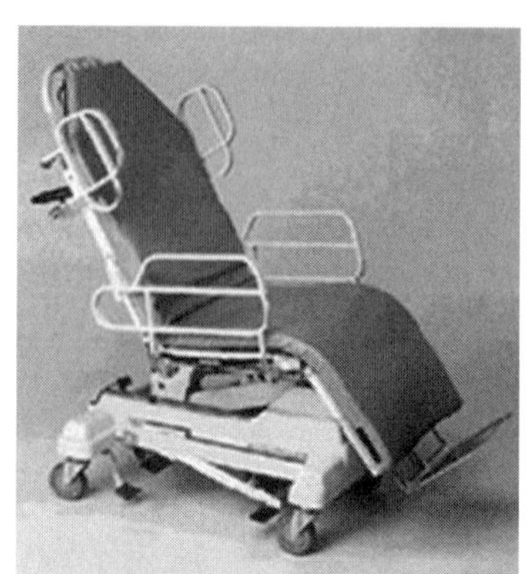

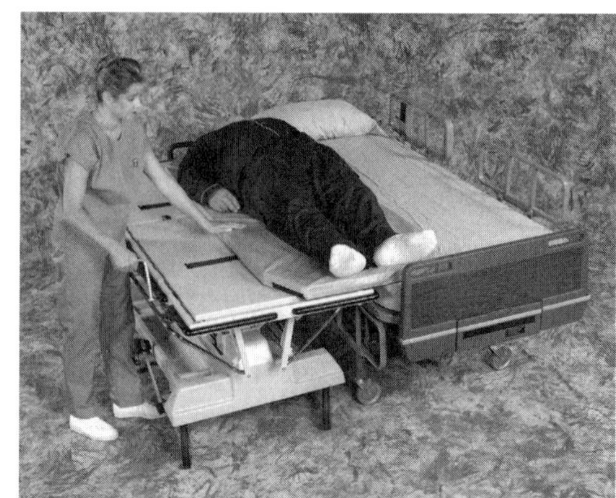

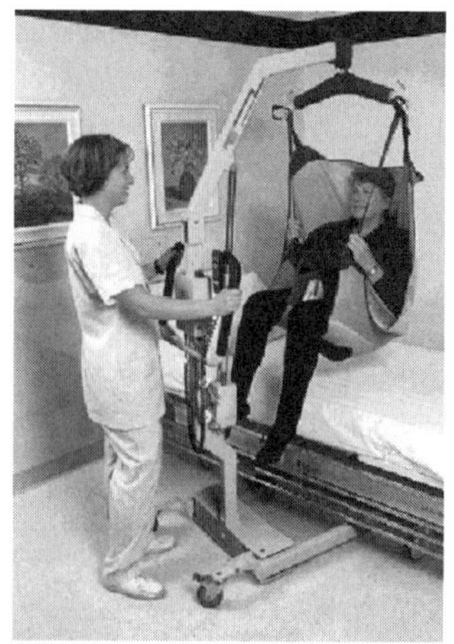

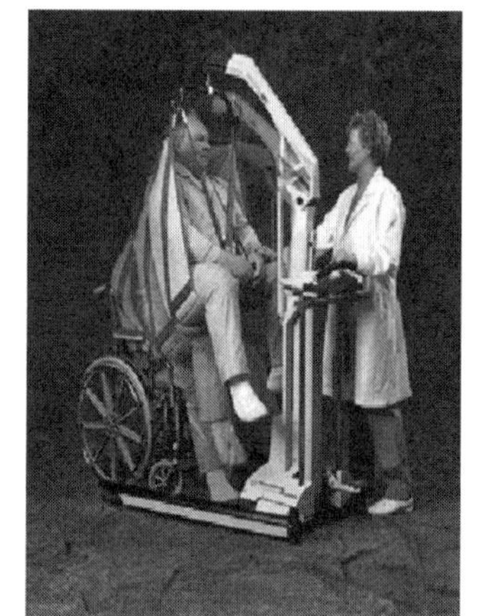

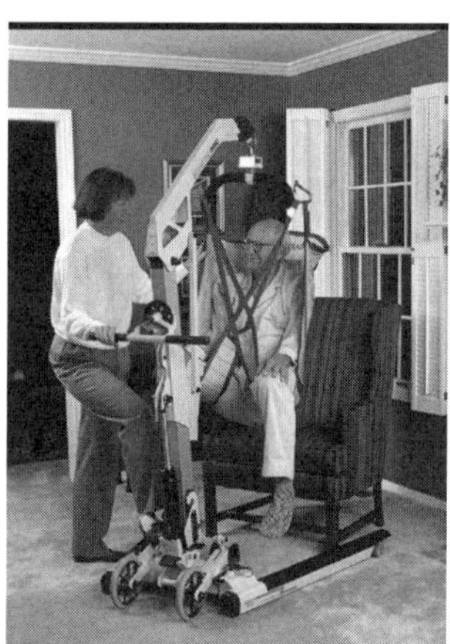

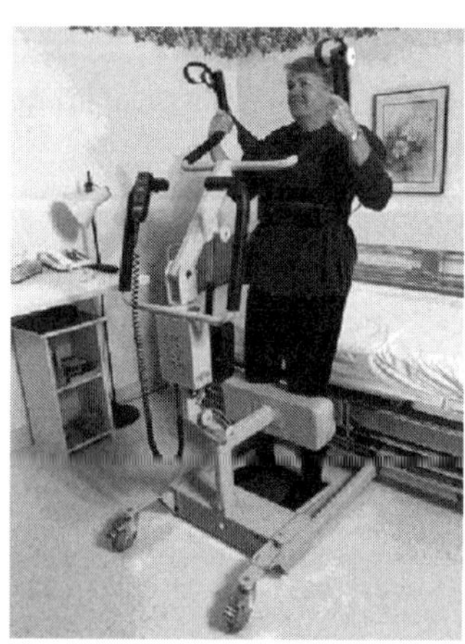

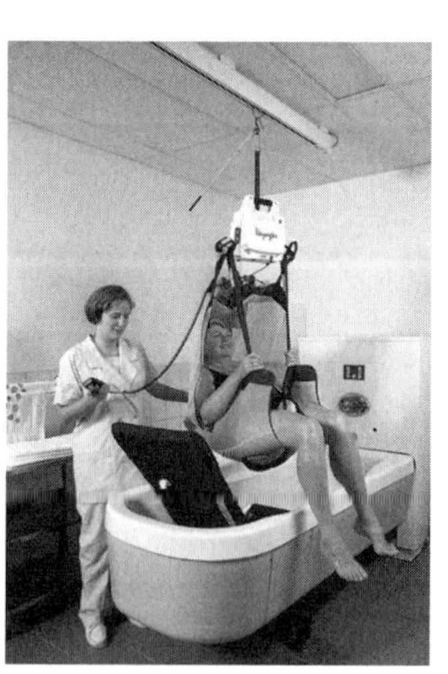

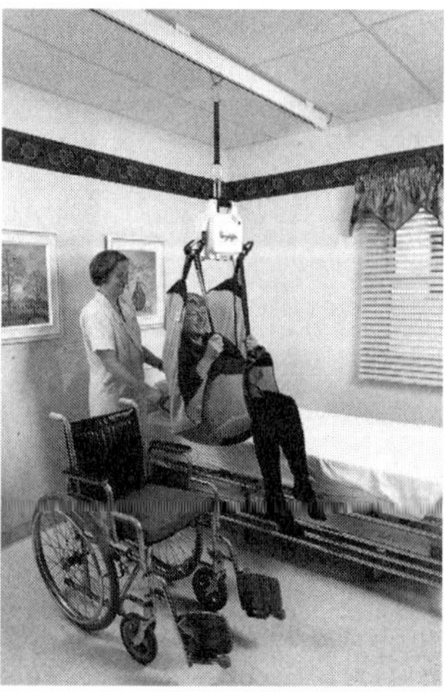

ARJO LIFTS

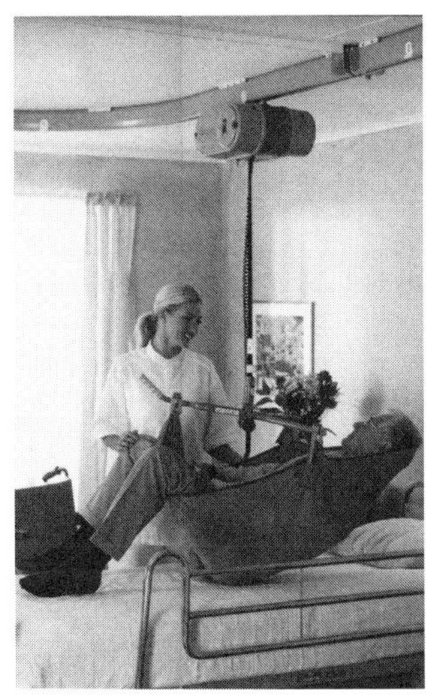

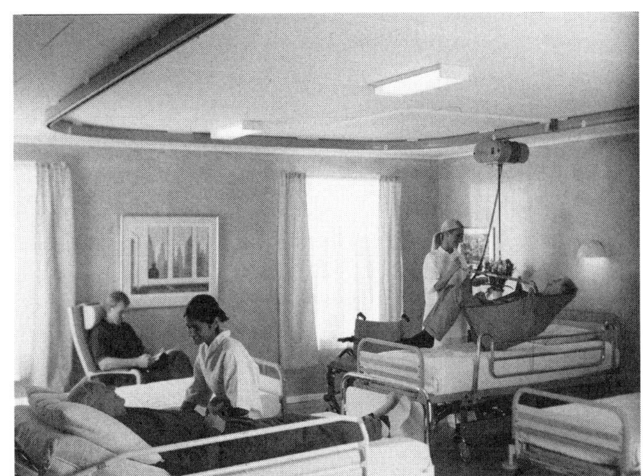

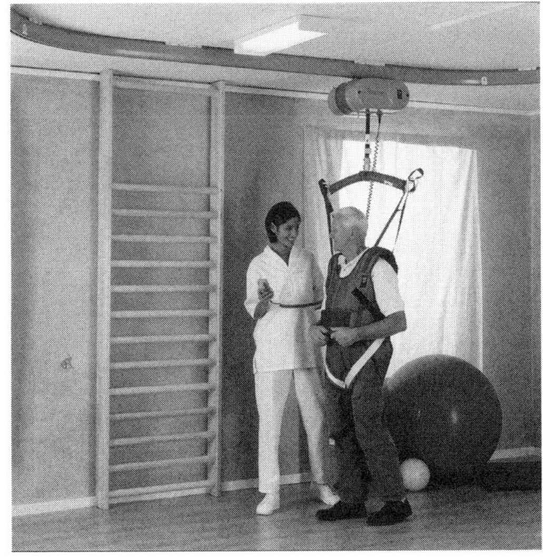

Index

I

J

L

P

Q

R

S

T

U

V

W

Z